The Discovery and Utility of Chemical Probes in Target Discovery

Chemical Biology

Editor-in-chief:
Kira J. Weissman, *Lorraine University, France*

Titles in the series:

1: High Throughput Screening Methods: Evolution and Refinement
2: Chemical Biology of Glycoproteins
3: Computational Tools for Chemical Biology
4: Mass Spectrometry in Chemical Biology: Evolving Applications
5: Mechanisms of Primary Energy Transduction in Biology
6: Cyclic Peptides: From Bioorganic Synthesis to Applications
7: DNA-targeting Molecules as Therapeutic Agents
8: Protein Crystallography: Challenges and Practical Solutions
9: Oxidative Folding of Proteins: Principles, Biological Regulation and Design
10: Chemical and Biological Synthesis: Enabling Approaches for Understanding Biology
11: Synthetic Glycomes
12: The Chemical Biology of Sulfur
13: The Chemical Biology of Phosphorus
14: DNA Damage, DNA Repair and Disease: Volume 1
15: DNA Damage, DNA Repair and Disease: Volume 2
16: The Discovery and Utility of Chemical Probes in Target Discovery

How to obtain future titles on publication:
A standing order plan is available for this series. A standing order will bring delivery of each new volume immediately on publication.

For further information please contact:
Book Sales Department, Royal Society of Chemistry, Thomas Graham House, Science Park, Milton Road, Cambridge CB4 0WF, UK
Telephone: +44 (0)1223 420066, Fax: +44 (0)1223 420247
Email: booksales@rsc.org
Visit our website at www.rsc.org/books

The Discovery and Utility of Chemical Probes in Target Discovery

Edited by

Paul Brennan
University of Oxford, UK
Email: paul.brennan@sgc.ox.ac.uk

and

Saleta Vazquez Rodriguez
University of Oxford, UK
Email: saleta.vazquezrodriguez@ndm.ox.ac.uk

ROYAL SOCIETY
OF **CHEMISTRY**

Chemical Biology No. 16

Print ISBN: 978-1-78801-589-9
PDF ISBN: 978-1-83916-074-5
EPUB ISBN: 978-1-83916-084-4
Print ISSN: 2055-1975
Electronic ISSN: 2055-1983

A catalogue record for this book is available from the British Library

The Royal Society of Chemistry is a charity, registered in England and Wales, Number 207890, and a company incorporated in England by Royal Charter (Registered No. RC000524), registered office: Burlington House, Piccadilly, London W1J 0BA, UK, Telephone: +44 (0) 20 7437 8656.

For further information see our web site at www.rsc.org

Printed in the United Kingdom by CPI Group (UK) Ltd, Croydon CR0 4YY, UK

Preface

Since the development of molecular biology and the ability to study single proteins on an atomic level, it has emerged that most small-molecule drugs and many pharmacologically active natural products exert their often profound phenotypic effects on cells and organisms by modulating the function of a single or small number of protein targets critical to cellular function. This knowledge sparked the modern era of (protein) target-based drug discovery and efforts to treat human disease have since paired molecular/cellular biology (MCB) and synthetic/medicinal chemistry (Med Chem) as a means of identifying disease-causing proteins in relevant cells and tissues and correct their function with bespoke small molecules, often antagonists or inhibitors. Over the following decades, the relationship between these experimentally very different disciplines has matured and produced scores of offspring: the medicines developed since the 1980s that have had such a huge impact on human health, increased lifespan and life quality.

However, the development of small molecules as modulators of protein function has not been limited in utility to the production of new medicines. Small molecules that show an effect on a specific target with certain potency and selectivity criteria do not necessarily need to fulfil all the requirements needed to become a drug, but instead they can be used as tool compounds, or what are now called chemical probes. These chemical probes can act as the keys to unlock the elusive and fractally complex secrets of cells and entire organisms.

A variety of compound sources have led to the discovery of chemical probes. In the biopharmaceutical industry, for example, where most small-molecule medicines are discovered, the intermediate forms in drug discovery – the hits, leads and candidates of pharma jargon – designed and made

Chemical Biology No. 16
The Discovery and Utility of Chemical Probes in Target Discovery
Edited by Paul Brennan and Saleta Vazquez Rodriguez

Published by the Royal Society of Chemistry, www.rsc.org

en route to the final medicine are extraordinarily useful in deciphering the complex disease biology in preclinical cellular and animal models. Historically, these proto-drugs were buried in patents and obscured by incomplete disclosure of biological potency and selectivity.

In parallel with mostly industry-led drug discovery, natural product chemists, primarily in academia, were discovering their own chemical probes. From the 1980s, despite the importance of natural product synthesis in training biopharma's army of medicinal chemists, it became more and more difficult to justify funding natural product synthesis as an end in itself. Hence synthetic organic chemists looked for other means to justify their obsession with Nature's own pharmacopoeia of complex secondary metabolites. More and more natural and semi-natural products were tested for their biological activity, with many showing cytotoxicity towards cancer and other cell types. What was more intriguing than their cytotoxicity was how these compounds killed cells, shedding light on the intricacies of the cell cycle, signalling and transport. Paclitaxel stabilizes microtubules and prevents cell-cycle progression; rapamycin forms a complex with FKBP12 and inhibits mTOR to prevent cell growth; epoxomicin selectively inhibits the proteasome; and brefeldin A prevents ARF1 GTPase reactivation and thereby vesicle transport. All of these natural products highlighted the importance of their respective targets and pathways on cellular physiology. Some of them led to the development of medicines (paclitaxel to Taxol, rapamycin to Rapamune, epoxomicin to Kyprolis), but they all showed unexpected mechanisms of action for small molecules on protein targets. Another extraordinary mechanism of action was shown by Craig Crews at Yale with his synthetic proteolysis-targeting chimeras (PROTACs), designed to induce protein degradation by tagging target proteins for destruction by the ubiquitin–proteasome system.

Seminal work from Tim Wilson (GlaxoSmithKline) and Stuart Schreiber (Harvard) in the late 1990s brought these nascent chemical probe strands together. Wilson's "reverse endocrinology" with nuclear hormone receptor ligands and Schreiber's "chemical genetics" with natural product-inspired scaffolds laid out the philosophy and groundwork of the field of chemical probes as it exists today. Essentially, potent and selective small molecules are not just medicines, failed medicines or substances with curious pharmacology, but are powerful and, critically, simple-to-use chemical probes for interrogating protein targets in complex biological systems. Expanding the haphazard collection of chemical probes discovered by chance and semi-rationally by natural product and medicinal chemists to the entire druggable genome, a goal still not yet realized, would allow us to interrogate and understand cell biology in extraordinary new ways. An extremely ambitious multi-collaborative project, Target 2035, is a global federation for developing and applying new technologies with the goal of creating chemical probes or antibodies for the entire proteome. Target 2035 is notable in its ambition, but also in the membership of the consortium. The mix of industry and academia shows that big pharma is no longer just burying priceless chemical probes in patents, but actively getting them into the laboratories of researchers worldwide, including their own competitors.

In this book, the authors review and exemplify the discovery and use of chemical probes as it exists today. Chapter 1 provides a background on the utility of chemical probes and presents the criteria for high-quality probes and different probe sources and databases available to the scientific community. Chapter 2 shows how the use of DNA-encoded libraries has expanded the breadth and depth of the protein targets available to chemical probe development. Chapter 3 reviews how computational tools are critical to the discovery of chemical probes. In Chapter 4, methodologies for the discovery and unique advantages of covalent chemical probes are presented. The rapidly developing field of peptidic chemical probes, which further expands chemical probe target scope, is reviewed in Chapter 5. In Chapter 6, an elegant case study of the derivatization of a natural product is used to untangle its unusual mechanism of cytotoxicity. Chapter 7 presents the principles and discovery of PROTACs, a field that grown exponentially in the last decade and has potential for enormous impact in new medicines and chemical probes. In Chapter 8, chemical probes and drugs for kinases, a protein family of enormous interest in drug discovery, are reviewed. Chapter 9 shows how small-molecule chemical probes can interrogate RNA as well as protein function. In Chapter 10, assays to understand chemical probes in native cellular environments are reviewed. Finally, in Chapter 11, the use of chemical probe libraries in phenotypic screening in cells is presented, with a focus on cellular models of neurodegeneration. Collectively, the principles, methods and examples in this book present a comprehensive picture of how chemical probes are used in biological and medical research today and offer a glimpse of how the field will evolve to achieve true chemical genetics when we have a chemical probe for every targetable protein encoded in the genome.

Paul Brennan and Saleta Vazquez Rodriguez

Contents

Chemical Biology No. 16
The Discovery and Utility of Chemical Probes in Target Discovery
Edited by Paul Brennan and Saleta Vazquez Rodriguez

Published by the Royal Society of Chemistry, www.rsc.org

CHAPTER 1

Introduction to Chemical Probes

BRIDGET K. WAGNER*

Chemical Biology and Therapeutics Science Program, Broad Institute, 415 Main Street, Cambridge, MA 02142, USA
*E-mail: bwagner@broadinstitute.org

1.1 Why Make and Use Chemical Probes?

The last 10 years have witnessed a rapid rise in enthusiasm for developing chemical probes as reagents in chemical biology research. Defined by Workman and Collins[1] as "small molecules to understand the function of genes and proteins and their role in physiology and pathology", the utility of chemical probes is without question. Frequently, small-molecule discovery is viewed as solely a drug discovery endeavour; however, chemical probes can provide unique laboratory reagents that fulfil a complementary, and often superior, role in understanding the underlying biology of a particular area.

Over 20 years ago, Stuart Schreiber raised the idea of developing a chemical probe for every gene.[2] He closed the perspective article with a call to arms that was well ahead of its time in the 1990s: "The original goal of the human genome project was 'to sequence every gene'. With that goal within sight, I suggest we consider a new goal for this project, one that can only be realized through the creative use of chemistry, 'to identify a small molecule partner for every gene product'." Years later, the field was finally on its way toward

Chemical Biology No. 16
The Discovery and Utility of Chemical Probes in Target Discovery
Edited by Paul Brennan and Saleta Vazquez Rodriguez

Published by the Royal Society of Chemistry, www.rsc.org

considering this idea, to the point that attention was refocused from whether this goal was even feasible to developing criteria for what constitutes a truly high-quality probe.

Small molecules provide temporal, spatial and concentration control that is not currently attainable by genetic modifications. Even better, when used in parallel with gene-based experiments, the use of chemical probes allows researchers to avoid being fooled by the off-target effects of either approach. In the case of genetic manipulation, examples exist where knockout mice often have a different, and sometimes the exact opposite, effect of chemical probes.[3] Further, genetic knockouts often result in the strong upregulation of related isoforms, which can confound the analysis of the role of proteins in cellular and organismal phenotypes. Thus, chemical probes serve a truly unique research purpose that cannot be accessed by genetic experiments. A challenge in the field has been to identify and use "good" chemical probes. The adage "garbage in, garbage out" applies here: developing a set of criteria for what constitutes a chemical probe, as opposed to, say, a lead candidate has been a major sources of discussion and debate over the last 10 years.

1.2 What Is a Chemical Probe?

Historically, within the research community, one often sees interchangeable use of the terms "compound", "small molecule", "drug", "tool", "lead" and "probe". What to make of these terms: how will we know if we even have a probe for research? Fortunately, many groups have made considerable intellectual contributions to this area and several key papers published in 2009 and 2010 helped propose what constitutes a high-quality probe.[4–7] Importantly (and sensibly), rather than establishing a set of hard-and-fast rules for probes, which can introduce an unhelpful rigidity to the definition, these authors have suggested sets of guidelines for the properties that probes should, for the most part, fulfil.

An important indication that not all probes are created equal was provided by Oprea and colleagues.[7] This work evaluated the 64 chemical probes emerging between 2004 and 2008 from the pilot phase of the NIH Molecular Libraries and Imaging (MLI) Initiative. The MLI Initiative funded several high-throughput screening centres, which ran 691 assays assembled from investigators throughout the USA. These assays represented 171 biological targets and 29 phenotypic assays. The authors then assembled a set of 11 experts in drug discovery and chemical biology and asked them to rate their levels of confidence in the 64 probes (on a scale of 0 to 10, with 0 representing the highest confidence). Overall, 16 compounds (25%) were deemed low confidence (score 5–6), an additional 16 compounds (25%) were of medium confidence (score 3–4) and 32 compounds (50%) were considered high confidence (score 0–2). In this study, much of the confidence score appeared to be driven by perceived chemical liabilities. Importantly, this paper lays out the evolution of the definition of chemical probes, going from the low bar of "adequate potency and solubility to be useful ... *in vitro*" to starting to incorporate notions of selectivity, toxicity, mode of action and availability to the community. A later analysis of the 375 total probes developed through 2015 in the Molecular Libraries Program highlighted some of the more sophisticated probes developed and emphasized the breadth and novelty of mechanisms discovered through the Program.[8]

Contemporaneous work by Edwards *et al.*[5] advocated for open collaboration between academics and industry, in a spirit of "open-access chemical biology". It was particularly noted that avoidance of duplicative studies and the resulting generation of higher quality probes would be a major outcome of such an effort. Some of the key historical lessons cited in the article that were learned from industry included the supremacy of chemical tractability, the success of clever experimental design over brute force, the importance of well-characterized probes and the involvement of the scientific community. As we shall see later, the fact that Target 2035 was launched only in 2019 is a testament to the scope and challenges of this goal.[9]

Another perspective on chemical probe criteria, focusing on protein kinase inhibitors, further helped enhance the stringency, and thus the quality, of probes emerging from studies.[4] These suggested guidelines included several essential criteria, including (1) that the compound must be tested on a panel of at least 50, but preferably 100, protein kinases; (2) at least two structurally unrelated inhibitors should work in cells; (3) the cellular effects should be observed at a reasonable concentration at which the kinase is inhibited; (4) analogues (when available) should rank-order in the same fashion for biochemical activity and cell potency; and (5) publications must include the structure, otherwise results cannot be replicated by other laboratories. Two other desirable criteria, but of much higher stringency, were that (1) the compound should be inactive in cells containing a drug-resistant mutation of the kinase and (2) the compound should be inactive in cells in which the target and related isoforms are knocked out. As we shall see later, the more recent failure of some notable drug candidates on this last "desirable" criterion has the potential to create havoc in cancer drug development.[10]

In 2010, Frye wrote about "the art of the chemical probe", with the clear implication that defining a probe was at that point still an art.[6] He specified five clear principles for chemical probes to follow; these values helped lay the foundation for later efforts to provide more precise definitions. First, there must be sufficient potency and selectivity to link up biochemical and cellular activity. Second, there should be sufficient mechanistic data on the probe's mechanism of action (MoA) to allow biological interpretation of cellular results. Third, a clear identity of the active species must be in place. Fourth, the probe must be useful and be able to address at least one hypothesis of the target's role in cells. Finally, the probe must be available to academic researchers without restriction. Together, this commentary helped advance the conversation about high-quality chemical probes.

A truly comprehensive treatment of probe quality was advanced in 2010 by Workman and Collins,[1] who proposed "fitness factors" for chemical probes, spanning across chemical, biological and contextual properties. Their paper is an essential resource for those interested in developing or using chemical probes, not only because all of the factors are listed in one place, but also because an extensive rationale for their inclusion is provided for the reader. First, the chemical properties of a probe include that the structure is not only revealed, but also can be reproducibly generated, the compound must be stable and membrane permeable and there should be good solubility, generally >100 μM. Importantly, the compound must not be an aggregator, which, as we shall see later, can interfere with many biochemical and cellular assay readouts.

Workman and Collins[1] went on to discuss the biological features that are important for a high-quality chemical probe. The biochemical potency should be <100 nM, with activity in cells in the range 1–10 μM. Of course, analogues with activity are enormously helpful, but perhaps even more useful are inactive analogues, whose lack of cellular activity helps determine selectivity towards the ostensible target. The appropriate pharmacokinetic profile for testing *in vivo* efficacy in the appropriate tissue is also an important component, but can take additional time to develop. Similarly, having additional chemotypes that show a correlative relationship between biochemical and cellular activity would provide the best evidence when using a chemical probe, but this is not always possible at an early point in probe development. In terms of selectivity, obtaining a target profile against the class of related proteins is nearly a must, especially in the protein kinase world. Venturing outside the target class is even better, but this feature is considered desirable, not essential. Cheminformatics to analyse the chemotype for other activities is also important. Finally, the context in which the probe will be used is a key consideration. For example, these parameters include understanding the biological hypothesis to be tested, the cellular context of the target and whether complementary genetic methods are available for comparison. Perhaps the most important consideration is compound availability, without restriction and in amounts (15–20 mg) to enable follow-up studies. Together, these guidelines are challenging to achieve, but it is better to lay out the highest quality probe characteristics and try to reach them than to easily attain sub-par goals.

The challenges involved in developing chemical probes was revisited in 2015, when a large consortium reflected on the perils and promise of chemical probes.[11] As the definition of a probe was becoming more accepted in the field, the next hurdle to be recognized was the enormous cost and effort required to make a high-quality probe. This challenge underscores the need for academic and industry researchers to work together on these probes. Importantly, the authors distinguished the prioritized properties for probes (*e.g.* MoA, selectivity, availability) from those for drugs (*e.g.* physicochemical properties), helping to emphasize the fact that the bar for quality is in some ways higher for a probe. As a cautionary note, the authors also provided examples of low-quality probes, including valproic acid and LiCl (both used at too high a concentration), flavones (promiscuous interference compounds) and resveratrol (an assay artefact).[12] Finally, this group introduced the Chemical Probes Portal, an online resource for sharing data and reviews of various compounds. More on this topic will be discussed in the last section of this Introduction.

1.3 How "Good" Must Chemical Probes Be?

As discussed, the work of Oprea and colleagues[7] provided the field with a good indication of some of the major challenges posed in developing high-quality probes, with one-quarter of the 64 reported probes from the NIH MLI Initiative programme considered highly dubious. With such a high number of questionable probes, what are the modes of failure? One of the most common shortcomings of low-quality probes is promiscuity. Frequently, a compound may be quite potent towards its intended target, but if it inhibits a whole swath of targets, cellular results cannot be reliably interpreted. This phenomenon has been observed for nearly 20 years in the protein kinase field. Philip Cohen and colleagues published some of the first papers describing non-radioactive profiling of compounds across many kinases.[13] Their work provided some of the first red flags that a number of commonly used compounds, such as rottlerin, quercetin and LY294002, inhibited other targets, sometimes with greater potency than the ostensible target. Despite this important work, poor tool compounds such as LY294002 are still used in research and result in publications;[11] that particular compound returns over 10 000 results in PubMed, including papers that have been published at the time of the preparation of this book. A stronger spotlight must be shone on these kinds of results, so that the continued use of these shoddy probes is halted.

Physicochemical properties comprise another major source of poor-quality probes. Of course, poor solubility and permeability will confound any cell-based research performed on a probe candidate, no matter how potent it is *in vitro*. Chemical reactivity, such as the inclusion of Michael acceptors, can cause instability and promiscuity. The problematic compounds with which most researchers are familiar are Pan Assay INterfering compoundS or PAINS.[14,15] The chemotypes associated with PAINS can cause false-positive results in many assays, due to covalent reactivity, redox activity and aggregation, among other mechanisms. Aggregation in particular is likely to be the major source of bad compound behaviour in cell-free experiments. An important study found that of 50 approved drugs tested, four demonstrated significant promiscuous, aggregation-dependent enzyme inhibition.[16] On that note, the Editors-in-Chief of several American Chemical Society (ACS) journals sounded an alarm that proper benchmarks need to be used to control for artefactual assay activity.[17] For PAINS themselves, many *in silico* tools exist to help predict whether compounds might cause a problem. PAINS are typically associated with synthetic compounds; the natural products that show several non-specific activities that cannot be explained by common failure modes of synthetic molecules have been termed invalid metabolic panaceas (IMPs).[18]

Several cautionary tales are worth exploring here; these stories involve the use of poor-quality probes, false-positive results and lack of reproducibility for probe activities. Curcumin, a natural product found in turmeric, is the poster child of a problem compound that has muddied the literature. A mini-perspective exploring the extent of curcumin's use in the literature tells the story of troublesome chemistry and just plain bad logic.[19] Multiple lines of evidence show that curcumin is an unstable, reactive and non-bioavailable compound; nonetheless, more than 120 clinical trials of curcuminoids have been launched, with none of them demonstrating efficacy. The concentrations at which curcumin displays its alleged activities, including inhibition of the histone acetyltransferase p300, histone deacetylase 8, glycogen synthase kinase-3β, tau fibril formation and cannabinoid receptor 1, are in the same range at which aggregation could occur. To add to the confusion, counterscreens were frequently not performed and compound stability was not always considered. These troubling findings result in a waste of time and money and plague the literature with results that are not likely to be real, perpetuating the reproducibility crisis with which biomedical science has been blighted. Importantly, the authors of this overview do not rule out that *crude turmeric extract* could have activity of interest, which appears to drive the continued fascination with this molecule; it is just not the activity of curcumin alone.

Aside from problematic compounds, the correct assignment of MoA for compounds is an extremely critical part of understanding disease biology. A recent study from Cold Spring Harbor Laboratory and Stony Brook University asked the question of why so many oncology drug candidates fail to advance to approval by the US Food and Drug Administration (FDA).[10] Amazingly, the authors found that the annotated target of 10 compounds tested (spanning five inhibited targets: HDAC6, MAPK14, PAK4, PBK and PIM1) were not essential to cancer-cell proliferation. Using CRISPR-Cas9 to knock out the putative targets, they found that there was no loss of viability of the knockout cells, which was not explainable by upregulation of a homologue. Further, in each case, the knockout cells retained sensitivity to the compounds, proving that the protein in question was not the target. Thus, off-target toxicity is a common MoA for these compounds. The targets had primarily been identified by siRNA or shRNA knockdown, which may explain the discrepancy: the known seed effects of siRNA experiments[20] make it all the more important to use well-validated methods to identify new targets. What is all the more remarkable, however, is that these are compounds in clinical trials, meaning they had advanced through many stages of development without these inconvenient facts rearing their heads. We can conclude from these results that prudent science, not just chemical liability filters, must be exercised when developing chemical probes and drugs.

A similar effort was performed to evaluate systematically probes focused on the expanding chemical matter to inhibit the interaction between Nrf2 and Keap1.[21] The authors collected the 19 small molecules reported in the literature to inhibit this protein–protein interaction (PPI). Six compounds were purchased, four compounds were easily synthesized and nine were synthesized with more difficulty. They then performed a side-by-side comparison of each compound, testing the inhibition by fluorescence polarization, thermal shift assays and surface plasmon resonance. Remarkably, half of the compounds were inactive or much less active than reported and 10 of the 19 compounds showed no evidence of binding at all; these results were independent of the source of the compound. This report should serve as a massive wake-up call to the chemistry and chemical biology communities that reproducibility *by other research groups* needs to be a key step in truly validating a chemical probe. Even more troubling, practitioners of basic cell and molecular biology will likely not even see this body of literature, resulting in the propagation of research using essentially useless reagents.

1.4 How Can One Discover and Use Chemical Probes?

Chemical probes have been discovered from a wide variety of chemical resources, including natural products, commercially available screening collections, drug-like libraries and diversity-oriented synthesis (DOS).[22] In the last case, over a decade of research activity has resulted in probes relevant to diabetes, cancer and, perhaps most promisingly, infectious disease.[23] A useful component of the DOS chemical strategy is to synthesize a smaller number of compounds across a larger number of scaffolds. Sampling across these scaffolds, using "Informer Sets", has yielded relative enrichment in activity in particular chemistries, including in projects focused on identifying inhibitors of insulin-degrading enzyme[24] and Cas9.[25] These insights, which can be achieved rapidly through pilot screening, enable researchers to screen more deeply in active scaffolds, enhancing the efficiency with which probe candidates are identified.

Of course, the use of chemical probes shows its greatest utility in demonstrating the role of a particular protein target in a cellular phenotype. Activity of a probe *in vivo* is even better, but sometimes pharmacokinetic properties are not yet sufficient to study without significant medicinal chemistry efforts. With the advent of DNA-encoded libraries (DELs) and the discovery of molecular glues that induce proximity between, for example, enzymes and substrates, exciting new probe opportunities arise. A popular example of this MoA is targeted protein degradation (*e.g.* PROTAC, degronimid) technologies.[26,27] This increasing focus on small molecules that bind targets, rather than inhibit enzymatic activity,[28] may yield probes that enable novel activities, previously accessible only by using genetic fusion proteins. These protein-binding compounds can also be used to exploit modulation of the protein interactome of the target, which can often be more useful in the case of, for example, non-enzymatic targets or allosteric binding sites.

Several groups have also made creative use of high-quality chemical probes as a pilot library itself, for understanding the types of targets that should emerge from phenotypic screening. For example, an overview by Wassermann and colleagues[29] discusses a number of collections with well-annotated and characterized compounds that exist for this kind of work (*e.g.* LOPAC, Oxford collection, Repurposing Hub). They also describe the approach developed at Novartis to a biodiverse screening deck. Similarly, cancer cell-line profiling benefits from the use of a focused library, which must therefore be composed of high-quality probes. For example, a study in 2015 examining the sensitivity of ~500 compounds across nearly 1000 cell lines focused on compounds that perturb key nodes in cellular signalling.[30] Finally, a call for contributions to a comprehensive kinase chemogenomic set (KCGS) aimed to develop a common set that the entire research community could use.[31] The first such step toward this goal was the assembly of the GlaxoSmithKline (GSK) published kinase inhibitor set (PKIS), a collection of 367 ATP-competitive inhibitors; kinase profiles were posted online and the physical set was distributed to hundreds of investigators.[32] These types of community efforts will improve data quality and enhance the efficiency of generating novel biological insights.

1.5 Where Can One Find Chemical Probes?

Garbaccio and Parmee[33] provided a perspective from the pharmaceutical industry to highlight the need for greater collaborative efforts between academia and industry. They acknowledge that such efforts will require huge investments, but make the argument that good probes help validate targets and show translatability in drug discovery. These advantages place chemical probes at a unique nexus between understanding fundamental biology and providing new avenues for therapeutic intervention. Overall, these authors exhort the community to provide continued investment and publication of high-quality probes for research use. Some pioneering groups have been at the vanguard of providing, free to the community, important high-quality probes. For example, Jay Bradner and Jun Qi's work to distribute the BET bromodomain inhibitor JQ1 has resulted in an explosion of research, with a more than fivefold increase in the number of patents for bromodomains four years after its first publication.[34] More such efforts are required to expand the biological space accessed by chemistry.

On that note, an excellent brief overview of chemical probe quality, literature and resources was provided in the paper "Revisiting the art of the chemical probe".[35] This overview highlights several key online resources, in which information about the chemistry, biology and availability of chemical probes dramatically lowers the barriers to using these probes to validate cellular targets. It is important for researchers to be aware of these resources and use the recommended probes available to the community, rather than perpetuating irreproducible research using shoddy probes.

Probe Miner (https://probeminer.icr.ac.uk) contains a database of over 1.8 million small molecules and provides binding and bioactivity data against 2220 human targets, information on structure–activity relationships and data on whether inactive analogues are available and their structures, in addition to including filters for common liabilities, such as PAINS.[36] This first large-scale resource also enables users to evaluate the quality of probes, by providing expert curation and judgement of whether small molecules fulfil minimum chemical probe criteria. Notably, the authors found that of the >350 000 active compounds in their database, only 2558 (0.7%) fulfil these minimal criteria: potency <100 nM, 10-fold selectivity and cell activity <10 μM. Although over half of the compounds possess good potency, the largest drop-off of compounds appears to be in cell activity, with only ~21 000 (6%) of compounds with cell activity, and even half of those compounds show lower potency than desired for a good probe. From a target perspective, there is a similarly sharp drop-off: of the 2220 targets represented, 1664 (75%) have potent compounds available, but 795 (36%) with selective compounds and only 250 (11%) with high-quality probes. These observations indicate that there does not appear to be much of a problem generating potent biochemical probes, but cellular activity for most probes has lagged. Perhaps the community should take heed of this knowledge and make cellular activity a more upstream requirement for probe generation.

The philosophy of expert curation has been taken to the next level by the Chemical Probes Portal (http://www.chemicalprobes.org), a resource that reports on fewer compounds but aims at high-quality review of each probe.[11] This database uses a principle of selective crowdsourcing: there are many invited probe reviewers from all over the world in the chemical biology community. Among the emphases of this resource is experimental reproducibility, which is hampered by poor-quality probes. To that aim, the authors also flag "historic" compounds, which are those that have previously been used as probes but should not be in the future. A comparative analysis of these historic compounds and probes showed that, unsurprisingly, historic compounds were more promiscuous than probes.[37] Overall, about half of the compounds were considered target selective, which the authors consider "an encouraging finding", which speaks to the unsatisfyingly low bar for reporting on active small molecules.

Two resources from the Structural Genomics Consortium (SGC) provide valuable information and, importantly, distribution of chemical probes for the research community to use. First, the SGC Chemical Probes site (https://www.thesgc.org/chemical-probes) provides open-access reagents, with detailed characterization and recommendations on which probes to use in cell-based assays and at what concentrations. The 73 probes on this site, largely focused on epigenetic targets and kinases, all fulfil the following criteria: biochemical potency <100 nM, >30-fold selectivity *versus* other subfamilies and significant on-target cell activity at 1 μM. This important resource is complemented by the Donated Chemical Probes resource (https://www.sgc-ffm.uni-frankfurt.de) run by SGC Frankfurt.[38] An initial consortium of seven pharmaceutical companies have opened up their set of undisclosed probes, which meet a strict set of criteria and are available for use by anyone in the research community. These criteria are similar to those discussed above in earlier literature, but this database also includes proof of target engagement, an increasingly important consideration in cell and *in vivo* treatment that is not always considered. Inactive control compounds are available for each probe and a variety of profiling data (kinome and GPCR data and select phenotypic assays, among others) is available for each compound. Although this set of compounds is still primarily focused on enzymatic targets, this framework for enabling open-access research to be performed on high-quality probes is a model for how chemical probes will advance our knowledge of target biology in a more rapid fashion.

Another recent contribution to the probes arsenal is the Probes & Drugs (P&D) database (https://www.probes-drugs.org/), developed by CZ-OPENSCREEN.[39] This resource contains information on >4000 chemical probes and >12 500 drugs, in addition to tens of thousands of other compounds (*e.g.* analogues). Further, the P&D database has the most sophisticated filtering, visualization and chemically intelligent tools available. Unlike some of the other databases, however, these probes are not distributed through this group. Therefore, consumers of these data should be cognizant of the advantages and limitations of each resource, with a combination of tools likely to provide the best direction for research.

Finally, a recent call to arms has been raised in the guise of "Target 2035", proposed by Carter *et al.*[9] This global federation aims to harness technological advances to generate reagents for the entire proteome, from chemical probes to selective antibodies to entire chemical libraries. Proceeding from the notion that the natural tendency is for scientists to study genes that are already under study, the authors argue that pharmacological modulators are "the most important reagent for any protein". It is remarkable to note several lessons that emerge from this effort: (1) the estimated cost for developing a high-quality probe is more than US$2 million, a cost sufficiently large that achieving Target 2035 will require enormous investment from diverse sources around the world; (2) the idea of generating proteome-wide probes is now, 20 years after originally being proposed,[2] finally sufficiently ripened to result in such a global effort; (3) the Target 2035 team wisely envisions two phases of the project over time: phase 1 (planned for 2020–2025) entails creating the necessary infrastructure, developing new technologies and self-organizing among chemical biologists, and phase 2 (anticipated to be 2025–2035) involves applying phase 1 resources to create a set of comprehensive modulators for most human protein-encoding genes. Each of these insights reflects the increasingly realistic expectations of the cost, effort and organization required to achieve such a worthwhile goal. An increasing partnership between public and private research organizations is absolutely required for the community to realize the promise of high-quality chemical probes to understand biology, identify and validate targets for therapeutic intervention and generate lead candidates for drug development. Further, greater incentive to study (and publish) the quality of chemical probes needs to be provided to the community, lest the community be saddled with poor probes and irreproducible results.

References

1. P. Workman and I. Collins, *Chem. Biol.*, 2010, **17**, 561–577.
2. S. L. Schreiber, *Bioorg. Med. Chem.*, 1998, **6**, 1127–1152.
3. Z. A. Knight and K. M. Shokat, *Cell*, 2007, **128**, 425–430.
4. P. Cohen, *Biochem. J.*, 2009, **425**, 53–54.
5. A. M. Edwards, C. Bountra, D. J. Kerr and T. M. Willson, *Nat. Chem. Biol.*, 2009, **5**, 436–440.
6. S. V. Frye, *Nat. Chem. Biol.*, 2010, **6**, 159–161.
7. T. I. Oprea, C. G. Bologa, S. Boyer, R. F. Curpan, R. C. Glen, A. L. Hopkins, C. A. Lipinski, G. R. Marshall, Y. C. Martin, L. Ostopovici-Halip, G. Rishton, O. Ursu, R. J. Vaz, C. Waller, H. Waldmann and L. A. Sklar, *Nat. Chem. Biol.*, 2009, **5**, 441–447.
8. S. L. Schreiber, J. D. Kotz, M. Li, J. Aubé, C. P. Austin, J. C. Reed, H. Rosen, E. L. White, L. A. Sklar, C. W. Lindsley, B. R. Alexander, J. A. Bittker, P. A. Clemons, A. de Souza, M. A. Foley, M. Palmer, A. F. Shamji, M. J. Wawer, O. McManus, M. Wu, B. Zou, H. Yu, J. E. Golden, F. J. Schoenen, A. Simeonov, A. Jadhav, M. R. Jackson, A. B. Pinkerton, T. D. Y. Chung, P. R. Griffin, B. F. Cravatt, P. S. Hodder, W. R. Roush, E. Roberts, D.-H. Chung, C. B. Jonsson, J. W. Noah, W. E. Severson, S. Ananthan, B. Edwards, T. I. Oprea, P. J. Conn, C. R. Hopkins, M. R. Wood, S. R. Stauffer, K. A. Emmitte, L. S. Brady, J. Driscoll, I. Y. Li, C. R. Loomis, R. N. Margolis, E. Michelotti, M. E. Perry, A. Pillai and Y. Yao, *Cell*, 2015, **161**, 1252–1265.
9. A. J. Carter, O. Kraemer, M. Zwick, A. Mueller-Fahrnow, C. H. Arrowsmith and A. M. Edwards, *Drug Discovery Today*, 2019, **24**, 2111–2115.
10. A. Lin, C. J. Giuliano, A. Palladino, K. M. John, C. Abramowicz, M. L. Yuan, E. L. Sausville, D. A. Lukow, L. Liu, A. R. Chait, Z. C. Galluzzo, C. Tucker and J. M. Sheltzer, *Sci. Transl. Med.*, 2019, **11**, eaaw8412.
11. C. H. Arrowsmith, J. E. Audia, C. Austin, J. Baell, J. Bennett, J. Blagg, C. Bountra, P. E. Brennan, P. J. Brown, M. E. Bunnage, C. Buser-Doepner, R. M. Campbell, A. J. Carter, P. Cohen, R. A. Copeland, B. Cravatt, J. L. Dahlin, D. Dhanak, A. M. Edwards, M. Frederiksen, S. V. Frye, N. Gray, C. E. Grimshaw, D. Hepworth, T. Howe, K. V. Huber, J. Jin, S. Knapp, J. D. Kotz, R. G. Kruger, D. Lowe, M. M. Mader, B. Marsden, A. Mueller-Fahrnow, S. Muller, R. C. O'Hagan, J. P. Overington, D. R. Owen, S. H. Rosenberg, B. Roth, R. Ross, M. Schapira, S. L. Schreiber, B. Shoichet, M. Sundstrom, G. Superti-Furga, J. Taunton, L. Toledo-Sherman, C. Walpole, M. A. Walters, T. M. Willson, P. Workman, R. N. Young and W. J. Zuercher, *Nat. Chem. Biol.*, 2015, **11**, 536–541.
12. M. Pacholec, J. E. Bleasdale, B. Chrunyk, D. Cunningham, D. Flynn, R. S. Garofalo, D. Griffith, M. Griffor, P. Loulakis, B. Pabst, X. Qiu, B. Stockman, V. Thanabal, A. Varghese, J. Ward, J. Withka and K. Ahn, *J. Biol. Chem.*, 2010, **285**, 8340–8351.

13. S. P. Davies, H. Reddy, M. Caivano and P. Cohen, *Biochem. J.*, 2000, **351**, 95–105.
14. J. B. Baell and G. A. Holloway, *J. Med. Chem.*, 2010, **53**, 2719–2740.
15. J. L. Dahlin, J. W. Nissink, J. M. Strasser, S. Francis, L. Higgins, H. Zhou, Z. Zhang and M. A. Walters, *J. Med. Chem.*, 2015, **58**, 2091–2113.
16. J. Seidler, S. L. McGovern, T. N. Doman and B. K. Shoichet, *J. Med. Chem.*, 2003, **46**, 4477–4486.
17. C. Aldrich, C. Bertozzi, G. I. Georg, L. Kiessling, C. Lindsley, D. Liotta, K. M. Merz Jr, A. Schepartz and S. Wang, *ACS Med. Chem. Lett.*, 2017, **8**, 379–382.
18. J. Bisson, J. B. McAlpine, J. B. Friesen, S. N. Chen, J. Graham and G. F. Pauli, *J. Med. Chem.*, 2016, **59**, 1671–1690.
19. K. M. Nelson, J. L. Dahlin, J. Bisson, J. Graham, G. F. Pauli and M. A. Walters, *J. Med. Chem.*, 2017, **60**, 1620–1637.
20. A. L. Jackson, J. Burchard, J. Schelter, B. N. Chau, M. Cleary, L. Lim and P. S. Linsley, *RNA*, 2006, **12**, 1179–1187.
21. K. T. Tran, J. S. Pallesen, S. M. O. Solbak, D. Narayanan, A. Baig, J. Zang, A. Aguayo-Orozco, R. M. C. Carmona, A. D. Garcia and A. Bach, *J. Med. Chem.*, 2019, **62**, 8028–8052.
22. S. L. Schreiber, *Science*, 2000, **287**, 1964–1969.
23. C. J. Gerry and S. L. Schreiber, *Nat. Rev. Drug Discovery*, 2018, **17**, 333–352.
24. J. P. Maianti, G. A. Tan, A. Vetere, A. J. Welsh, B. K. Wagner, M. A. Seeliger and D. R. Liu, *Nat. Chem. Biol.*, 2019, **15**, 565–574.
25. B. Maji, S. A. Gangopadhyay, M. Lee, M. Shi, P. Wu, R. Heler, B. Mok, D. Lim, S. U. Siriwardena, B. Paul, V. Dancik, A. Vetere, M. F. Mesleh, L. A. Marraffini D. R. Liu, P. A. Clemons, B. K. Wagner and A. Choudhary, *Cell*, 2019, **177**, 1067–1079 e1019.
26. Y. Che, A. M. Gilbert, V. Shanmugasundaram and M. C. Noe, *Bioorg. Med. Chem. Lett* , 2018, **28**, 2585–2592.
27. Y. Zhang, C. Loh, J. Chen and N. Mainolfi, *Drug Discovery Today: Technol.*, 2019, **31**, 53–60.
28. S. L. Schreiber, *Isr. J. Chem.*, 2019, **59**, 52–59.
29. A. M. Wassermann, L. M. Camargo and D. S. Auld, *Front. Pharmacol.*, 2014, **5**, 164.
30. B. Seashore-Ludlow, M. G. Rees, J. H. Cheah, M. Cokol, E. V. Price, M. E. Coletti, V. Jones, N. E. Bodycombe, C. K. Soule, J. Gould, B. Alexander, A. Li, P. Montgomery, M. J. Wawer, N. Kuru, J. D. Kotz, C. S. Hon, B. Munoz, T. Liefeld, V. Dancik, J. A. Bittker, M. Palmer, J. E. Bradner, A. F. Shamji, P. A. Clemons and S. L. Schreiber, *Cancer Discovery*, 2015, **5**, 1210–1223.
31. D. H. Drewry, C. I. Wells, D. M. Andrews, R. Angell, H. Al-Ali, A. D. Axtman, S. J. Capuzzi, J. M. Elkins, P. Ettmayer, M. Frederiksen, O. Gileadi, N. Gray, A Hooper, S. Knapp, S. Laufer, U. Luecking, M. Michaelides, S. Muller, E. Muratov, R. A. Denny, K. S. Saikatendu, D. K. Treiber, W. J. Zuercher and T. M. Willson, *PLoS One*, 2017, **12**, e0181585.
32. P. Dranchak, R. MacArthur, R. Guha, W. J. Zuercher, D. H. Drewry, D. S. Auld and J. Inglese, *PLoS One*, 2013, **8**, e57888.

33. R. M. Garbaccio and E. R. Parmee, *Cell Chem. Biol.*, 2016, **23**, 10–17.
34. A. R. Scott, *Nature*, 2016, **533**, S60–S61.
35. D. M. C. Schwarz and J. E. Gestwicki, *ACS Chem. Biol.*, 2018, **13**, 1109–1110.
36. A. A. Antolin, J. E. Tym, A. Komianou, I. Collins, P. Workman and B. Al-Lazikani, *Cell Chem. Biol.*, 2018, **25**, 194–205 e195.
37. F. Miljković and J. Bajorath, *Molecules*, 2018, **23**, 2434.
38. S. Muller, S. Ackloo, C. H. Arrowsmith, M. Bauser, J. L. Baryza, J. Blagg, J. Bottcher, C. Bountra, P. J. Brown, M. E. Bunnage, A. J. Carter, D. Damerell, V. Dotsch, D. H. Drewry, A. M. Edwards, J. Edwards, J. M. Elkins, C. Fischer, S. V. Frye, A. Gollner, C. E. Grimshaw, A. IJzerman, T. Hanke, I. V. Hartung, S. Hitchcock, T. Howe, T. V. Hughes, S. Laufer, V. M. Li, S. Liras, B. D. Marsden, H. Matsui, J. Mathias, R. C. O'Hagan, D. R. Owen, V. Pande, D. Rauh, S. H. Rosenberg, B. L. Roth, N. S. Schneider, C. Scholten, K. Singh Saikatendu, A. Simeonov, M. Takizawa, C. Tse, P. R. Thompson, D. K. Treiber, A. Y. Viana, C. I. Wells, T. M. Willson, W. J. Zuercher, S. Knapp and A. Mueller-Fahrnow, *eLife*, 2018, **7**, e34311.
39. C. Skuta, M. Popr, T. Muller, J. Jindrich, M. Kahle, D. Sedlak, D. Svozil and P. Bartunek, *Nat. Methods*, 2017, **14**, 759–760.

CHAPTER 2

DNA-encoded Chemistry and Its Use in Discovering Chemical Probes

ANTHONY D. KEEFE*

X-Chem Pharmaceuticals, 100 Beaver Street, Waltham, MA 02453, USA
*E-mail: akeefe@x-chemrx.com

2.1 Introduction to DNA-encoded Chemical Synthesis

DNA-encoded libraries are comprised of a plurality of individual chimeric molecules each containing both a chemically synthesized structure and an oligonucleotide with a sequence that encodes that structure. Each member of a DNA-encoded chemical library contains a combination of chemical building blocks covalently bonded to each other according to a defined chemical scheme that is also covalently attached to an encoding oligonucleotide. Individual compounds within such libraries may be identified even when they are mixed together. This is accomplished by the determination of the sequences of the oligonucleotides to which they are conjugated.

DNA-encoded chemical libraries contain vast mixed collections of small molecules each comprised of building block combinations reacted together in defined chemical schemes. Each individual library member is displayed upon an encoding DNA oligonucleotide. The successive introduction of

Chemical Biology No. 16
The Discovery and Utility of Chemical Probes in Target Discovery
Edited by Paul Brennan and Saleta Vazquez Rodriguez

Published by the Royal Society of Chemistry, www.rsc.org

individual oligonucleotides by ligation permits the encoding of building block and chemical scheme information directly into the sequence of the DNA attached to the small molecule. This approach renders the chemical structures of individual members of such libraries discoverable by the cloning and subsequent sequencing of the encoding DNA. This in turn allows the discovery of rare ligands, hits or probes contained within these libraries by the processes of affinity-mediated selection for binding to an arbitrary target of interest, usually a protein, followed by cloning and sequencing. Amplification of the encoding DNA by PCR is also supported and greatly facilitates the handling of the small amounts of DNA that exist in screen outputs. Figure 2.1 shows a representation of a single DNA-encoded small molecule representing a single member of such a library and indicates the relationship between the various concatenated sequence elements and the small molecule that they encode. For this generic individual member of a DNA-encoded chemical library, three successive chemical steps introduce three different building blocks that together comprise the encoded small molecule.

The successively ligated individual oligonucleotide tags are also indicated along with the type of information that each encodes. Terminal constant sequences allow for primer binding during PCR amplification. The library identifier permits different libraries to be combined for affinity-mediated enrichment together. The experiment identifier permits different experiments to be pooled and sequenced together to take advantage of the depth of sequencing offered by modern sequencing platforms. The degenerate sequence tag permits the identification of amplification duplicates to normalize for amplification biases.

DNA-encoded chemical libraries may be generated using a range of different methods, but there are many commonalities between all of these methods. In all cases the chemical history of each individual library member is recorded in the sequence of the DNA to which it is covalently attached. Additional information such as experiment identity may also be recorded to permit efficient utilization of the sequencing platform. Descriptions of various methods of obtaining DNA-encoded chemical libraries are provided

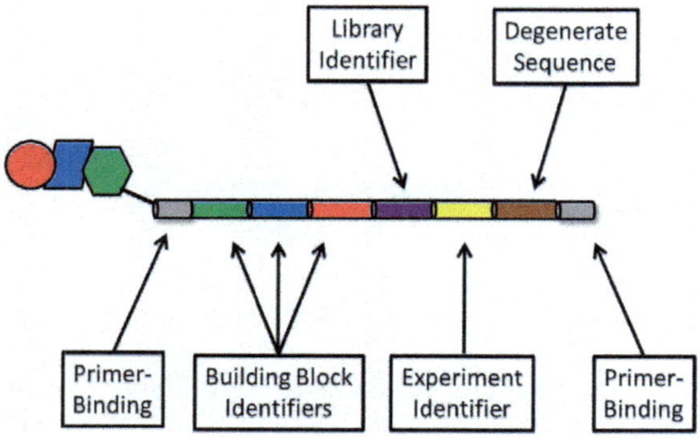

Figure 2.1 Generic individual member of a DNA-encoded chemical library.

in Figures 2.2 and 2.3. DNA-encoded chemistry takes advantage of DNA cloning and the depth of DNA sequencing available using platforms such as Roche/454[1] and Illumina/Solexa.[2] Cloning is a process whereby single DNA molecules are spatially isolated and amplified to give multiple copies of the same sequence, thereby permitting signal amplification during the process of determining the sequence of one DNA molecule. This process can be achieved by a number of different technical means and most importantly permits the determination of the sequences of individual DNA molecules even when they exist in a mixed population. When DNA is used to encode chemistry, this enables delivery upon the early promises of combinatorial chemistry[3] by rendering complex mixtures of small molecules deconvolutable and understandable. When cloning and sequencing are conducted at scale, large numbers of individual sequence reads can be determined and large populations can begin to be understood through statistical methods. This approach can then be used to understand the very large numbers of compounds that emerge from affinity-mediated screens with libraries of DNA-encoded small molecules, thereby supporting downstream decision-making processes, including choosing compounds for resynthesis and subsequent optimization.

Different methods exist for the generation of DNA-encoded small-molecule libraries and these may be broadly divided into two classes, DNA-recorded synthesis and DNA-directed synthesis, of which DNA-recorded synthesis is the most widely practiced variant.[4] In both approaches, combinatorial synthesis schemes are designed and tested for their applicability to on-DNA synthesis in that conditions are found which permit the successive installation of building blocks onto DNA to generate assemblages of multiple covalently joined building blocks without causing DNA damage or modification. Any chemical scheme that supports the covalent reaction of building blocks with each other and does not damage or modify DNA can be used to generate a library of DNA-encoded small molecules. Such a scheme can then be adapted to DNA-recorded synthesis by sourcing large numbers of building blocks for incorporation into the library. Efforts should then be invested to optimize the yield of each individual chemical step that will be used to synthesize the library and to

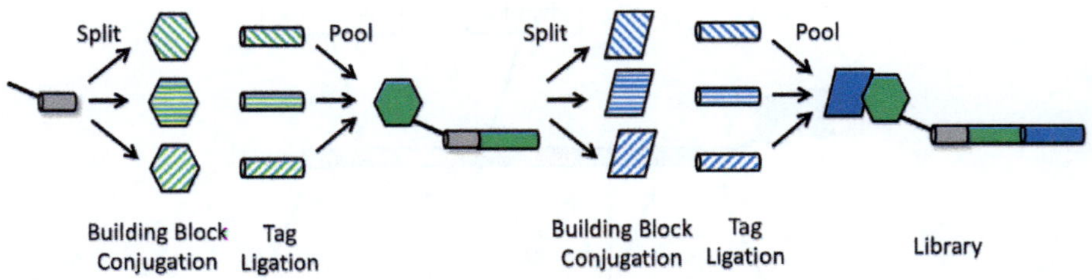

Figure 2.2 DNA-recorded chemical synthesis methodology. Sequence of events for the synthesis of a two-cycle DNA-recorded chemical library with chemistry and encoding steps performed upon individual "splits" before subsequent pooling.

validate individually every building block that will be incorporated and ensure that the yield for each surpasses a minimum acceptable threshold, for example 50%. Once this has been accomplished and the tag architecture and tag design process has been completed and suitable oligonucleotide reagents have been acquired, the synthesis of the DNA-encoded library can begin.

For DNA-recorded library synthesis, the encoding strategy is designed such that individual encoding tags can be successively installed by either enzymatic[5] or chemical[6] ligation methodologies. Library synthesis is then undertaken using a "split-and-mix" strategy that is initiated by placing aliquots of a chemically functionalized DNA oligonucleotide, the "headpiece", into individual wells of multi-well plates, followed by two successive operations within each well. Individual building blocks for the first cycle of chemistry are added and reacted under the previously established conditions and individual encoding oligonucleotides are added and ligated. When both reactions are complete, the contents of the wells are pooled and aliquots are again distributed within the wells of multi-well plates and a second chemical reaction and oligonucleotide ligation step is conducted to install the second set of building blocks and encoding tags. Purification or buffer exchange activities may need to be conducted both within the individual wells and with the pool in order to permit the successive steps to be accomplished and the fidelity of the library construction process to be maintained at a high level. DNA-recorded libraries are generally constructed with two or three successive chemical steps such that the final library contains encoded chemical structures comprised of two or three building blocks covalently attached to each other. Larger numbers of chemical steps have also been reported to have been used to construct DNA-recorded chemical libraries.[7] In all cases, at the conclusion of the library synthesis process the library will contain all combinations of the individual building blocks joined together in a defined chemical scheme with each building block combination encoded by a unique combination of oligonucleotide tags. Figure 2.2 outlines this overall approach. It should be acknowledged that the vast majority of DNA-recorded library experiments will use pre-existing libraries rather than require new ones to be synthesized because individual DNA-recorded library synthesis projects will generate enough physical library material to support thousands of screens.

An alternative set of methods for the creation of DNA-encoded chemical libraries exist that rely upon the hybridization of DNA sequences to each other in order that encoding relationships are established between tag sequences and building block identities (DNA-directed synthesis of DNA-encoded chemical libraries). There are several variants on this theme, many of which are depicted in Figure 2.3. Most of these methods are initiated by the synthesis of a library of templates that is generated by a combinatorial process that produces all combinations of the defined tag sequences within it. For example, individual tags may be synthesized with distinct invariant flanking sequences that will define their location relative to other tags. These are then combined and allowed to hybridize to each other before ligation or polymerase-mediated extension with thermal cycling yields the full-length template flanked at each terminus by invariant primer-binding regions for amplification by PCR. Templates created in such a fashion have a number of "codons" distributed throughout the template at defined positions and with defined variation permitted. An example of such a template is shown in Figure 2.3A. Templates with this general design may then be used to enable the synthesis of encoded chemical libraries either by routing them to specific compartments where specified chemistry steps may be performed, as in the case of DNA-routed synthesis,[8] as shown in Figure 2.3B, or complementary oligonucleotides displaying encoded building blocks may hybridize and thereby permit chemical reaction to occur based upon proximity to the relevant template, as occurs in DNA-templated synthesis as shown in Figure 2.3A.[9] The YoctoReactor®, as developed and utilized by Vipergen, is a similar hybridization-mediated proximity-driven template-based system except that the proximity of the species that are intended to react with each other is driven by the three-dimensional architecture of the system rather than sequence adjacency on the template.[10] A related library synthesis methodology that was originated in the Neri group at ETH is termed encoded self-assembling chemistry (ESAC), in which pairs of encoded building blocks are each displayed on separate single-stranded oligonucleotides and held in close proximity by hybridization of a non-encoding region as shown in Figure 2.3C. Each oligonucleotide separately contains an encoding sequence that identifies the building block that it displays. Polymerase-mediated extension of one strand using its hybridization partner as a template transfers the encoding information from one strand to the other and results in both encoding sequences being present in the extended strand, thereby permitting identification of each proximal pair by sequencing.[11]

Although each of these approaches can be used to generate libraries of DNA-encoded small molecules, each approach has its advantages and disadvantages. In-solution hybridization-mediated approaches such as ESAC, DNA-templated synthesis and the YoctoReactor® allow library synthesis to be achieved in a single compartment without the need to perform different chemical steps in spatially separated compartments, but each requires that individual building blocks must themselves first be conjugated to individual encoding oligonucleotides, in distinct compartments. Strategies that utilize capture methods such as DNA-routed synthesis generate very small amounts of material because of the limitations of operating in two-dimensional space. DNA-recorded synthesis requires the use of separate compartments for each and every synthesis step with each and every building block, but generates sufficient material for thousands of discovery campaigns. Many hybridization-mediated approaches permit amplification of the chemically synthesized entity, but reported library sizes for therapeutic or probe-relevant chemical space are not large enough to require that this be utilized. Lastly, hybridization-mediated approaches support limited split sizes because of the need to limit encoding tag sequences to a mutually orthogonal set of hybridizing sequences that inappropriately hybridize at a low rate, whereas DNA-recorded synthesis can utilize tag sequences that need only be distinguished by sequencing, a much lower bar that permits individual split sizes as high as in the multiple thousands.[12] Advantages and disadvantages of these different approaches to DNA-encoded chemistry tagging strategies are also listed in Table 2.1.

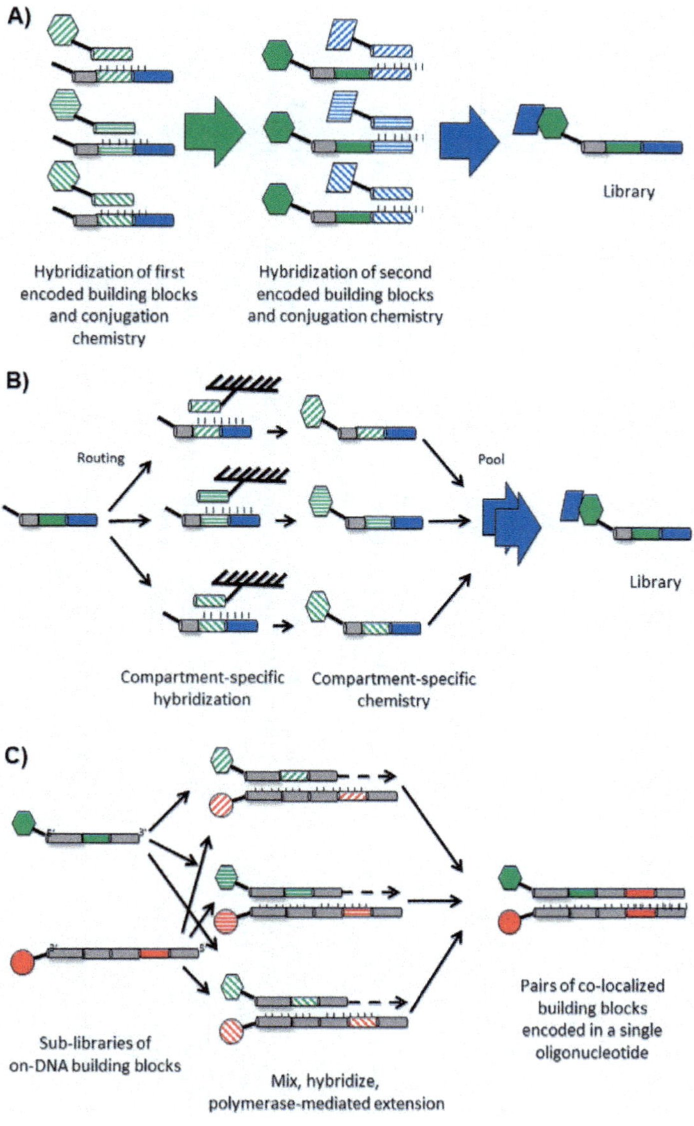

Figure 2.3 Various distinct methodologies are depicted for the generation of DNA-encoded chemical libraries. (A) Template-directed chemistry in which on-DNA building blocks are aligned for encoded chemical reaction by hybridization in a single-pot reaction. (B) Template-mediated routing in which pre-existing templates are routed to distinct compartments in which different chemical conditions occur that result in a succession of encoded chemical steps. (C) Template-mediated co-localization (alternatively, ESAC) in which pairs of on-DNA building blocks are co-localized according to hybridization preferences; enzymatic extension of one DNA strand using the other strand as a template transfers the encoding information of the first building blocks into a single strand already encoding the second building block, thereby recording the co-localization event.

Table 2.1 Different methodologies for the generation of DNA-encoded chemical libraries and their relative advantages and disadvantages.

Tagging strategy	Tagging method	Advantages	Disadvantages
DNA-recorded chemical synthesis	Ligation of tags and synthesis in separate compartments	Large quantities of library generated. Large split sizes. Building blocks utilized without pre-conjugation to DNA	Synthesis occurs in as many as 10 000 compartments. Does not support amplification of chemistry
DNA-templated chemical synthesis	Chemistry occurs owing to hybridization-mediated proximity of DNA-conjugated building blocks	Synthesis occurs in a single compartment. Supports amplification of chemistry	Each building block needs to be pre-conjugated to DNA. Small split sizes
DNA-directed chemical synthesis	Chemistry occurs owing to hybridization-mediated routing of DNA templates to specific compartments	Compartmentalization is automatic. Supports amplification of chemistry	Small quantities of library generated. Small split sizes
Encoded self-assembly	Programmed spatial proximity owing to hybridization of DNA-conjugated building blocks	Library spontaneously associates in a single compartment	Building blocks are not covalently attached to each other. Each building block needs to be pre-conjugated to DNA

The successful outcome of a campaign to discover new probes from one or more DNA-encoded chemical libraries is highly dependent upon the chemical schemes and building blocks that are utilized in their construction. Ideally, discovery campaigns will be conducted with a large number of libraries each with a different DNA-encoded chemical scheme and each utilizing a large number of diverse building blocks. This will provide a wide range of different chemical structures displayed in a range of topologies. Attractive hits or probes will derive from libraries that contain large numbers of individual compounds exhibiting attractive properties. Productive library schemes will utilize low-mass building blocks that are available in large numbers and can participate in chemical schemes that lead to a diversity of spatial presentations within defined chemical products while maintaining low numbers of invariant atoms, hydrogen-bond acceptors and hydrogen-bond donors and a low topological polar surface area. These last three parameters increase the chances that individual library members are cell permeable. The optimal design of such libraries is highly challenging and is best left to expert practitioners; a range of successfully utilized on-DNA library chemistries may be found in Salamon *et al.*,[13] but this is an area of active research and new on-DNA chemistries are currently appearing in the literature on an ongoing basis. DNA-encoded libraries fall into a range of overall architectures that are outlined in Figure 2.4, but the effectiveness of each depends upon the ability

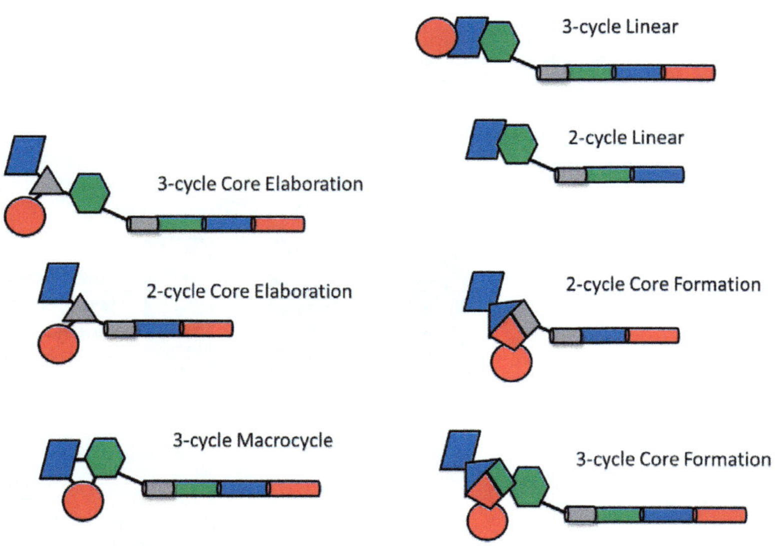

Figure 2.4 Library architectures supported by DNA-encoded chemistry. Each strategy is defined by its topology and this is in turn supported by a combination of mono-, bi- and trifunctional building blocks. Each topological scheme can be accessed by multiple different chemistries subject to acquisition of appropriately functionalized building blocks in suitable numbers and the design of conditions under which they can be efficiently reacted.

to design efficient chemical schemes to create them and to source and utilize large numbers of the mono-, di- and trivalent building blocks required for their synthesis.

Because DNA-encoded chemical libraries are comprised of large numbers of different chemical entities mixed together, they can only be used in screening approaches that result in the spatial separation of library members from each other. This is most usually accomplished by affinity-mediated methods in which the target is immobilized on a solid surface in the presence of the library and subsequent washing steps remove most non-binders while retaining binders. Although members of the library that bind to the target are also lost during this process, they are enriched compared with the non-binding members and thereby rendered identifiable. The library may be incubated with the protein in solution prior to the capture of the target or directly with a pre-immobilized target. A wide range of capture methodologies have been used, including with affinity tags designed into recombinant proteins[5] or by using chemically or enzymatically biotinylated proteins or by covalent immobilization.[14] After the washing steps have been completed, the retained library members are eluted, most usually by the thermal denaturation of the target protein, and these may then be further enriched by additional cycles of affinity-mediated screening with fresh aliquots of target protein. These steps are depicted in Figure 2.5.

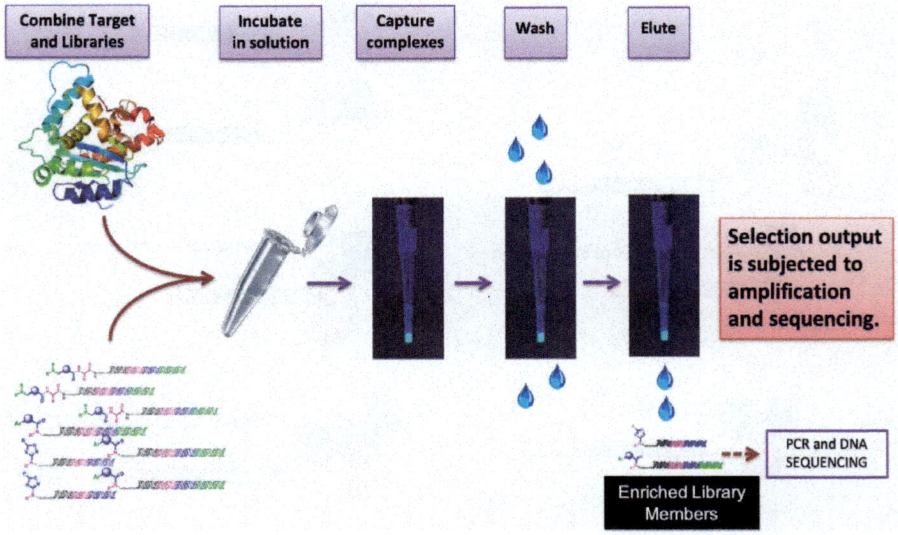

Figure 2.5 Affinity-mediated selection of encoded libraries. Sequence of events for the affinity-mediated selection of DNA-encoded libraries in which the protein target and libraries are combined and incubated together in solution followed by capture on an affinity matrix, stringent washing and elution. The eluted enriched library may be subject to additional rounds of selection prior to amplification and sequencing.

The design of the screen should include experimentally determined parameters that maintain the proper conformation and function of the protein target while also supporting the stringent washing away of unbound library members. Standard biophysical techniques routinely applied to the assessment of protein function should be utilized such as SEC, DLS, TSA and SDS-PAGE.

Subsequent to elution of the retained library from the final round of affinity-mediated selection, the library is amplified and a sample is prepared for deep sequencing. The Illumina platform currently offers the most economical approach with a single lane on a HiSeq 2000 or 2500 instrument, providing around 250 million 100-base single-end reads for around US$2000. Once sequence data have been received, each read will need to be parsed into its individual encoding tags, as defined by the original tag architecture, and these are then assigned to individual building block combinations within individual chemical schemes along with experiment identification. Typically, data are also filtered for amplification duplicates and possibly other parameters such as enrichments observed in a parallel no-target control experiment. Statistical parameters are then calculated for all building block combinations, ideally including normalization to the naïve (unselected) library. Owing to the scale of this endeavour, these steps may need to be automated.

Once the output has been translated back into the chemical structures that the sequences represent, then enriched sets of structurally related chemical compounds may be identified along with an assessment of the variability that is tolerated at different parts of the chemotype. Compounds may then be designed for resynthesis and testing in biochemical, biophysical and cell-based assays depending on the goals of the project. An overview of different methods for the presentation and interpretation of enrichment data to support this process was presented by Goodnow *et al.*[4]

Discovery campaigns for hits or probes that utilize DNA-encoded chemistry have modest requirements for amounts of protein reagents because the entire screen can be conducted in less than 1 mL of solution in total, even for multiple rounds of selections conducted under multiple conditions. Also, DNA-recorded chemistry in particular results in the generation of amounts of libraries sufficient to support thousands of screens. As such, this technique appears particularly applicable to proposed schemes to generate probes to many of the proteins that exist in the human proteome[15]

Lastly, it is worth pointing out that small molecules capable of specifically engaging protein targets discovered from DNA-encoded chemical libraries are discovered in a linked form. During the screen, the small molecule is covalently attached to, and displayed on, a linker moiety that is in turn covalently attached to the encoding DNA. As such, applications that require linking of a probe to another functional element are readily supported. Examples include the conversion of probes into fluorescent reagents that can be used for homogeneous probe-displacement assays such as FRET[16] or direct imaging.[17] Other applications that require linkage of a target-engaging reagent to a second functional element also exist, for example the generation of bifunctional PROTAC® molecules that are able to engage simultaneously both a target of interest and an E3 ligase engager, thereby eliciting target ubiquitination and subsequent proteasome-mediated degradation. Applications such as these are supported by the knowledge of where a linker may be attached, that a particular design of linker is tolerated by the target of interest and that a synthetic methodology has already been developed to generate the relevant linked form efficiently.

2.2 Comparison with Other Methods

Chemical probes have been identified using the same broad range of methods that are utilized for hit finding to initiate drug discovery projects. The use of DNA-encoded chemical libraries in affinity-mediated screening may therefore be considered an alternative discovery method for chemical probes when compared with these. Different methods draw upon different kinds of compound collections and these are considered separately below.

Fragment-based discovery methods utilize small collections of low molecular weight compounds that are screened at very high concentration.[18,19] Typical fragment libraries have molecular weights below 230 Da and, because of their low molecular weight, generally yield hits of low affinity. These will then require significant medicinal chemistry efforts to be applied in order that larger molecules with the specificity and potency required to support their intended applications be discovered. Fragment libraries generally number in the tens to hundreds depending upon the molecular weight cut-off value applied to them. Fragment libraries must be screened using techniques that are capable of detecting weak interactions and at concentrations of target and/or fragment that enable weak interactions to manifest themselves. Interaction detection techniques suitable for the discovery of weak interactions include NMR,[20] SPR,[21] ITC,[22] TSA,[23] X-ray structure determination[24] and covalent tethering.[25] Fragment-based approaches have resulted in over 20 molecules that have reached clinical trials and some of these have become marketed products.[18,19] From a scalability perspective, the small library sizes mean that some of these approaches might be considered suitable for the discovery of probes to large numbers of proteins, but these screens tend to require large quantities of protein and significant optimization efforts may be needed to advance each hit fragment into a potent and selective probe.

High-throughput screening is perhaps the most widely used method for the discovery of hits for the initiation of drug discovery projects. High-throughput screening utilizes large collections of small molecules with higher molecular weights than are commonly utilized for fragment-based approaches. Because of their higher molecular weights, these collections may number into the millions. High-throughput screening compound collections are often proprietary to large pharmaceutical companies, but academic and accessible commercial collections also exist. High-throughput screening has also been adapted to probe discovery with a wide range of example methodologies.[26] This list includes a range of homogeneous luminescence- and fluorescence-based approaches along with scintillation proximity, protein complementation, direct absorbance and electrophysiology. Considerable effort needs to be invested in screen robustness and miniaturization in view of the large numbers of compounds that need to be screened in parallel. The undertaking of a large high-throughput screen requires a considerable investment of resources and, although suitable for the discovery of probes to individual targets, this approach may be difficult to use to support the discovery of probes to large numbers of targets. Phenotypic screening has also been applied to probe discovery,[27] as has virtual screening,[28] but neither appears to have been broadly utilized for this purpose.

DNA-encoded chemistry provides an alternative strategy to discover chemical probes compared with the methodologies outlined above. An alternative viewpoint is that DNA-encoded chemistry may complement other strategies. One approach that exemplifies this latter point is a report by GlaxoSmithKline in which different representations of the same chemotype for WIP1 phosphatase were discovered independently using high-throughput screening with a biochemical assay with a fluorescent readout and affinity-mediated selection of a DNA-encoded chemical library.[29]

Because DNA-encoded libraries are utilized in a mixed format and DNA-recorded chemical libraries may be synthesized in sufficient quantities to support thousands of screens, DNA-recorded chemistry appears to be much more readily scaled up to support projects directed at discovering probes to hundreds or thousands of different proteins. From a workflow perspective, it requires less resources to conduct the screen than it does to either source the protein or resynthesize and test the resultant compounds, so these latter processes become the rate-determining steps.

2.3 Examples of Challenging Targets

Defined narrowly, chemical probes are small organic compounds that can be used to assess the therapeutic potential of individual targets in a biologically relevant environment. The extent to which individual compounds are suitable for use as probes will depend on their ability to access and inhibit specifically and fully the desired target without affecting other targets or cellular processes. In short, useful probes must be cell permeable, specific or at least highly selective and able to inhibit fully the function of the target at moderate concentrations (see Chapter 1 for more details on chemical probe selection and validation). Evidence should exist that supports these assertions with a high degree of certainty and ideally a negative control should also be available – a molecule that is structurally related to the probe but that does not bind to or modulate the activity of the target. The exploration of selectivity (for example) in the medicinal chemistry literature appears alarmingly limited for most disclosed molecules with target modulation activities.[30] Probes need not fulfil other parameters, for example those associated with drug likeness, and generally need not be orally available or have well-characterized DMPK parameters. The majority of human proteins have no known ligands, as reported by Antolin *et al.* that only 11% of the proteome has been liganded.[30] Also, there is little consensus on what fraction of these ligands are suitable as probes except that the number is low. As discussed in Chapter 1, less than 4% of the proteome has probes and one-third of the proteome is protein "dark matter" about which nothing is known.[31] Antolin *et al.*[30] concluded that only 0.7% of reported active compounds for human targets fulfil the minimum probe requirements of potency, selectivity and permeability. As a consequence, the scenario in which a therapeutic target protein is known but has no known probe is commonly encountered.

For probe identification technologies to be of true utility, they will need to demonstrate the ability to discover probes for targets for which probes do not yet exist and demonstrate that the probes thereby discovered satisfy the definition as discussed above and listed in Table 2.2. Over the last few years, a wide range of successes have been reported for the discovery of new chemical matter that modulates target proteins using DNA-encoded chemistry. Lists of ligands discovered using DNA-encoded chemistry have been reported.[4,13,32] In common with reports of ligands or hits discovered using alternative methodologies to DNA-encoded chemistry, most of these reported ligands do not have sufficiently detailed accompanying reported characterization to determine whether they satisfy generally accepted criteria for use as chemical probes.

One example of a compound that fulfils the declared minimum criteria for a chemical probe that was found in a DNA-encoded library using affinity-mediated selection against a challenging target can be found listed as BAY-850 in the SGC chemical probes catalogue (https://www.thesgc.org/chemical-probes/BAY-850). The discovery and subsequent characterization of this compound were reported in 2017[33] as part of a collaboration between X-Chem Pharmaceuticals, where the original hit compound was identified, and Bayer, where the compound series was characterized and optimized along with the identification of the negative control compound, BAY-460. An account of the discovery and characterization of this probe is described below.

ATPase family AAA domain-containing protein 2 (ATAD2) is an epigenetic reader that binds to acetyllysine using its bromodomain.[34] High expression levels of ATAD2 have been reported to be correlated with poor patient prognosis for multiple different tumour types, although the mechanism by which

Table 2.2 Chemical probe requirements according to the Structural Genomics Consortium (SGC).

Chemical probe requirement
In vitro potency of <100 nM
>30-fold selectivity *versus* other subfamilies
Significant on-target cell activity at 1 μM

these effects occur is not well understood.[35,36] At the time when this project was initiated and the first hit identified in late 2013, there were no reported small-molecule modulators for this target and bromodomain-containing targets have been reported to be a challenging target class for any kind of ligand discovery effort.[37] The ATAD2 protein was expressed in *Escherichia coli* as a recombinant GST fusion and was shown to interact with acetylated peptide substrates both with a single acetylated lysine and with four acetylated lysines, and TR-FRET assays were established using labelled versions of these peptides. These assays were used to qualify the reagent in solution and when immobilized on a glutathione matrix as well as a primary assay for putative inhibitors by assessing the ability of small molecules to inhibit the bromodomain from interacting with the peptide. The protein was characterized by a range of biophysical techniques, including SEC, TSA and NMR using a ^{15}N-labelled version, and shown to be a soluble dimer, as expected for a GST-tagged protein, with a single thermal transition when cleaved and purified from the GST. The protein was demonstrated to capture efficiently on an immobilized glutathione matrix and to exhibit activity in this state as assessed using the TR-FRET assay.[33]

Affinity-mediated selection with a combined set of 11 different DNA-recorded chemical libraries containing over 65 billion distinct unique chemical compounds was initiated by incubating the combined libraries in solution with the target followed by capture on a glutathione affinity matrix, washing and thermal elution to liberate bound library members. After a second cycle of selection initiated with fresh protein, the DNA encoding the eluted library members was amplified using PCR, clustered and sequenced using an Illumina 2500 HiSeq instrument in high-output mode. The identified sequences were translated back into building block, library scheme and experiment identifiers and statistical parameters were calculated, thereby allowing individual enriched compounds to be identified.

The DNA-recorded chemical library from which BAY-850 was derived was synthesized using three distinct chemical steps. Initially, 300 amino acids were installed upon a primary amine-functionalized synthetic DNA oligonucleotide, followed by acetylation with 150 formyl acids, followed by reductive amination with 2300 amines to yield 103.5 million uniquely encoded individual compounds. Each step was individually encoded by ligation of encoding tag oligonucleotides ultimately followed by the installation of a library-encoding tag and terminal constant sequence to permit co-amplification and subsequent identification when combined with other libraries.

The sequence information derived from the selection output was parsed into individual libraries and the individual building blocks within them and statistical information was calculated for all building block combinations. This process included steps directed at the removal of error-containing sequences, the removal of PCR amplification duplicates and the calculation of enrichment metrics with normalization to the naïve (unselected) library. Analysis of the output of this library showed the co-enrichment of a set of structurally related compounds defined by a single formyl acid containing the biaryl moiety

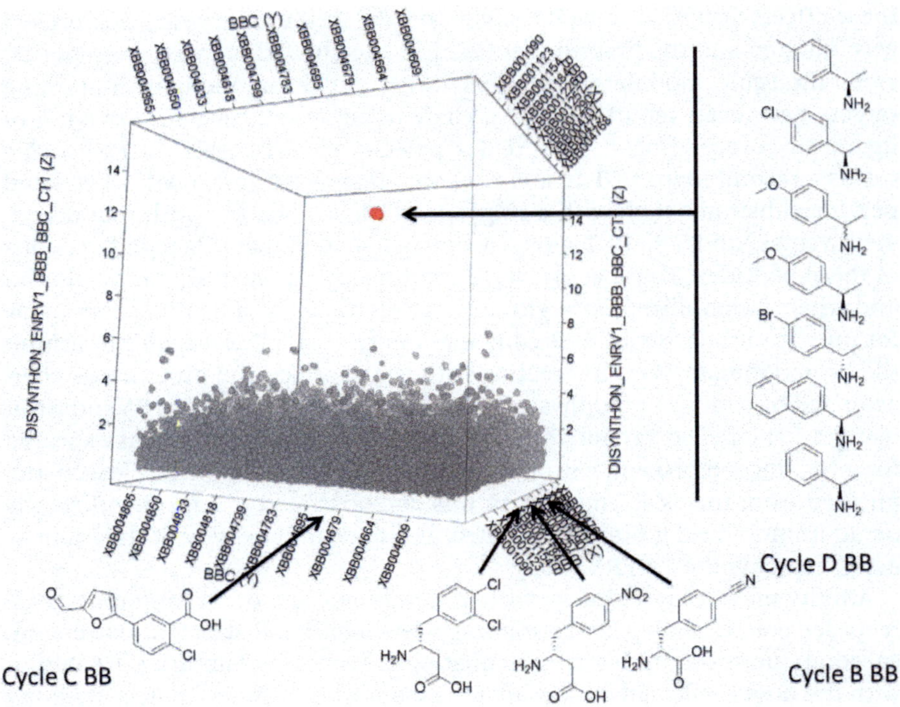

Figure 2.6 Cubic scatter plot showing the co-enrichment of individual building block combinations derived from a three-cycle DNA-recorded chemical library; 300 cycle B amino acids and 150 cycle C formyl acids are represented on the *x*- and *y*-axes, respectively, and the *z*-axis represents a statistical enrichment metric for each of the 45 000 pairs calculated using the negative $\log_{10}$ of the asymptotic significance value and normalization to the naïve library. Co-enriched building block combinations show structural relatedness, thereby exemplifying individual variants of the chemotype that binds to the target. Individual building blocks existing within the enriched individual compounds are indicated.

depicted in Figure 2.6 as the cycle C building block. Enriched compounds displayed ranges of chemically related building blocks at both the proximal (B) and distal (D) chemical cycles with a set of mostly *para*-substituted α-methylbenzylamines at the former and *para*-substituted phenylalanines and homophenylalanines at the latter, as depicted in Figure 2.6.

The synthesis of the most highly enriched individual building block combination from this cluster gave compound 1 shown in Figure 2.7. Compound 1 was synthesized and determined to have an IC_{50} in the single-digit μM range using the TR-FRET assay. Subsequent efforts at Bayer to improve the potency of this hit were undertaken and resulted in the discovery of the probe BAY-850 and also a structurally related inactive compound BAY-460. BAY-850 displaces a tetraacetylated peptide from ATAD2 with an IC_{50} value of 22 nM as assessed by TR-FRET, whereas BAY-460 was inactive when assessed in the same assay at concentrations of up to the low μM range. Similar results were observed for BAY-850 in a range of orthogonal assays including probe

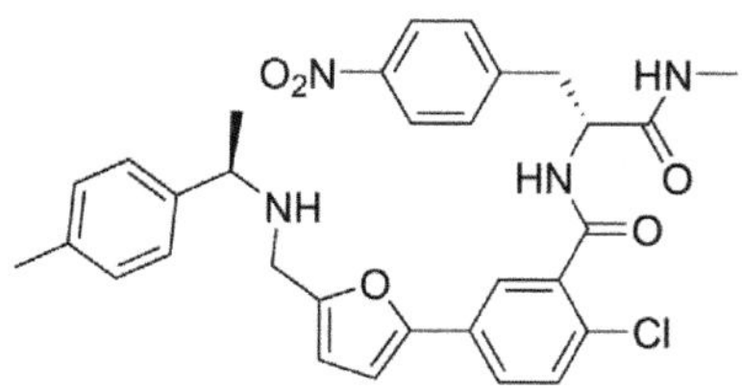

Compound 1
Initial hit

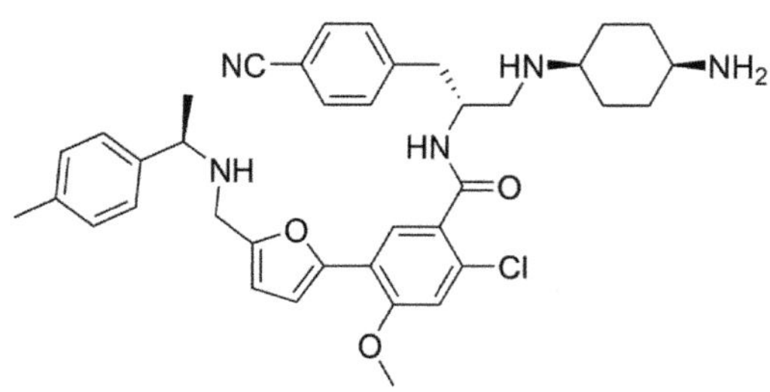

BAY-850

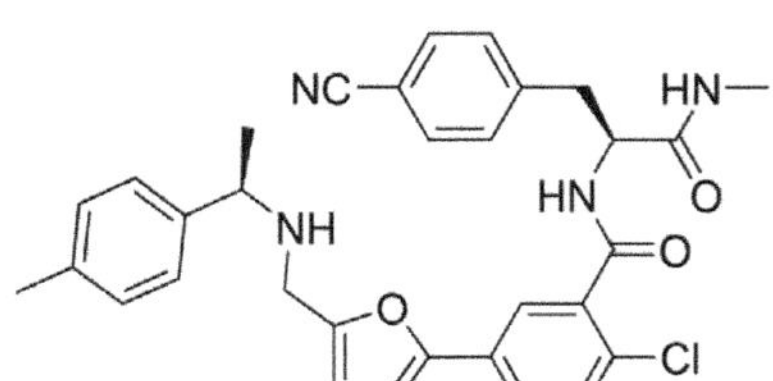

BAY-460
Negative Control

Figure 2.7 Chemical structures of the hit (compound 1), probe (BAY-850) and negative control probe (BAY-460) for ATAD2 discovered using DNA-recorded chemistry.

displacement assays including BROMOscan® and AlphaScreen® and direct binding assays including NMR, MST and TSA. The TSA and BROMOscan® assays were also used to assess the specificity of BAY-850 with regard to a range of close homologues, and in both cases indicated apparent complete specificity for all proteins tested at 10 μM, including the most homologous such as ATAD2B.[33] The TSA data generated using untagged ATAD2 also confirmed that the interaction was not mediated by the GST. Further testing revealed no inhibitory activity against any of a panel of 354 kinases. The specificity of this interaction, unprecedented for a bromodomain inhibitor, led the authors to consider the possibility of a novel mode of action. Both SEC and native MS conducted with untagged ATAD2 indicated that ATAD2 is exclusively dimeric in the presence of BAY-850 and exclusively monomeric in its absence. Native MS was also used to show that no further multimerization of dimeric GST-ATAD2 was observed in the presence of BAY-850 and that interaction with the acetylated peptide was inhibited by dimerization. No dimerization-inducing effects were observed for the control molecule, BAY-460.

Both BAY-850 and BAY-460 were determined to be sufficiently cell permeable to support the assessment of target engagement in a cell assay and the fluorescence recovery after photobleaching (FRAP) assay was used to measure the ability of these compounds to displace full-length ATAD2 from chromatin in a human tumour cell line (MCF7). Using this technique, BAY-850 maximally displaces ATAD2 from chromatin at a concentration of 1 μM compared with a mutated form of the target that is unable to engage chromatin. As such, this compound satisfies the SGC definition of a chemical probe. BAY-460 is inactive in the same assay. In an effort to support the further study of this target, BAY-850 and BAY-460 are available as specific well-characterized probes to the scientific community from the SGC and elsewhere.

This account of the discovery and characterization of BAY-850 as a chemical probe for ATAD2 is presented here as an example of the application of DNA-recorded chemistry to the discovery of chemical probes for targets for which no ligands are known. In this example, a simple library was created by installing 300 amino acids upon a chemically functionalized DNA oligonucleotide followed by further successive elaborations with acetylation using 150 formyl acids and reductive amination using 2300 amines yielding 103.5 million individually DNA-encoded distinct compounds. This mixture of compounds was then rapidly interrogated in parallel using an affinity-mediated screen followed by the amplification of the encoding DNA of target-associated library members and deep sequencing on the Illumina platform. These steps directly led to the identification of this series of novel highly selective inhibitors of ATAD2 that elicit their effects by the induction of a dimeric form of the target with a reduced propensity to associate with acetylated lysine substrates. This ATAD2 inhibitor fully satisfies the generally accepted criteria for a chemical probe and does so for a challenging target for which there were no known inhibitors at the time at which the screen was conducted. Other reported examples of DNA-encoded chemistry-derived compounds where the extent of reported characterization is less extensive or the target is not considered challenging also exist. Select examples of these are described below. Also shown in Table 2.3 are the structures of three compounds to three distinct protein targets that were discovered using DNA-encoded chemistry and were subsequently progressed to the point where they were approved for evaluation in humans as potential therapeutics.

GPCRs are membrane-embedded proteins for which a wide range of small-molecule agonists and antagonists are known, including many that are utilized for therapeutic purposes. Nonetheless, GPCRs are considered inherently challenging targets and especially so for discovery approaches that utilize the preparation of functional purified recombinant protein reagents such as affinity-mediated selection with DNA-encoded chemical libraries. Three different successful DNA-encoded chemistry GPCR discovery projects have been reported. These were conducted by three independent groups, each of which used a different approach to reagent generation and they are described in turn below.

GlaxoSmithKline used a strategy in which affinity-mediated selections were conducted directly against the NK3 GPCR displayed upon living cells.[38] In this study, NK3 was highly overexpressed and was determined to have a

Table 2.3 Structures of compounds discovered using DNA-encoded chemistry and subsequently progressed to the point where they were approved for evaluation in humans as potential therapeutics.

Target name	DECL hit structure	Clinical candidate structure	Identifier or publication
Soluble epoxide hydrolase	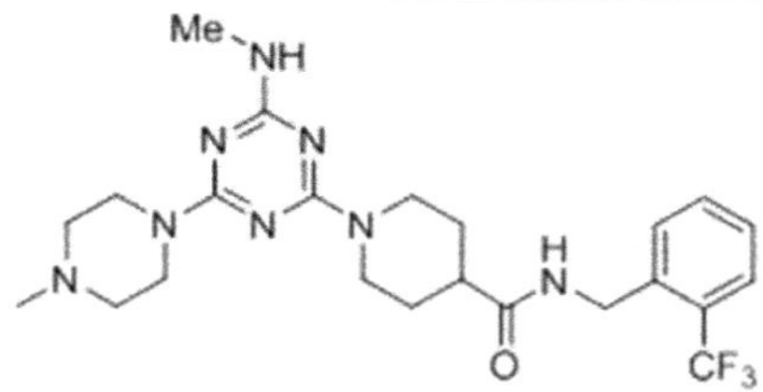	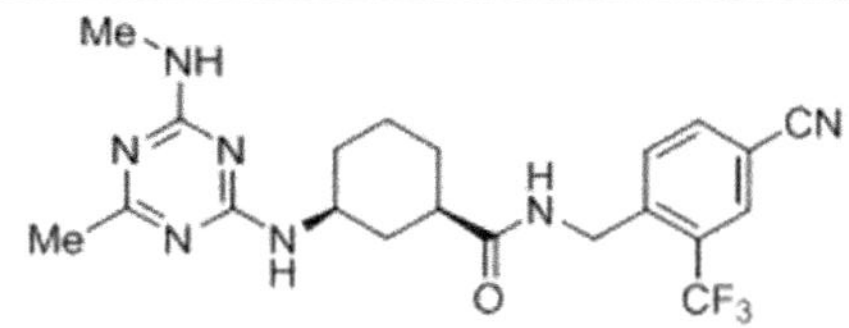	GSK2256294, NCT01762774
RIP1K	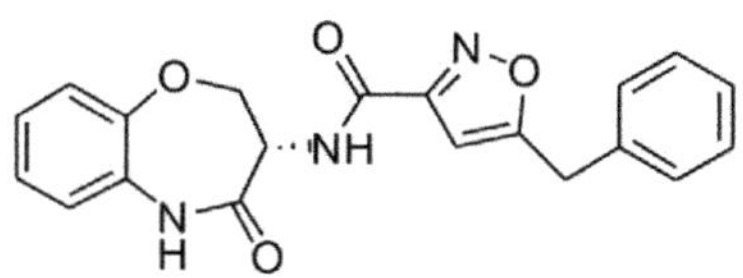	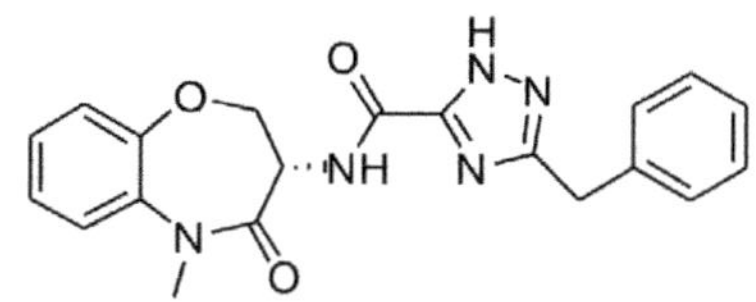	GSK2982772, NCT02776033
Autotaxin	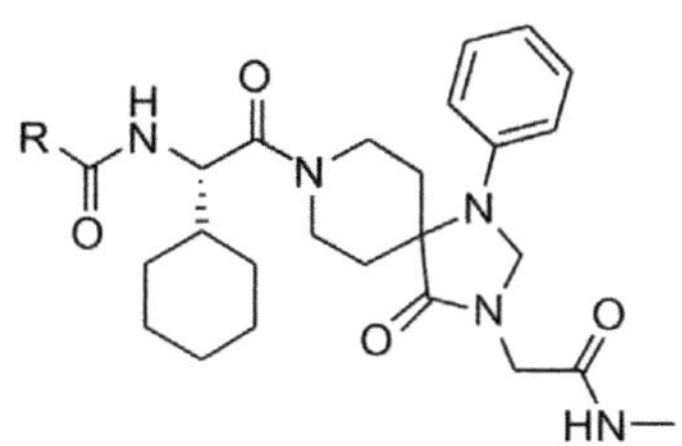	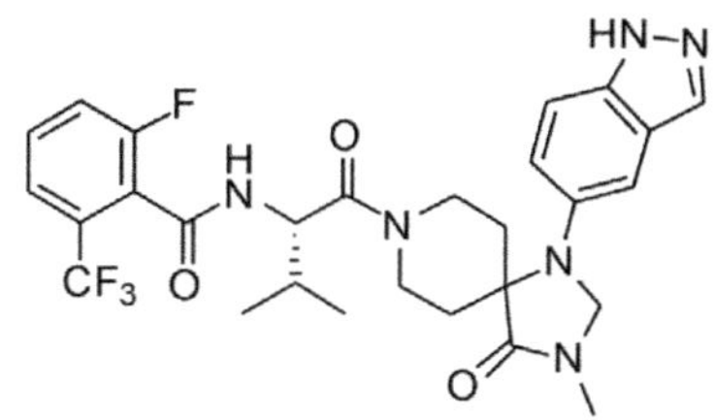	Ref. 39

copy number of approximately 500 000 upon each HEK293 cell utilized in the selection experiment. It is likely that overexpression levels of this magnitude are important for cell-based selections. A cell line not expressing NK3 was utilized in a parallel selection as a negative control. This strategy ultimately resulted in a range of antagonists for NK3 with IC_{50} values as low as 1 nM and at least 1000-fold selectivity over each of NK1 and NK2, satisfying the generally accepted criteria for use as a chemical probe.

X-Chem Pharmaceuticals, working in collaboration with both AstraZeneca and Heptares, used a StaR® strategy to generate the PAR2 GPCR in a thermally stabilized mutant form in which the protein is sufficiently stable that it can achieve an active conformation in a detergent-stabilized environment. This enabled PAR2 to be utilized in a soluble form independent of an extended lipid bilayer. In this study, the authors reported both agonists and antagonists of PAR2 and showed them to be selective over both PAR1 and PAR4.[40,41] This compound also satisfies generally accepted criteria for use as a chemical probe. Notably, the antagonist was found to bind to a novel allosteric site within a membrane-embedded part of the target.

Nuevolution, working in collaboration with Duke University, used a nanodisc strategy in which the recombinant GPCR was embedded in a stabilized lipid raft that renders it soluble while retaining the GPCR in its membrane-embedded immediate environment. In this study, the authors reported a compound that binds to the GPCR target β_2-adrenergic receptor (β_2AR).[42] Cell-based functional assays showed inhibition of β_2AR agonist-promoted Gs-mediated cAMP generation and also β-arrestin recruitment to the receptor. No selectivity information was provided.

A range of compounds against various antibacterial targets have also been reported. Machutta *et al.*[43] reported parallel selections with a range of DNA-recorded chemical libraries against a range of antibacterial targets and some cell activity data for some of these was reported, including a sub-nM active (biochemical IC_{50} value) against MRS for *Staphylococcus aureus* (MIC 0.5 μg mL^{-1}). Selectivity data were not provided. Soutter *et al.*[44] reported parallel selections against the TB target InhA in a range of forms, including apo, NAD-bound and NADH-bound with actives found in the nM range for biochemical assay and MIC values as low as single-digit μM; again, selectivity data were not provided.

Lastly, several reports of the use of DNA-encoded chemistry to find highly potent cell-active inhibitors to a range of targets have been reported for targets that are not generally considered challenging. Cuozzo *et al.*[45] reported a BTK inhibitor discovered from a DNA-recorded chemical library with a sub-nM IC_{50} in biochemical assay and a 2 nM IC_{50} in Ramos cell assay. Some selectivity information was also provided in that 15-fold selectivity was observed for Jurkat cells for which T-cell activation is not BTK-driven. This compound may therefore be suitable as a chemical probe. GlaxoSmithKline have reported several inhibitors discovered through DNA-encoded libraries, such as RIPK3, she and PDE12 inhibitors. RIP3K inhibitors showed[46] a biochemical IC_{50} value of 300 pM that was active in cell assay and when tested at 1 μM failed to inhibit most of 300 kinases tested in a kinase panel. Also, an sEH inhibitor with a biochemical IC_{50} value in the pM range[47,48] inhibits the conversion of 14,15-EET to 14,15-DHET in human, rat and mouse whole blood *in vitro*, is active in animal models of lung inflammation and has entered human clinical trials. PDE12 inhibitors that exhibit inhibition of PDE12 in a biochemical assay with a K_i value of 18 nM, >500-fold selectivity over the related target CNOT6 and antiviral activity comparable to that observed using gene inactivation have also been reported.[49] GlaxoSmithKline have also reported inhibitors of RIP1 kinase that satisfy the generally accepted criteria for chemical probes.[50] The reported compounds are highly potent in cell assays with IC_{50} values as low as 10 nM in U937 cells and a biochemical IC_{50} value of 1 nM and demonstrated complete specificity for RIP1 kinase when tested against 456 kinases at 10 μM.

Vipergen have reported highly potent P38α inhibitors discovered using their YoctoReactor® and BinderTrap® technologies.[51] These exhibit IC_{50} values of 7 nM in cell assays and in a kinase panel were observed to inhibit significantly only PI3Kα, PI3Kβ and KIT when tested against 99 kinases at 2 μM in biochemical assays. As such, this molecule satisfies the generally accepted criteria for chemical probes with regard to both cell assay potency and characterized biochemical selectivity.

In conclusion, DNA-encoded chemical library outputs have been sufficiently characterized in multiple instances for use as chemical probes including for challenging targets and at least one target for which no prior matter existed. Projects specifically designed to discover probes only require access to the libraries themselves and relatively small amounts of the protein targets. It is also relatively easy to envisage scaling this approach to access probes to large numbers of different targets should this be desired.

Abbreviations

ATAD2	ATPase family AAA domain-containing protein 2
β_2AR	β_2-Adrenergic Receptor
BTK	Bruton tyrosine kinase
cAMP	Cyclic adenosine monophosphate
CNOT6	CCR4-NOT transcription complex subunit 6
DECL	DNA-encoded chemical library
14,15-DHET	14,15-Dihydroxyeicosatrienoic acid
DLS	Dynamic light scattering
DMPK	Drug metabolism and pharmacokinetics
DNA	Deoxyribonucleic acid
E3	E3 Ubiquitin ligase
14,15-EET	14,15-Epoxyeicosatrienoic acid
ESAC	Encoded self-assembling chemistry
FRAP	Fluorescence recovery after photobleaching
FRET	Fluorescence resonance energy transfer
GPCR	G Protein-coupled receptor
GST	Glutathione *S*-transferase
ITC	Isothermal titration calorimetry

MIC	Mean inhibitory concentration
MRS	Methicillin-resistant *Staphylococcus*
MS	Mass spectrometry
MST	Microscale thermophoresis
NAD	Nicotinamide adenine dinucleotide
NADH	Nicotinamide adenine dinucleotide, reduced form
NK3	Neurokinin 3
NMR	Nuclear magnetic resonance
PAR2	Protease-activated receptor 2
PDE12	Phosphodiesterase 12
PCR	Polymerase chain reaction
PROTAC	Proteolysis-targeting chimera
RIP3K	Receptor-interacting serine/threonine-protein kinase 3
SDS-PAGE	Sodium dodecyl sulfate polyacrylamide gel electrophoresis
SEC	Size-exclusion chromatography
SGC	Structural Genomics Consortium
sEH	Soluble epoxide hydrolase
SPR	Surface plasmon resonance
TB	Tuberculosis
TR-FRET	Time-resolved fluorescence resonance energy transfer
TSA	Thermal shift assay
WIP1	Wild-type p53-induced (phosphatase)

Acknowledgements

I would like to acknowledge X-Chem Pharmaceuticals for providing support for the writing of this chapter.

References

1. M. Margulies, M. Egholm, W. E. Altman, S. Attiya, J. S. Bader, L. A. Bemben, J. Berka, M. S. Braverman, Y.-J. Chen, Z. Chen, S. B. Dewell, L. Du, J. M. Fierro, X. V. Gomes, B. C. Godwin, W. He, S. Helgesen, C. H. Ho, G. P. Irzyk, S. C. Jando, M. L. I. Alenquer, T. P. Jarvie, K. B. Jirage, J.-B. Kim, J. R. Knight, J. R. Lanza, J. H. Leamon, S. M. Lefkowitz, M. Lei, J. Li, K. L. Lohman, H. Lu, V. B. Makhijani, K. E. McDade, M. P. McKenna, E. W. Myers, E. Nickerson, J. R. Nobile, R. Plant, B. P. Puc, M. T. Ronan, G. T. Roth, G. J. Sarkis, J. F. Simons, J. W. Simpson, M. Srinivasan, K. R. Tartaro, A. Tomasz, K. A. Vogt, G. A. Volkmer, S. H. Wang, Y. Wang, M. P. Weiner, P. Yu, R. F. Begley and J. M. Rothberg, *Nature*, 2005, **437**, 376–380.
2. D. R. Bentley, S. Balasubramanian, H. P. Swerdlow, G. P. Smith, J. Milton, C. G. Brown, K. P. Hall, D. J. Evers, C. L. Barnes, H. R. Bignell, J. M. Boutell, J. Bryant, R. J. Carter, R. Keira Cheetham, A. J. Cox, D. J. Ellis, M. R. Flatbush, N. A. Gormley, S. J. Humphray, L. J. Irving, M. S. Karbelashvili, S. M. Kirk, H. Li, X. Liu, K. S. Maisinger, L. J. Murray, B. Obradovic, T. Ost, M. L. Parkinson, M. R. Pratt, I. M. Rasolonjatovo, M. T. Reed, R.

Rigatti, C. Rodighiero, M. T. Ross, A. Sabot, S. V. Sankar, A. Scally, G. P. Schroth, M. E. Smith, V. P. Smith, A. Spiridou, P. E. Torrance, S. S. Tzonev, E. H. Vermaas, K. Walter, X. Wu, L. Zhang, M. D. Alam, C. Anastasi, I. C. Aniebo, D. M. Bailey, I. R. Bancarz, S. Banerjee, S. G. Barbour, P. A. Baybayan, V. A. Benoit, K. F. Benson, C. Bevis, P. J. Black, A. Boodhun, J. S. Brennan, J. A. Bridgham, R. C. Brown, A. A. Brown, D. H. Buermann, A. A. Bundu, J. C. Burrows, N. P. Carter, N. Castillo, E. C. M. Chiara, S. Chang, R. N. Cooley, N. R. Crake, O. O. Dada, K. D. Diakoumakos, B. Dominguez-Fernandez, D. J. Earnshaw, U. C. Egbujor, D. W. Elmore, S. S. Etchin, M. R. Ewan, M. Fedurco, L. J. Fraser, K. V. Fuentes Fajardo, W. S. Furey, D. George, K. J. Gietzen, C. P. Goddard, G. S. Golda, P. A. Granieri, D. E. Green, D. L. Gustafson, N. F. Hansen, K. Harnish, C. D. Haudenschild, N. I. Heyer, M. M. Hims, J. T. Ho, A. M. Horgan, K. Hoschler, S. Hurwitz, D. V. Ivanov, M. Q. Johnson, T. James, T. A. H. Jones, G. D. Kang, T. H. Kerelska, A. D. Kersey, I. Khrebtukova, A. P. Kindwall, Z. Kingsbury, P. I. Kokko-Gonzales, A. Kumar, M. A. Laurent, C. T. Lawley, S. E. Lee, X. Lee, A. K. Liao, J. A. Loch, M. Lok, S. Luo, R. M. Mammen, J. W. Martin, P. G. McCauley, P. McNitt, P. Mehta, K. W. Moon, J. W. Mullens, T. Newington, Z. Ning, B. L. Ng, S. M. Novo, M. J. O'Neill, M. A. Osborne, A. Osnowski, O. Ostadan, L. L. Paraschos, L. Pickering, A. C. Pike, A. C. Pike, D. C. Pinkard, D. P. Pliskin, J. Podhasky, V. J. Quijano, C. Raczy, V. H. Rae, S. R. Rawlings, A. C. Rodriguez, P. M. Roe, J. Rogers, M. C. Rogert Bacigalupo, N. Romanov, A. Romieu, R. K. Roth, N. J. Rourke, S. T. Ruediger, E. Rusman, R. M. Sanches-Kuiper, M. R. Schenker, J. M. Seoane, R. J. Shaw, M. K. Shiver, S. W. Short, N. L. Sizto, J. P. Sluis, M. A. Smith, J. E. S. Sohna, E. J. Spence, K. Stevens, N. Sutton, L. Szajkowski, C. L. Tregidgo, G. Turcatti, S. Vandevondele, Y. Verhovsky, S. M. Virk, S. Wakelin, G. C. Walcott, J. Wang, G. J. Worsley, J. Yan, L. Yau, M. Zuerlein, J. Rogers, J. C. Mullikin, M. E. Hurles, N. J. McCooke, J. S. West, F. L. Oaks, P. L. Lundberg, D. Klenerman, R. Durbin and A. J. Smith, *Nature*, 2008, **456**, 53–59.

3. A. Furka, F. Sebestyen, M. Asgedom and G. Dibo, *Int. J. Pept. Protein Res.*, 1991, **37**, 487–493.
4. R. A. Goodnow Jr, C. E. Dumelin and A. D. Keefe, *Nat. Rev. Drug Discovery*, 2017, **16**, 131–147.
5. M. A. Clark, R. A. Acharya, C. C. Arico-Muendel, S. L. Belyanskaya, D. R. Benjamin, N. R. Carlson, P. A. Centrella, C. H. Chiu, S. P. Creaser, J. W. Cuozzo, C. P. Davie, Y. Ding, G. J. Franklin, K. D. Franzen, M. L. Gefter, S. P. Hale, N. J. Hansen, D. I. Israel, J. Jiang, M. J. Kavarana, M. S. Kelley, C. S. Kollmann, F. Li, K. Lind, S. Mataruse, P. F. Medeiros, J. A. Messer, P. Myers, H. O'Keefe, M. C. Oliff, C. E. Rise, A. L. Satz, S. R. Skinner, J. L. Svendsen, L. Tang, K. van Vloten, R. W. Wagner, G. Yao, B. Zhao and B. A. Morgan, *Nat. Chem. Biol.*, 2009, **5**, 647–654.
6. A. D. Keefe, M. A. Clark, C. D. Hupp, A. Litovchick and Y. Zhang, *Curr. Opin. Chem. Biol.*, 2015, **26**, 80–88.
7. Z. Zhu, A. Shaginian, L. C. Grady, T. O'Keeffe, X. E. Shi, C. P. Davie, G. L. Simpson, J. A. Messer, G. Evindar, R. N. Bream, P. P. Thansandote, N. R. Prentice, A. M. Mason and S. Pal, *ACS Chem. Biol.*, 2018, **13**, 53–59.

8. D. R. Halpin and P. B. Harbury, *PLoS Biol.*, 2004, **2**, E173.
9. B. N. Tse, T. M. Snyder, Y. Shen and D. R. Liu, *J. Am. Chem. Soc.*, 2008, **130**, 15611–15626.
10. M. H. Hansen, P. Blakskjaer, L. K. Petersen, T. H. Hansen, J. W. Hojfeldt, K. V. Gothelf and N. J. Hansen, *J. Am. Chem. Soc.*, 2009, **131**, 1322–1327.
11. S. Melkko, J. Scheuermann, C. E. Dumelin and D. Neri, *Nat. Biotechnol.*, 2004, **22**, 568–574.
12. A. Litovchick, X. Tian, M. I. Monteiro, K. M. Kennedy, M. A. Guie, P. Centrella, Y. Zhang, M. A. Clark and A. D. Keefe, *Molecules*, 2019, **24**, 2026.
13. H. Salamon, M. Klika Skopic, K. Jung, O. Bugain and A. Brunschweiger, *ACS Chem. Biol.*, 2016, **11**, 296–307.
14. C. E. Dumelin, S. Trussel, F. Buller, E. Trachsel, F. Bootz, Y. Zhang, L. Mannocci, S. C. Beck, M. Drumea-Mirancea, M. W. Seeliger, C. Baltes, T. Muggler, F. Kranz, M. Rudin, S. Melkko, J. Scheuermann and D. Neri, *Angew. Chem., Int. Ed.*, 2008, **47**, 3196–3201.
15. A. Mullard, *Nat. Rev. Drug Discovery*, 2019, **18**, 733–736.
16. L. Yuan, W. Lin, K. Zheng and S. Zhu, *Acc. Chem. Res.*, 2013, **46**, 1462–1473.
17. L. M. Wysocki and L. D. Lavis, *Curr. Opin. Chem. Biol.*, 2011, **15**, 752–759.
18. D. A. Erlanson, S. W. Fesik, R. E. Hubbard, W. Jahnke and H. Jhoti, *Nat. Rev. Drug Discovery*, 2016, **15**, 605–619.
19. C. W. Murray and D. C. Rees, *Angew. Chem., Int. Ed. Engl.*, 2016, **55**, 488–492.
20. M. J. Harner, A.O. Frank and S. W. Fesik, *Journal of Biomolecular NMR*, 2016, **56**, 65–75.
21. A. Chavanieu and M. Pugnière, *Expert Opinion on Drug Discovery*, 2016, **11**, 489–99.
22. E. Ruhmann, M. Betz, M. Fricke, A. Heine, M. Schafer and G. Klebe, *Biochim. Biophys. Acta*, 2015, **1850**, 647–656.
23. J. K. Kranz and C. Schalk-Hihi, in *Methods in Enzymology*, 2011, **493**, 277–298.
24. J. Schiebel, S. G. Krimmer, K. Rower, A. Knorlein, X. Wang, A. Y. Park, M. Stieler, F. R. Ehrmann, K. Fu, N. Radeva, M. Krug, F. U. Huschmann, S. Glockner, M. S. Weiss, U. Mueller, G. Klebe and A. Heine, *Structure*, 2016, **24**, 1398–1409.
25. S. G. Kathman and A. V. Statsyuk, *MedChemComm*, 2016, **7**, 576–585.
26. J. Inglese, R. L. Johnson, A. Simeonov, M. Xia, W. Zheng, C. P. Austin and D. S. Auld, *Nat. Chem. Biol.*, 2007, **3**, 466–479.
27. I. M. Wallace, M. L. Urbanus, G. M. Luciani, A. R. Burns, M. K. Han, H. Wang, K. Arora, L. E. Heisler, M. Proctor, R. P. St Onge, T. Roemer, P. J. Roy, C. L. Cummins, G. D. Bader, C. Nislow and G. Giaever, *Chem. Biol.*, 2011, **18**, 1273–1283.
28. V. A. Kincaid, N. London, K. Wangkanont, D. A. Wesener, S. A. Marcus, A. Heroux, L. Nedyalkova, A. M. Talaat, K. T. Forest, B. K. Shoichet and L. L. Kiessling, *ACS Chem. Biol.*, 2015, **10**, 2209–2218.
29. A. G. Gilmartin, T. H. Faitg, M. Richter, A. Groy, M. A. Seefeld, M. G. Darcy, X. Peng, K. Federowicz, J. Yang, S. Y. Zhang, E. Minthorn, J. P. Jaworski, M. Schaber, S. Martens, D. E. McNulty, R. H. Sinnamon, H. Zhang, R. B. Kirkpatrick, N. Nevins, G. Cui, B. Pietrak, E. Diaz, A. Jones, M. Brandt, B. Schwartz, D. A. Heerding and R. Kumar, *Nat. Chem. Biol.*, 2014, **10**, 181–187.

30. A. A. Antolin, J. E. Tym, A. Komianou, I. Collins, P. Workman and B. Al-Lazikani, *Cell Chem. Biol.*, 2018, **25**, 194–205 e195.
31. T. I. Oprea, C. G. Bologa, S. Brunak, A. Campbell, G. N. Gan, A. Gaulton, S. M. Gomez, R. Guha, A. Hersey, J. Holmes, A. Jadhav, L. J. Jensen, G. L. Johnson, A. Karlson, A. R. Leach, A. Ma'ayan, A. Malovannaya, S. Mani, S. L. Mathias, M. T. McManus, T. F. Meehan, C. von Mering, D. Muthas, D. T. Nguyen, J. P. Overington, G. Papadatos, J. Qin, C. Reich, B. L. Roth, S. C. Schurer, A. Simeonov, L. A. Sklar, N. Southall, S. Tomita, I. Tudose, O. Ursu, D. Vidovic, A. Waller, D. Westergaard, J. J. Yang and G. Zahoranszky-Kohalmi, *Nat. Rev. Drug Discovery*, 2018, **17**, 317–332.
32. A. D. Keefe, *A Handbook for DNA-encoded Chemistry*, 2014, pp. 45–66.
33. A. E. Fernández-Montalván, M. Berger, B. Kuropka, S. J. Koo, V. Badock, J. Weiske, V. Puetter, S. J. Holton, D. Stöckigt, A. ter Laak, P. A. Centrella, M. A. Clark, C. E. Dumelin, E. A. Sigel, H. H. Soutter, D. M. Troast, Y. Zhang, J. W. Cuozzo, A. D. Keefe, D. Roche, V. Rodeschini, A. Chaikuad, L. Díaz-Sáez, J. M. Bennett, O. Fedorov, K. V. M. Huber, J. Hübner, H. Weinmann, I. V. Hartung, and M. Gorjánácz, *ACS Chemical Biology*, 2017, **12**, 2730–2736.
34. P. Filippakopoulos, S. Picaud, M. Mangos, T. Keates, J. P. Lambert, D. Barsyte-Lovejoy, I. Felletar, R. Volkmer, S. Muller, T. Pawson, A. C. Gingras, C. H. Arrowsmith and S. Knapp, *Cell*, 2012, **149**, 214–231.
35. F. Boussouar, M. Jamshidikia, Y. Morozumi, S. Rousseaux and S. Khochbin, *Biochim. Biophys. Acta*, 2013, **1829**, 1010–1014.
36. J. T. Lloyd and K. C. Glass, *J. Cell. Physiol.*, 2018, **233**, 1877–1886.
37. P. Filippakopoulos and S. Knapp, *Nat. Rev. Drug Discovery*, 2014, **13**, 337–356.
38. Z. Wu, T. L. Graybill, X. Zeng, M. Platchek, J. Zhang, V. Q. Bodmer, D. D. Wisnoski, J. Deng, F. T. Coppo, G. Yao, A. Tamburino, G. Scavello, G. J. Franklin, S. Mataruse, K. L. Bedard, Y. Ding, J. Chai, J. Summerfield, P. A. Centrella, J. A. Messer, A. J. Pope and D. I. Israel, *ACS Comb. Sci.*, 2015, **17**, 722–731.
39. J. W. Cuozzo, M. A. Clark, A. D. Keefe, A. Kohlmann, M. Mulvihill, H. Ni, L. M. Renzetti, D. I. Resnicow, F. Ruebsam, E. A. Sigel, H. A. Thomson, H. A. Thomson, C. Wang, Z. Xie and Y. Zhang, *J. Med. Chem.*, 2020, 7840–7856.
40. R. K. Y. Cheng, C. Fiez-Vandal, O. Schlenker, K. Edman, B. Aggeler, D. G. Brown, G. A. Brown, R. M. Cooke, C. E. Dumelin, A. S. Dore, S. Geschwindner, C. Grebner, N. O. Hermansson, A. Jazayeri, P. Johansson, L. Leong, R. Prihandoko, M. Rappas, H. Soutter, A. Snijder, L. Sundstrom, B. Tehan, P. Thornton, D. Troast, G. Wiggin, A. Zhukov, F. H. Marshall and N. Dekker, *Nature*, 2017, **545**, 112–115.
41. D. G. Brown, G. A. Brown, P. Centrella, K. Certel, R. M. Cooke, J. W. Cuozzo, N. Dekker, C. E. Dumelin, A. Ferguson, C. Fiez-Vandal, S. Geschwindner, M. A. Guie, S. Habeshian, A. D. Keefe, O. Schlenker, E. A. Sigel, A. Snijder, H. T. Soutter, L. Sundstrom, D. M. Troast, G. Wiggin, J. Zhang, Y. Zhang and M. A. Clark, *SLAS Discovery*, 2018, **23**, 429–436.
42. S. Ahn, A. W. Kahsai, B. Pani, Q. T. Wang, S. Zhao, A. L. Wall, R. T. Strachan, D. P. Staus, L. M. Wingler, L. D. Sun, J. Sinnaeve, M. Choi, T. Cho, T. T. Xu, G. M. Hansen, M. B. Burnett, J. E. Lamerdin, D. L. Bassoni, B. J. Gavino, G. Husemoen, E. K. Olsen, T. Franch, S. Costanzi, X. Chen and R. J. Lefkowitz, *Proc. Natl. Acad. Sci. U. S. A.*, 2017, **114**, 1708–1713.

43. C. A. Machutta, C. S. Kollmann, K. E. Lind, X. Bai, P. F. Chan, J. Huang, L. Ballell, S. Belyanskaya, G. S. Besra, D. Barros-Aguirre, R. H. Bates, P. A. Centrella, S. S. Chang, J. Chai, A. E. Choudhry, A. Coffin, C. P. Davie, H. Deng, J. Deng, Y. Ding, J. W. Dodson, D. T. Fosbenner, E. N. Gao, T. L. Graham, T. L. Graybill, K. Ingraham, W. P. Johnson, B. W. King, C. R. Kwiatkowski, J. Lelievre, Y. Li, X. Liu, Q. Lu, R. Lehr, A. Mendoza-Losana, J. Martin, L. McCloskey, P. McCormick, H. P. O'Keefe, T. O'Keeffe, C. Pao, C. B. Phelps, H. Qi, K. Rafferty, G. S. Scavello, M. S. Steiginga, F. S. Sundersingh, S. M. Sweitzer, L. M. Szewczuk, A. Taylor, M. F. Toh, J. Wang, M. Wang, D. J. Wilkins, B. Xia, G. Yao, J. Zhang, J. Zhou, C. P. Donahue, J. A. Messer, D. Holmes, C. C. Arico-Muendel, A. J. Pope, J. W. Gross and G. Evindar, *Nat. Commun.*, 2017, **8**, 16081.
44. H. H. Soutter, P. Centrella, M. A. Clark, J. W. Cuozzo, C. E. Dumelin, M. A. Guie, S. Habeshian, A. D. Keefe, K. M. Kennedy, E. A. Sigel, D. M. Troast, Y. Zhang, A. D. Ferguson, G. Davies, E. R. Stead, J. Breed, P. Madhavapeddi and J. A. Read, *Proc. Natl. Acad. Sci. U. S. A.*, 2016, **113**, E7880–E7889.
45. J. W. Cuozzo, P. A. Centrella, D. Gikunju, S. Habeshian, C. D. Hupp, A. D. Keefe, E. A. Sigel, H. H. Soutter, H. A. Thomson, Y. Zhang and M. A. Clark, *ChemBioChem*, 2017, **18**, 864–871.
46. P. Mandal, S. B. Berger, S. Pillay, K. Moriwaki, C. Huang, H. Guo, J. D. Lich, J. Finger, V. Kasparcova, B. Votta, M. Ouellette, B. W. King, D. Wisnoski, A. S. Lakdawala, M. P. DeMartino, L. N. Casillas, P. A. Haile, C. A. Sehon, R. W. Marquis, J. Upton, L. P. Daley-Bauer, L. Roback, N. Ramia, C. M. Dovey, J. E. Carette, F. K. Chan, J. Bertin, P. J. Gough, E. S. Mocarski and W. J. Kaiser, *Mol. Cell*, 2014, **56**, 481–495.
47. P. L. Podolin, B. J. Bolognese, J. F. Foley, E. Long 3rd, B. Peck, S. Umbrecht, X. Zhang, P. Zhu, B. Schwartz, W. Xie, C. Quinn, H. Qi, S. Sweitzer, S. Chen, M. Galop, Y. Ding, S. L. Belyanskaya, D. I. Israel, B. A. Morgan, D. J. Behm, J. P. Marino Jr, E. Kurali, M. S. Barnette, R. J. Mayer, C. L. Booth-Genthe and J. F. Callahan, *Prostaglandins Other Lipid Mediators*, 2013, **104–105**, 25–31.
48. S. L. Belyanskaya, Y. Ding, J. F. Callahan, A. L. Lazaar and D. I. Israel, *Chembiochem*, 2017, **18**, 837–842.
49. E. R. Wood, R. Bledsoe, J. Chai, P. Daka, H. Deng, Y. Ding, S. Harris-Gurley, L. H. Kryn, E. Nartey, J. Nichols, R. T. Nolte, N. Prabhu, C. Rise, T. Sheahan, J. B. Shotwell, D. Smith, V. Tai, J. D. Taylor, G. Tomberlin, L. Wang, B. Wisely, S. You, B. Xia and H. Dickson, *J. Biol. Chem.*, 2015, **290**, 19681–19696.
50. P. A. Harris, B. W. King, D. Bandyopadhyay, S. B. Berger, N. Campobasso, C. A. Capriotti, J. A. Cox, L. Dare, X. Dong, J. N. Finger, L. C. Grady, S. J. Hoffman, J. U. Jeong, J. Kang, V. Kasparcova, A. S. Lakdawala, R. Lehr, D. E. McNulty, R. Nagilla, M. T. Ouellette, C. S. Pao, A. R. Rendina, M. C. Schaeffer, J. D. Summerfield, B. A. Swift, R. D. Totoritis, P. Ward, A. Zhang, D. Zhang, R. W. Marquis, J. Bertin and P. J. Gough, *J. Med. Chem.*, 2016, **59**, 2163–2178.
51. L. K. Petersen, P. Blakskjær, A. Chaikuad, A. B. Christensen, J. Dietvorst, J. Holmkvist, S. Knapp, M. Kořínek, L. K. Larsen, A. E. Pedersen, S. Röhm, F. A. Sløk and N. J. V. Hansen, *MedChemComm*, 2016, **7**, 1332–1339.

CHAPTER 3

Computational Methods for the Discovery of Chemical Probes

G. PODA*[a,b] AND V. TANCHUK[c]

[a]Drug Discovery, Ontario Institute for Cancer Research, Toronto, Ontario, Canada; [b]Leslie Dan Faculty of Pharmacy, University of Toronto, Toronto, Ontario, Canada; [c]V. P. Kukhar Institute of Bioorganic Chemistry and Petrochemistry, National Academy of Sciences of Ukraine, Kyiv, Ukraine
*E-mail: gennady.poda@oicr.on.ca

3.1 Introduction to Computational Methods in Drug Discovery and Discovery of Chemical Probes

In the field of chemical biology, a chemical probe can be defined as a small-molecule modulator of a protein target. Ideally, chemical probes should be potent (K_i <100 nM), selective (>30-fold selectivity against proteins of the same protein family) and potent in cells, which allows interrogation of changes in their phenotype and, thus, determination of the function of the protein with which it interacts (see Chapter 1 for an introduction to chemical probes).

The holy grail of computational drug discovery consists in the prediction of how small-molecule modulators bind to their protein targets and prediction of their binding affinities: dissociation (K_D) or inhibitory constants (K_i). The same could be applied to the computational discovery of chemical probes. Hence most of the computational methods used in the discovery of chemical probes are the same as the methods used in drug design. The most important and widely used among them is molecular docking.

Chemical Biology No. 16
The Discovery and Utility of Chemical Probes in Target Discovery
Edited by Paul Brennan and Saleta Vazquez Rodriguez

Published by the Royal Society of Chemistry, www.rsc.org

3.2 Molecular Docking

In the field of molecular modelling, docking is a method that predicts the preferred orientation of one molecule relative to a second when bound to each other to form a stable complex.[1] The prediction of binding poses and binding affinities of protein–ligand interactions is a very important problem in computational structure-based drug discovery and also in the development of molecular probes, hence molecular docking is widely used in drug design.

It is known on a molecular level that many diseases are caused by the protein dysfunctions and the development of therapies is most commonly focused on the inhibition or activation of their functions. Molecular docking was primarily designed to investigate the binding of small drug-like molecules to target proteins. Nowadays, molecular docking is one of the most frequently used methods in structure-based drug design, because of its ability to predict the binding conformation of small-molecule ligands and find the appropriate target binding site. Investigation of the binding of small molecules plays an important role in the rational design of drugs and probes and it also helps in understanding the mechanisms of fundamental biochemical processes.

3.2.1 Molecular Docking Background

Docking methods have been in use since the 1980s and many molecular docking programs have been developed (Table 3.1). Examples of today's popular molecular docking programs include DOCK,[2,3] AutoDock,[4] GOLD,[5] Glide,[6,7] ICM Pro,[8,9] Surflex-Dock,[10,11] AutoDock Vina[12] and MOE-Dock.[13] AutoDock is one of the first and most widely used docking packages that has been distributed to more than 29 000 academic, governmental, non-profit and commercial users around the world and cited in more than 2700 publications (http://autodock.scripps.edu/#WHERE).

In order to find the preferred binding poses of a ligand with respect to a given protein target, computational docking methods sample possible ligand binding poses and rank them using various scoring functions, which are critical to the interpretation of the docking results. In addition to the scoring functions being embedded in the docking software, standalone scoring functions, such as X-Score 9[14] and DrugScore,[15] have also been developed that are used to rank compounds after the docking software has generated the binding poses. The most common uses of docking and scoring in structure-based drug design are as follows:[16]

1. hit identification *via* high-throughput virtual screening;
2. hit-to-lead and lead optimization (drug discovery);
3. prediction of binding affinities (pK_D);
4. binding site prediction (blind docking);

5. protein–protein and protein–nucleic acid interactions;
6. structure–function studies;
7. mechanisms of enzymatic reactions.

Most of the above uses are also applicable in the case of chemical probes. For example, docking and scoring may be used to virtually screen large compound libraries to identify hit molecules that fit into the binding site of a given target. Only the highest scoring virtual screening hits will be tested in laboratory-based experiments, saving cost and labour. After confirming true binders, molecular docking can be repeated to find their binding modes. This knowledge may be used in turn to improve or tune scoring functions to predict ligand binding affinities to the specific target. Even though experimental methods such as X-ray crystal diffraction, NMR spectroscopy and cryogenic electron microscopy (cryoEM) are able to determine real binding modes, they all have certain technical bottlenecks (such as designing protein constructs, protein expression and purification, crystallization, achieving sufficient solubility of the protein, preventing protein aggregation, assigning chemical shifts, poor resolution or poor diffraction of crystals), so they often fail to provide ligand binding modes in a timely manner. In such cases, hypotheses produced by molecular docking serve as valuable guidance for rational, structure-guided optimization of lead molecules. Nevertheless, more reliable and more efficient docking and scoring methods are still needed in structure-based drug design.

In general, molecular docking is a "rough and fast" method used for screening as many compounds as possible. It uses only simple types of scoring functions, such as molecular mechanics (MM) or knowledge-based.[17] However, a more detailed description of binding is sometimes required. In such cases, more accurate prediction of binding modes is achieved using the methods of quantum mechanics coupled with molecular mechanics (QM/MM). This allows a more accurate representation of the electronic structure of the docking site and ligand while still taking into account the electrostatic contributions of the rest of the protein environment at the MM level.[18,19] The large number of compounds screened would normally make QM/MM techniques computationally forbiddingly expensive; however, this more computationally expensive method could be used for challenging docking problems and has been demonstrated in recent studies that give greater agreement with experimental results.[20,21]

3.2.2 Rigid and Ensemble Docking

Sampling algorithms for protein–ligand docking can be classified in two categories, rigid-body docking and flexible docking, depending on the degree to which the methods consider both ligand and protein flexibility during the docking process.[22] Rigid-body methods do not change the conformations of the ligand and protein and take into account only their geometric or physicochemical complementarities. Other approaches consider several discrete conformations of the ligand and protein.[23] The most practical approach uses a rigid protein and a flexible ligand, where the protein is represented by a single or small group (an ensemble) of conformations. These conformations may be generated by different modelling software (*i.e.* molecular dynamics simulations) or may be taken from existing X-ray or NMR structural data. In some cases, there are hundreds of available enzyme conformations in the PDB database for a single protein target. It was proposed that in such cases, such as PTP1B or thrombin, where a substantial amount of crystallographic data exists, it is possible and necessary to use ensemble docking. On the other hand, it is not desirable to dock to all conformations of the enzyme. All existing active centre conformations should be clustered by their similarity and cluster centroids used to generated a smaller ensemble of structures.[24] Likewise, during an ensemble docking procedure, the conformations of the ligand are pre-computed and the rigid ensemble of ligands is docked in the rigid ensemble of protein conformations. Rigid docking simulations are generally preferred when time is critical, *e.g.* when a large number of compounds are to be docked during an initial virtual high-throughput screening (vHTS). On the other hand, rigid docking may not give accurate results as the binding conformation of a ligand may be substantially different from the selected conformations. There may also be an induced fit of the enzyme when the protein changes conformation for better interaction with the ligand. Flexible docking methods are needed for the refinement and optimization of poses obtained from an initial rigid docking procedure.

3.2.3 Flexible Docking

Flexible docking requires some strategy for sampling of protein and ligand conformations and also for their subsequent optimization. Some of the most popular approaches include systematic enumeration of conformations,[7,10] molecular dynamic (MD) simulations,[75] Monte Carlo search algorithms with the Metropolis criterion (MCM)[25,26] and genetic algorithms (GA).[3–5] Comprehensive exploration of ligand flexibility using all of a molecule's degrees of freedom is the basis of systematic algorithms. Starting from the same exact state and same set of parameters, systematic methods will yield exactly the same final state. Systematic methods can be categorized into exhaustive search algorithms and fragmentation algorithms.

Exhaustive searches elucidate ligand conformations by systematically moving all possible rotatable bonds by a given rotational interval.[27] A large number of free rotating bonds makes the ligand conformational space too large for an exhaustive systematic search. Some heuristics are used to focus on regions that are most important for ligand binding. This approach is used in Glide.[28] Glide creates an initial set of low-energy ligand conformations and finds initial favourable ligand poses for them by approximate positioning and scoring methods (shape and geometric complementarities). A precomputed grid representation of the docking target is used for this. After this initial screening that finds the most favourable docking regions and poses, high-resolution screening is applied. A high-resolution search uses standard molecular mechanics energy functions. It combines the minimization of the ligand with a Monte Carlo procedure for examining nearby torsional minima.[29]

An alternative approach is called fragmentation methods. In this case, a ligand is divided into computational fragments, each of which has far fewer rotatable bonds than the parent ligand. Fragments are docked separately in the binding site and grown incrementally to give the full ligand. FlexX[30] uses this "anchor and grow method" for ligand conformational sampling. A base fragment has to be interactively selected by the user, which is followed by automatic determination of placements for the fragment that maximize favourable interactions with the target protein. The base fragment is grown incrementally by adding new fragments in all possible conformations and the extended fragment is selected if no significant steric clashes are observed between ligand and target atoms. Extended ligands are optimized if new interactions are found and if minor steric interactions exist.[31] Fully automated "anchor and grow" methods have been implemented in several methods, such as FLOG,[32] SURFLEX[33] and SEED.[34] Alternatively, all fragments are docked into the binding site and then linked by appropriate bonds to give the full ligand. This method of separately docking fragments into the binding site and subsequently joining them is used in a variation of the DOCK algorithm to allow for ligand flexibility.[35]

Genetic algorithms, or similar approaches called evolution strategy, simulate the evolutionary process to find the best ligand binding pose. This is a modern stochastic optimization that replaces older Monte Carlo methods in many cases. A modified Monte Carlo method called biased probability Monte Carlo (BPMC) is used by ICM-Pro.[8] Initially, a set of state variables are defined that describe the ligand pose in the protein: translations, rotations, torsion angles and hydrogen bonding groups. Each set of values for the state variables is called an individual or genotype and the resulting structural model of the ligand in the protein corresponds to the phenotype. Analogously, the binding energy or the value of some other scoring function corresponds to the fitness of the individual. An initial population is generated randomly and the most fit individuals, according to the scoring functions, are chosen to survive. They produce the next generation of individuals by recombination of parent parameters to give the children. The rules of recombination may be different, but each child combines the features of its parents. This next generation is also evaluated, the survivors are selected and the process repeats many times. The standard number of generations produced in AutoDock is 27 000. Sometimes mutations, which are random changes in a standalone individual, are used to increase the diversity of the individuals who are sampled.

Genetic Optimization for Ligand Docking (GOLD)[5] explores full ligand flexibility with partial target flexibility using a genetic algorithm. The GOLD algorithm optimizes rotatable dihedrals and ligand–target hydrogen bonds. The fitness of a generation is evaluated based on a maximization of intermolecular hydrogen bonds. The fitness function is the sum of a hydrogen bonding term, a term for steric energy interaction between the protein and the ligand and a Lennard-Jones potential for internal energy of the ligand. AutoDock[4] uses the Lamarckian genetic algorithm, which allows favourable phenotypic characteristics to become inheritable. GOLD has demonstrated better accuracy than most docking algorithms in various benchmark studies.[36–38] Sampling methods optimize docking poses of ligands and also sometimes flexible parts of receptors such as connecting loops and/or selected key side chains. This can be achieved by optimizing either the energy of binding or the scoring function.

In order to approximate better the reality of ligand–protein interactions, it is also desirable to treat not only the ligand as flexible but also the protein or at least the binding site or side chains of key residues. Most modern docking packages allow some parts of receptors also to be flexible, but unfortunately it is not always computationally feasible to do this for a large number of potential ligands, as would be required for a virtual screening approximation. In addition, allowing full protein flexibility can lead to unrealistic conformations, so this approach must be used with caution.[39]

Sometimes the interaction between ligand and receptor involves water molecules or metal ions that are present in active centres of some enzymes. This can also be considered during docking, although in many cases it is not. Docking of covalent compounds creates an additional level of complexity that is often ignored, although there have been some attempts to take it into account and there are covalent docking capabilities in AutoDock, DOCK and Glide.

3.2.4 Scoring Functions

Much depends on scoring functions, *i.e.* the functions used to evaluate the strength of the obtained ligand–receptor complex. Accurate scoring of the binding pose and ranking of docked compounds is a crucial step in virtual screening. Frequently used scoring functions, as reported in several papers,[40,41] are divided into three categories: (a) force-field based functions that calculate binding affinity derived from physical atomic contacts in the target–ligand complex where both solvation and entropy terms can also be evaluated, (b) empirical-based functions that employ simpler energy terms, such as that for hydrogen bonding and hydrophobic interactions, fitted to experimental binding affinity data, and (c) knowledge-based functions that derive binding energy from statistical analyses taken from a training set containing protein–ligand atom pair frequencies. Numerous studies have compared different docking programs and their scoring functions to evaluate their ability to correctly score and rank ligands that bind to a protein target.[42–44]

Even though current scoring functions have greatly advanced, more development is necessary to increase accuracy in hit ranking. The performance of scoring functions varies depending on the purpose (scoring or ranking) and on the protein families and ligand series being considered.[45,46] Combining the results from two or more scoring functions and forming a consensus have proved to be more beneficial in scoring and ranking compounds. Consensus scoring has superior accuracy, thereby improving the probability of determining true binding poses and affinities.[47]

Although the methods of molecular docking are routinely applied in both academia and the pharmaceutical industry nowadays, and many successful applications have been publicly reported, it is also well known that the performance of current docking and scoring methods is still below one's expectation in many respects.[48] Although the pioneering studies in this field were published almost three decades ago,[49] a strong motivation for improvement makes the development of docking and scoring methods continue to be an active field.

The scoring function measures and ranks the binding of a ligand–receptor complex. It is therefore not only important to have an efficient, accurate scoring function that gives the best rank to true bound structures, but it must give the correct relative rank of each ligand in the input dataset. Ideally, the score should directly correspond to the binding affinity of the ligand for the protein so that the top scoring compounds are also the best binders. There are three main types of scoring functions that are used by various docking programs (Table 3.1): (1) force-field based – GOLD, ICM Pro/VLS, DOCK, MOE-Dock, AutoDOCK derived from AMBER and CHARMM; (2) empirical scoring – PLP, ChemScore, Glide SP/XP and PLANTSchemplp/PLANTSplp; and (3) knowledge-based – PMF, DrugScore and its derivatives DrugScore CSD and Astex statistical potential.

There are several groups that have compared different docking methods and reported results on a regular basis.[50] Among them there is a CSAR (Community Structure–Activity Resource) team (http://www.csardock.org)[51] and the group of Prof. Renxiao Wang at Shanghai Institute of Organic Chemistry.[52,53] Despite the fact that during the past 20 years or so significant progress in the development of the scoring functions for molecular docking has been made, correlations between the experimental inhibitory constants (expressed as a negative logarithm of K_i or pK_i) and docking scores remain mediocre at best. Therefore, every step in the direction of improving the accuracy of the scoring functions should be taken into account. Accurately predicting the binding affinities of large sets of diverse protein–ligand complexes is a very difficult and important task. As solving the problem would make virtual screening of potential binders much more practical, many different approaches have been tried and many scoring functions created. Some of them use a very sophisticated description of molecular structures and the others rely on complex mathematical algorithms.

The methods of scoring function evaluation are also evolving.[38] The comparative assessment of scoring functions (CASF) is designed as a "scoring benchmark" dataset, where the scoring process is decoupled from the docking process to depict the performance of the scoring function more precisely. The latest update of this benchmark, CASF-2016, uses 285 protein–ligand complexes with high-quality crystal structures and reliable binding constants to evaluate scoring function by four metrics that are correlated with each other but represent different uses of scoring functions, including "scoring power", "ranking power", "docking power" and "screening power".[38] Screening power refers to the ability of a scoring function to identify the true binders to a given target protein among a pool of random molecules. Scoring power refers to the ability of a scoring function to produce binding scores in a linear correlation with experimental binding data. Ranking power refers to the ability of a scoring function to rank correctly the known ligands of a certain target protein by their binding affinities when the precise binding poses of those ligands are given. Finally, docking power refers to the ability of a scoring function to identify the native ligand binding pose among computer-generated decoys. The results reveal that the performance of the scoring functions is more promising in terms of docking power than scoring, ranking and screening power. The results obtained with CASF-2016 may provide valuable guidance for end users to make smart choices among available scoring functions depending on their desired use. Moreover, CASF was created as an open-access benchmark so that other researchers can utilize it to test a wider range of scoring functions (http://www.pdbbind-cn.org/casf.asp/).

Scoring functions are very different in their nature, and sometimes they are built on a very simple basis. For example, the RF-Score scoring function was described by Morris *et al.*[4] They thought that complex machine learning-based regression techniques can be used to learn the non-linear relationship between the structure of the protein–ligand complex and its binding affinity. At the same, time they decided to use a very simple approach to describe the binding between ligand and receptor. They used the number of occurrences of a particular protein–ligand atom type pair interacting within a certain distance range. Their main criterion for the selection of atom types was to generate features that are as dense as possible, while considering all the heavy atoms that are commonly observed in PDB complexes. Only nine types of atoms were considered (C, N, O, F, P, S, Cl, Br, I). More complex differentiation by atom subtypes or by hybridization or protonation state was not used. Furthermore, only four of these heavy atom types (C, N, O, S) are usually found in proteins. Atoms types are denoted by their atom numbers. The cutoff distance for interatomic interactions is 12 Å, which is considered large enough to implicitly capture solvation effects.[54]

The interaction between protein and ligand is described by a vector X, which has 36 components (four protein atom types times nine ligand atom types). Each element of the vector is denoted as $x_{i,j}$, where *i* and *j* are atomic numbers. For example, $x_{7,8}$ is the number of occurrences of protein nitrogen interacting with a ligand oxygen within 12 Å. This simple set of descriptors has been used to build a scoring function with the help of a modern non-linear learning algorithm known as Random Forest (RF).[55] RF is an ensemble learning method for classification, regression and other tasks, that operates by constructing a multitude of decision trees at training time and outputting the class that is the mode of the classes (classification) or mean prediction (regression) of the individual trees. RF corrects for decision trees' habit of overfitting to their training set. As has already been mentioned, such a complex and sophisticated technique as RF is able to generalize the data and build a robust model even on a simple set of descriptors suggested by the authors of the RF-Score. As follows from the paper,[56] RF-Score performs better than 15 other scoring functions. Strikingly good results were reported by another group.[57] They retrained RF-Score on the PDBbind v. 2007 refined set (1300 compounds)[58] and expanded the test set to the much larger PDBbind v. 2012 refined set (2897 compounds). The predictions were very good ($R = 0.765$, $RMSE = 1.26$, where R is the non-parametric Pearson correlation coefficient and $RMSE$ is the root mean square error of atomic positions between the docked ligand and its crystal structure). The entire PDBbind v. 2012 refined set (2897 compounds) was also used to rebuild the RF-Score. Although the descriptors used to build the RF-Score seemed oversimplified, especially in comparison with other scoring functions, it appeared that they might be sufficient to build a reasonably well-performing, simpler scoring function. Thus a simple multiple linear regression analysis (MLRA) equation built on the same data appeared to be only marginally worse than the RF-Score model itself. Both scoring functions performed much better than scoring functions of the popular docking packages AutoDock and AutoDock Vina.[59] MolSoft's ICM software that uses a force-field type of scoring function ranked in first place for average RMSD docking accuracy in the 2016–17 Drug Design challenge competition.[60,61] Sometimes a fairly big set of scoring functions is used in the same docking package, as in MOE,[62] LigandFit[63] and QXP.[64]

In general, the speed and low cost of docking represent a huge advantage when it comes to the study of protein–ligand interactions compared with laboratory-based techniques. There are millions of compounds, ligands, drugs and receptors the 3D structures of which have been crystallized. Virtual screening of all commercially available compounds (which is currently about 7.6 million unique structures, *e.g.* MolPort, eMolecules and ZINC databases) or compounds made on demand (*e.g.* the Enamine Chemspace REAL Library, which currently consists of 13 billion compounds) can be performed to identify potential binders for protein targets.

Basic characteristics of current protein–ligand docking tools are summarized in Table 3.1.

Table 3.1 Basic characteristics of current protein–ligand docking tools.

Name	Authors	License type	Supported OS	Docking algorithm	Scoring function
AutoDock[4]	D. S. Goodsell and A. J. Olson, Scripps Research Institute	Free for academic use	Unix, Mac OS X, Linux, SGI, Windows	Lamarckian genetic algorithm	AutoDock, force-field type
DOCK[3]	I. Kuntz, University of California, San Francisco	Free for academic use	Unix, Linux, Sun, IBM AIX, Mac OS X, Windows	Shape fitting	Chem Score, GB/SA solvation scoring, other
FlexX[7]	T. Lengauer and M. Rarey	Commercial, free evaluation (6 weeks)	Unix, Linux, SGI, Sun, Windows	Incremental construction	FlexX Score, PLP, Screen Score, Drug Score
FRED[8]	Open Eye Scientific Software	Free for academic use	Unix, Linux, SGI, Mac OS X, IBM AIX, Windows	Shape fitting (Gaussian)	Screen Score, PLP, Gaussian shape score, user defined
Glide[6]	Schrödinger Inc.	Commercial	Unix, Linux, SGI, IBM AIX, Mac OS X, Windows	Monte Carlo Sampling	Glide vHTVS, SP and XP Scores, MM/GBSA, InducedFit Docking Score
GOLD[5]	Cambridge Crystallographic	Commercial, free evaluation (2 months)	Unix, Linux, SGI, Sun, Windows	Genetic Algorithm	Gold Score, Chem Score, user defined
LigandFit[63]	Accelrys Inc	Commercial	Linux, SGI, IBM AIX	Monte Carlo sampling	Lig Score, PLP, PMF
AutoDock Vina[10]	O. Trott and A. J. Olson, The Scripps Research Institute	Free for academic use	Unix, Mac OS X, Linux, SGI, Windows	Stochastic optimization	AutoDock Vina, empirical type
QXP[64]		Free for academic use	Unix, Linux, SGI	Multi-dimensional optimization	Several force-field type functions
ICM-Pro[8,61]	Molsoft LLC	Commercial	Windows, Linux, Mac OS X	Biased probability Monte Carlo (BPMC)	ICM VLS Scoring, force-field type[63]
MOE[62]	Chemical Computing Group	Commercial	Windows, Linux, Mac OS X	Several sampling algorithms with default "triangle matcher"	London-dG, ASE, Affinity, Alpha-HB – a set of empirical functions, GBVI/WSA dG – force-field type

3.3 Target Databases for Structure-based Drug Design

The knowledge of the structure of the target protein is required for structure-based design and there are a number of free and easily accessible databases that facilitate this. The Protein Data Bank (PDB, www.rcsb.org, www.wwpdb.org), established in 1971 at the Brookhaven National Laboratory and the Cambridge Crystallographic Data Centre, is among the most commonly used databases for protein structures. PDB currently houses more than 164 000 protein structures, the majority of which have been determined using X-ray crystallography, and a smaller set determined using NMR spectroscopy and electron microscopy. It is well known that proteins are encoded by genes and the most frequently used protein databases of protein sequences and other characteristics are UniProt (www.uniprot.org), which includes manually annotated and reviewed records, Swiss-Prot and TrEMBL, containing proteins awaiting full manual annotation, and Gene Cards (www.genecards.org), which provides comprehensive information on all annotated and predicted human genes. Other popular protein–ligand databases include Binding MOAD (Mother Of All Databases, www.bindingmoad.org), database of experimentally measured binding affinity data for biomolecular complexes (PDBbind, www.pdbbind.org, www.pdbbind-cn.org), binding database of measured binding affinities for small, drug-like molecules (BindingDB, www.bindingdb.org), AffinDB (pc1664.pharmazie.uni-marburg.de), PDSP K_i Database (https://pdsp.unc.edu/databases/kidb.php), Chemical Entities of Biological Interest (ChEBI, www.ebi.ac.uk/chebi/), a manually curated database of bioactive molecules with drug-like properties (ChEMBL, www.ebi.ac.uk/chembl/), an Open Patent Data resource (SureChEMBL, www.surechembl.org), Ligand–Protein Binding database (BioLiP, https://zhanglab.ccmb.med.umich.edu/BioLiP/), a free online repository of protein–ligand complexes structurally resolved and thermodynamically characterized, SCORPIO (Structure–Calorimetry Of Reported Protein Interactions Online, scorpio.biophysics.ismb.lon.ac.uk/scorpio.html), and many others.

Even though the DNA sequence determines the protein sequence, the function of the protein depends on a large number of other factors. Proteins undergo a number of important modifications following the protein biosynthesis, called post-translational modifications (PTMs). The major PTMs include protein folding, methylation and acetylation of lysines and/or arginines, phosphorylation of specific serines, threonines or tyrosines, glycosylation, lipidation, disulfide bond formation and cleavage of peptide bonds to remove some part(s) of the initiator methionine residue. In some cases, proteins may not be expressed at all or made in abundance (due to the copy number variations in their genes), or undergo carbonylation or oxidation of methionine and cysteine residues due to excessive oxidative stress. Some excellent resources are available to analyse protein sequences, identify homologous proteins and protein ontology, conduct multiple sequence alignments and a variety of other tools at the Swiss Institute of Bioinformatics (SIB, www.sib.swiss/research-infrastructure/database-software-tools/sib-resources), Protein Information Resources (PIR, https://proteininformationresource.org), European Bioinformatics Institute Tools and Data Resources (EBI, www.ebi.ac.uk/services/all), and many others.

When an experimentally determined structure of a protein is not available, it is often possible to create a comparative model based on the experimental structure of a structurally related protein. Most frequently the relation is based in evolution that introduced the term "homology model". Usually a protein with high (or reasonably high) sequence identity and experimental 3D structure is used as a template for comparative homology modelling. A set of closely related proteins could be identified by performing a BLAST search (*e.g.* www.ebi.ac.uk/Tools/sss/ncbiblast/) against the PDBe protein structure sequences database and the top candidates are reviewed. Usually the longest protein sequence with available 3D structure and highest sequence identity represents the best candidate for a template. Then a multiple sequence alignment (MSA) should be performed for the homologous and the target proteins (ideally including sequences of closely related species) and evaluated for gaps, deletions and insertions. Usually, the structurally conserved regions (the α-, 3/10- and π-helices, β-sheets) are modelled with good accuracy, whereas the connecting loops, varying in length, require special loop modelling tools to come up with a reasonable solution. During the homology modelling procedure, coordinates of the protein backbone of the structurally conserved regions are swapped and the side chains are placed in their conformationally preferred rotamers, avoiding steric clashes with other side chains and the protein backbone and favouring specific (H-bonding and ionic), π–π, positive charge (hetero-)aromatic and hydrophobic interactions. The resulted first-pass homology model could be further refined by conformational sampling of the protein side chain (*e.g.* using SCWRL4, dunbrack.fccc.edu/scwrl4, or Schrödinger's PRIME, www.schrodinger.com/prime), and finally energy minimized, either fully solvated (rare) or using a continuum solvation model (more common), ideally in the presence of a known active ligand at the binding site. The final model should be checked for errors, violation, inversion of chiral centres and overall stereochemical quality of the protein structure using one of the protein health tools, *e.g.* PROCHEK (www.ebi.ac.uk/thornton-srv/software/PROCHECK/) or ModEval (https://modbase.compbio.ucsf.edu/evaluation).

In case of a large number of insertions and deletions in MSA and, in particular proteins with large domain movements, the use of multiple templates is recommended, where different parts of the protein will be modelled after different templates. This usually improves the quality of the model. If a model is built to guide a structure-based analogue design effort, it is highly recommended to build multiple homology models and probe them by docking of known active inhibitors (binders). There are a large number standalone and Web-based software tools, both commercial and free to use, that allow the building of homology models. Every commercial software package, such as Schrödinger Drug Discovery and BioLuminate suites, MolSoft's ICM Homology and CCG MOE, has a homology modelling workflow built in. Free for academia tools include MODELLER (https://salilab.org/modeller), SWISS-MODEL (https://swissmodel.expasy.org), Robetta (www.robetta.org/), CPHmodels (www.cbs.dtu.dk/services/CPHmodels), HHpred (toolkit.tuebingen.mpg.de/tools/hhpred), MODBASE (https://salilab.org/modbase), Phyre^2 (www.sbg.bio.ic.ac.uk/phyre2), LOMETS (https://zhanglab.ccmb.med.umich.edu/LOMETS), MPACK (curie.utmb.edu/mpack) and many others. The SWISS-MODEL server is one of the most widely used Web-based tools for homology modelling and also provides access to a large number of pre-computed homology models that are stored on the server.[65] Other popular repositories for homology models are EBI BioModels (www.ebi.ac.uk/biomodels) and ModBase (https://modbase.compbio.ucsf.edu).

It is worth noting that even a high-resolution crystal structure of a protein is still a model that fits in the electron density [usually $(2F_o - F_c)$]. The protein is packed in a crystal and the data are usually collected using crystals frozen at a liquid nitrogen temperature of −196 °C and, as a result, the protein does not move much. In this way, a protein crystal structure is like an individual frame taken out of a movie episode. As has been demonstrated, X-ray crystallographic data collected at room temperature result in a conformational ensemble of structures where some loops or even well-packed core side chains repack and adopt different conformations.[66,67] One should think of a protein as a highly dynamic, constantly moving macromolecule. Its behaviour could be simulated by a long molecular dynamics trajectory of a fully solvated protein, ionized to a given pH and with added ions to mimic the ionic strength of the solvent at physiological conditions.

3.4 Molecular Dynamics Simulations

Molecular dynamics (MD) simulation calculates the trajectory of a system by the application of Newtonian mechanics in differential form using a numerical Verlet algorithm.[68] The first MD simulation of a biomolecule was accomplished in 1977 by McCammon *et al.*[69] It was a 9.2 ps simulation of a 58-residue bovine pancreatic trypsin inhibitor (BPTI), performed in vacuum. The most often used force fields in MD are CHARMM,[70] AMBER,[71] OPLS[72] and GROMOS.[73] The observation of a binding (or unbinding) event during an MD simulation event is very rare. In cases of very tight binders with slow on and off kinetics, the binding and unbinding event may take minutes and in some cases hours. Hence it is unrealistic to expect MD to simulate these kinds of events. As a result, standard MD simulations depend heavily on the starting conformation and are not readily appropriate for the simulation of ligand–target interactions. Because of its nature, MD is not able to cross high-energy barriers within the simulation time and is not efficient for traversing the rugged hypersurface of protein–ligand interactions. Strategies such as simulated annealing have been applied for more efficient use of MD in docking.[74]

Molecular dynamics starts where molecular docking finishes. It may be used for further in-depth investigations of the interaction between ligands and receptors. Docking poses are usually the starting points for the investigation of the behaviour of the obtained ligand–receptor complexes.[75] In the early 2000s, the use of MD required the utilization of supercomputers. Fortunately, at the end of the first decade of the twenty-first century scientists were able to develop code able to run on GPUs (graphics processing units), thus distributing calculations over thousands of cores. In more recent years, a new method, called supervised molecular dynamics (SuMD), was introduced to speed up the binding process.[76,77] An SuMD simulation is done by a series of small MD windows (SuMD steps) where step $n + 1$ is run after the evaluation of step n in terms of ligand–target approaching. During each MD step, the distance between the centres of mass of the ligand and of the protein binding site is gradually reduced (following a certain algorithm), thus the binding event can be observed in tens to hundreds of nanoseconds. In the future, MD techniques will allow the simulation of large conformational movements upon binding, thus giving a mechanistic interpretation of binding pathways.[75]

3.5 Quantitative Structure–Activity Relationship (QSAR)

Quantitative structure–activity relationship (QSAR) models, currently also known as machine learning or *in silico* models, are regression or classification models applied in the chemical and biological sciences and engineering and are used to predict ligand potency in the absence of ligand–protein structural information. QSAR implies that changes in bioactivity are associated with structural variations in a set of compounds.[40] The types of models may be very different, ranging from simple multiple linear regression analysis (MLRA) to sophisticated deep learning artificial neural networks.[78] Molecules are represented by a set of molecular descriptors and/or substructural fingerprints. As was defined by one of the greatest specialists in the field, Roberto Todeschini, "*The molecular descriptor is the final result of a logic and mathematical procedure which transforms chemical information encoded within a symbolic representation of a molecule into a useful number or the result of some standardized experiment*".[79]

The descriptors may be experimental or theoretical. There are many thousands of theoretical descriptors calculated by software packages such as Todeschini's DRAGON, which calculates more than 5000 descriptors,[80] and PaDEL-Descriptor,[81] which calculates 1875 descriptors. Another example is comparative molecular field analysis (CoMFA),[82] which was established more than three decades ago, but is still one of the most widely used 3D-QSAR methods. CoMFA calculates 3D grids around molecules (molecular fields), aligns them and finds the special regions that are most related to the activity. Since the number of parameters is much greater than the number of molecules, a special method of partial least squares (PLS) is used to score overlays. There are many modern derivatives of 3D-QSAR strategies that include Topomer CoMFA,[83] spectral structure–activity relationship (S-SAR),[84] adaptation of the fields for molecular comparison (AFMoC)[85] and comparative residue interaction analysis (CoRIA).[86] Despite its notable successes in the drug discovery field, 3D-QSAR still has numerous shortcomings that can be solved by more advanced multi-dimensional QSAR strategies in the form of 4D-, 5D- and 6D-QSAR. 4D-QSAR was developed to address ligand conformation and orientation in the target binding site, whereas 5D-QSAR incorporates issues such as receptor flexibility and induced fit effects. Finally, 6D-QSAR takes note of the solvation effects to include its critical role in receptor–ligand interactions.[87] Advances in computational power and software performance have also been applied in improving QSAR model development and validation through Discovery Bus[88] and AutoQSAR.[89] In both of these methods, hundreds of highly predictive statistical models can be objectively discovered, updated and validated by continuously integrating new machine learning agents and descriptors into the system.

Selected standalone and Web-based programs commonly used for QSAR studies are summarized in Table 3.2.

There are several considerations that one should keep in mind in order to generate a reliable QSAR model: (a) the bioactivity data should be of sufficient number (a minimum of 20 active and inactive compounds) and acquired from common or comparable experimental protocols, (b) proper selection of compounds for the training and test sets, including the size of the sets, (c) molecular descriptors for the ligands should have minimal internal correlation to avoid over-fitting, (d) the model should be validated using internal and/or external validation sets to determine its applicability and predictive power and (e) the applicability domain should be kept in mind while making predictions for new compounds.[90] An essential paper to read on how to build and validate quality machine learning models was written by Prof. Alexander Tropsha nearly two decades ago.[91] As a rule of thumb, more, good-quality data outperform a better, more sophisticated method. Frequently, the models are just as good as the data that were used to build them. Computer programs (*e.g.* DRAGON, PaDEL) calculate a very large number of descriptors and fingerprints. However, most of them (usually close to 90%) are not useful in building a given QSAR model, for two reasons: (1) some descriptors do not vary much [as could be seen from their coefficients of variation (CVs), which are standard deviations divided by their averages, expressed as percentage]; and (2) poor correlation of a given descriptor with the modelled endpoint (as seen by, *e.g.*, a non-parametric Pearson correlation coefficient). Hence it is very wise to exclude descriptors with CV <5% and Pearson correlation coefficient <0.1. This will also reduce the data noise and the chance of overfitting the data and will improve the prediction accuracy for novel compounds. Although the models usually predict very well compounds within the applicability domain, their prediction quality deteriorates quickly when trying to predict novel compounds that are outside the applicability domain.

Table 3.2 Selected standalone and Web-based programs commonly used for QSAR studies.

Program	Status	Description	Institution	Website
McQSAR	Free	Uses genetic function approximation paradigm	Abo Akademi University	http://users.abo.fi/mivainio/mcqsar/index.php
D360 Scientific Informatics Platform	Commercial	Discovery platform for data design, access and analysis	CERTARA	https://www.certara.com/data-and-informatics/d360-scientific-informatics-platform/
WEKA	Free	Workbench for machine learning	University of Waikato	https://www.cs.waikato.ac.nz/ml/weka/
Open3DQSAR	Free	Generates chemometric analysis of molecular interaction fields (MIFs)	University of Turin	http://open3dqsar.sourceforge.net
OCHEM, online chemical database with modelling environment	Free	Computes molecular descriptors for structure–activity or structure–property relationship studies		https://www.ochem.eu

Keeping in mind the recent advances and successes of deep neural networks in image (face and other objects) and speech recognition, one should be well aware that the same level of success in the chemical probe design, and in particular drug design, is still far away, if it could ever be achieved. Indeed, the triangulation of the human face or an object results in a limited set of input descriptors. The same is true about any language – it has a limited number of words and grammatical constructs. Even though the human genome has a limited number of genes that encode proteins (about 23 000), the size of chemical space is enormous and estimated to be 10^{60},[92] which is far more than the number of stars in the observable universe (estimated to be 10^{24})! During a chemical probe development, one should aim for self-evolving ML models when the newly generated data are fed right back into the model and improve the model prediction accuracy in this particular chemical sub-space. This could be achieved by using a central storage for all project-related data (such as a relational database or a flat file) and setting up cron jobs (Linux) or using other scheduling techniques to retrain the models using all combined data on regular basis, *e.g.* every weekend.

3.6 Pharmacophore Modelling

IUPAC defines a pharmacophore as *"an ensemble of steric and electronic features that is necessary to ensure the optimal supramolecular interactions with a specific biological target and to trigger (or block) its biological response"*.[93] The presence of pharmacophore for a specific target explains why structurally diverse molecules can have the same type of activity. Pharmacophore screening aims to identify compounds containing different scaffolds, but with a similar 3D arrangement of key interacting functional groups.[94] Binding site information can be incorporated into the pharmacophore model by exploiting the bioactive conformation of candidate compounds. Pharmacophore modelling is also often performed in the molecular alignment stage of QSAR modelling studies.[95] Several commonly used programs for automatic pharmacophore generation include Discovery Studio (Accelrys Inc., now BIOVIA/Dassault Systèmes), PHASE,[96] LigandScout[97] and CCG MOE. These software packages and other algorithms have been extensively reviewed.[50,51] It is worth noting that combining pharmacophore modelling and structure-based approaches (such as docking) significantly improves the chances of identifying true binders (or true actives) for a given target. Indeed, although a molecule could fit well to all defined pharmacophore features, it might have an extra part or a functional group that does not fit in the binding site. A very powerful approach that combines these two approaches was pioneered by Hans-Joachim Böhm in 1992[98] and was named LUDI. Based on structure-based derived pharmacophores, it allows the design of new substituents for enzyme inhibitors in an automated fashion. This was later used to derive a LUDI scoring function to estimate binding constants for protein–ligand interactions.[99,100] The LUDI scoring function was used to develop a very powerful structure-based focusing (SBF) approach that was a part of the Catalyst application in Cerius^2 (MSI, later Accelrys Inc.) and currently available in Discovery Studio as a part of DS Catalyst Structure Based Pharmacophore (SBP) (Accelrys Inc., now BIOVIA/Dassault Systèmes).

3.7 Screening in Virtual Chemical Space

Virtual screening (VS) is a very fast and efficient alternative to high-throughput screening (HTS) for finding hits. However, it is usually limited to available sources of compounds – in-house or databases of commercially available compounds. Some pharmaceutical companies, *e.g.* Boehringer Ingelheim, started to exploit virtual chemical space.[101] To achieve this expansion they created a global platform called the BI Comprehensive Library of Accessible and Innovative Molecules (BICLAIM).[102] It consists of putative library cores and reagents that were extracted computationally from the corporate compound databases – chemical space that they assume is accessible by mining electronic laboratory notebooks, as well as commercial sources. The current version of BICLAIM contains almost 90 000 cores and tens of thousands of reagents, giving a combinatorial virtual chemistry space of more than 10^{17} compounds. Searching such a huge space requires new approaches. Several new 2D and 3D approaches were developed. This includes a 2D feature tree search.[103] This was followed by 3D search using the rapid overlay of chemical structures (ROCS) algorithm from OpenEye (https://www.eyesopen.com/).[104] This is a powerful virtual screening tool that can rapidly identify potentially active compounds by shape comparison. ROCS is competitive with, and often superior to, structure-based approaches in virtual screening,[105,106] in terms of both overall performance and consistency.[107]

ROCS is a fast shape comparison application, based on the idea that molecules have similar shapes if their volumes overlay well and any volume mismatch is a measure of dissimilarity. It uses a smooth Gaussian function to represent the molecular volume,[108] so it is possible to routinely minimize to the best global match.

It is based on comparisons of molecular shapes. Traditionally, molecules were modelled as a set of rigid spheres corresponding to atoms. The spheres were overlapping and the volume was calculated as a volume of all spheres minus overlapping parts. Unfortunately, it was difficult to operate with spheres, but a breakthrough came when it was discovered that a sphere could be replaced by a Gaussian distribution.[108] This new approach is much simpler from a computational point of view and fairly accurate. The difference between sphere volume and volume calculated using Gaussians is less than 0.1%. In addition, the Gaussian contraction formula states that an overlap of two Gaussians is a Gaussian itself. This fact greatly simplifies the calculation of all overlaps of all orders. It also makes it possible to overlap two different structures very efficiently. Molecular shape and shape overlapping calculations became approximately 1000 times faster than calculations using spheres, although this speed enhancement was still not sufficient to screen a very large chemical space such as that of Boehringer Ingelheim's BICLAIM. A newer version that uses graphics processing units (GPUs) was reported to be 1000 times faster than the original CPU version and only 5 days were required to screen a large library whereas the CPU version required 11.3 computational years.[102] Novel and interesting molecular scaffolds have been identified using ROCS against targets often considered very difficult for computational techniques to address. The method is usually used to find compounds similar in shape to known inhibitors. The inhibitors are usually obtained from crystallographic PDB files.[109] Two successful examples where novel chemicals were identified using this approach are GPR119 agonists[110] and CDK2 inhibitors.[111] ROCS alignments have a number of applications: 3D-QSAR, SAR analysis, understanding of scaffold diversity and detection of common binding elements.[112] ROCS alignments to crystallographic conformations have also been useful in pose prediction in the absence of a protein structure.[113]

Another recent approach, called deep docking, developed by Prof. Artem Cherkasov's group at the University of British Columbia, is a very powerful technique to virtually screen very large databases of billions of compounds (e.g. ZINC, Chemspace REAL Library) in a fraction of the time compared with regular molecular docking. A deep neural network (DNN) is first trained on a smaller set of compounds to predict the docking scores accurately. The DNN trained in this way is then used to screen the whole dataset and pull out compounds with better docking scores without performing the actual docking. Likewise, this group virtually screened 1.36 billion compounds from the ZINC15 library against 12 target proteins and identified hits. This new approach demonstrated 100-fold chemical data reduction and 6000-fold enrichment for top hits, without a notable loss of well-docked entities.[114] Keeping in mind that computers are getting faster every year, novel methods to conduct VS will emerge.

3.8 *De Novo* Design and Big Data

The combination of big data and artificial intelligence (AI) was referred to by the World Economic Forum as the fourth industrial revolution that can radically transform the practice of scientific discovery.[115] AI is revolutionizing medicine,[116] including radiology, pathology and other medical specialties.[117] Deep learning (DL) technologies are beginning to find applications in drug discovery,[118,119] including areas of molecular docking,[120] reaction mechanism elucidation[121] and molecular energy prediction.[122,123] Deep learning is machine learning based on artificial neural networks (ANNs), usually with multiple layers of neurons between input and output layers. Although neural networks have been known since 1948, a revolution occurred starting in 2012 when GPU implementation of ANNs achieved for the first time superhuman performance in visual recognition.[124] It was followed by the Merck Molecular Activity Challenge, won by a group using multi-task deep learning to predict a biomolecular target through the use of special implementations of convolution neural networks (CNNs) with max-polling on GPUs.[125]

The formulation of a well-motivated hypothesis for new lead compound generation (*de novo* design) or compound selection from available or synthetically feasible chemical libraries based on the available structure–activity relationship (SAR) data is a crucial step in many new drug discovery projects. The design hypotheses are often biased towards preferred chemistry[126] or driven by model interpretation.[127] Research aimed at automating approaches to design compounds with the desired properties *de novo* has become an active field of research in the past 15 years.[128] It was estimated that there could be between 10^{30} and 10^{60} synthetically feasible drug-like molecules.[129] Despite great advances in computational algorithms,[130] hardware and high-throughput screening technologies, the size of this chemical space does not allow its exhaustive search. Although local optimization approaches have been proposed, they are not able to ensure the optimal solution. The design process is usually limited to a "practical" optimum or restricts the search to some predefined section of chemical space that can be screened exhaustively.[131]

A method based on continuous encodings of molecules was proposed recently for exploring chemical space.[132] It allows efficient, directed gradient-based searches through chemical space but does not involve biasing libraries towards special physical or biological properties. Another recent approach using recurrent neural networks (RNNs) for generating focused molecular libraries with the desired bioactivity was proposed.[133] However, it is not able to control the properties of the molecules produced well enough. An adversarial auto encoder was proposed[134] as a tool for selecting compounds with the desired properties from large libraries, but not by designing novel molecules.

A novel method for generating chemical compounds with desired physical, chemical and biological properties *de novo* that is based on deep reinforcement learning (RL) has been proposed.[135] RL is a subfield of AI that is used to solve dynamic decision problems. It involves the analysis of possible actions and estimation of the statistical relationship between the actions and their possible outcomes, followed by the determination of a treatment regime that attempts to find the most desirable outcome.[136] When RL is combined with deep neural networks, deep RL is the result. Also recently, RL was used to achieve superhuman performance in the game Go,[137] which was considered an impossible task given the theoretical complexity of more than 10^{140} possible solutions.[138] The scale of this problem resembles that of the complexity of searching chemical space.

The proposed ReLeaSE approach alleviates the deficiency of a small group of methodologically similar approaches discussed above. The most distinct innovative aspects of the approach proposed by Krakovsky[136] includes the representation of molecules by their simplified molecular-input line-entry system (SMILES) strings for both generative and predictive phases of the method and integration of these phases into a single workflow that includes an RL module. It was demonstrated that ReLeaSE allows the design of chemical libraries with the desired physicochemical and biological properties.[139]

One should keep in mind that there is major overhype in claims about what ML and AI methods can do. Indeed, even the most elaborate and sophisticated ML and AI models are only as good as the data on which they have been trained. When building models, one should not ignore simpler, faster methods (*e.g.* Random Forest and XGBoost). They require very little (if any) fine tuning of parameters, work much faster and produce models either similar to or marginally inferior in prediction accuracy than more complicated AI methods such as DNN, RL and others. AI cannot develop a chemical probe (or a drug) for you. Even though some companies or groups have claimed successes in retrospectively "designing" a drug or have come up with a drug candidate for a given (usually a well-known, easy to inhibit target), even chemical probes cannot be developed fully using *in silico* approaches. Chemical probes are designed to interrogate particular cellular functions though inhibition, activation or disruption of a protein target or a complex. Keeping in mind that cells within the same body or even organ may differ both genetically and, more importantly, epigenetically and functionally, the phenotypic effects of the chemical probes may be different in different cells, organs and individuals. Combining artificial and human intelligence (of experienced medicinal and computational chemists) is always the best strategy.

3.9 Selected Examples of Chemical Probe Development

As one would expect, chemical probe development starts from the identification of quality hits, without much in the way of liabilities (if any). Experienced medicinal chemists call them "developable hits", meaning that the compounds are usually easy to make, span across the binding site, are chemically stable and are without reactive functional groups (unless the development of a covalent probe is the goal), are soluble enough to be tested in a variety of enzymatic assays, do not undergo colloidal aggregation under the concentration and assay conditions, do not interfere with the assay readouts and inhibit the target by the expected mechanism. One could perform a virtual screening or experimental screening to identify developable hits.

Likewise, a collaborative group of scientists at the OICR Drug Discovery and Structural Genomics Consortium (SGC, University of Toronto) used a quality OICR 16 000 diversity set to identify a set of quality low-micromolar WDR5 inhibitors (17 primary hits at 50 μM).[140] Promptly, the scientists at SGC were able to solve a structure with the lead-like WDR5 inhibitor (PDB code 3SMR) and thus provided the required structure-based design support throughout the life of the project. Early SAR data were published elsewhere,[141] but the project continued. A first breakthrough in the project in terms of a significant improvement of potency came from the successful design of a focused virtual library of all in-house and commercially available amide coupling building blocks, docking this virtual library using Glide, evaluation by experienced computational chemists and selection of a small set of compounds to make. The best set of compounds was profiled in a variety of biochemical and cell-based assays, including the kinase selectivity screen (KSS) of 100 kinases at Eurofins for the best compound. However, the KSS revealed that the probe candidate inhibited 26 protein kinases, showing >50% inhibition at 10 μM, and thus would interfere with the probe phenotype in cell-based assays. A set of 19 best compounds in cell-based, biochemical and direct-binding assays were docked against four key protein kinases with available crystal structures (seven in total) in both DFG-in and DFG-out conformations (α-C-helix in), and it was hypothesized that the quinoline moiety is likely responsible for the off-target activity. A set of top four compounds were selected and profiled in KSS; none of them inhibited any protein kinase at 10 μM. OICR9429 was additionally profiled in a variety of assays and showed exclusive selectivity against 100 protein kinases, nine other WD repeat domain proteins,

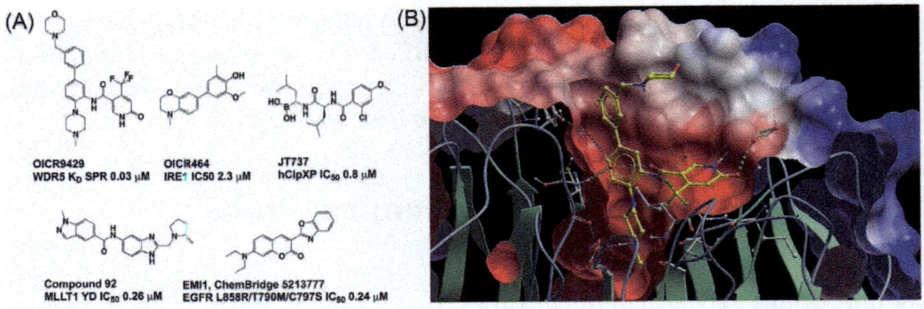

Figure 3.1 Selected examples of successful chemical probe developments. (A) Chemical structures of probes along with their activities. (B) Crystal structure of WDR5 chemical probe OICR9429 (PDB code 4QL1).

22 methyltransferases, 150 ion channels and GPCRs and was declared as a small-molecule chemical probe for WDR5[142] (PDB code 4QL1; see Figure 3.1). OICR9429 is bioavailable when administered *via* the intraperitoneal (IP) route and showed IP C_{max} = 6.8 μM at 30 mg kg^{-1} and IV $T_{½}$ = 4.7 h at 5 mg kg^{-1}. More details and SAR and phenotypic profiling are available.[143,144] Once more evidence on the important role of WDR5 in development of leukaemia emerged, OICR Drug Discovery pursued WDR5 as a drug discovery project and further improved the biochemical potency by >600-fold and cellular activity by >200-fold and achieved a good level of oral exposure. This effort led to partnership with Celgene/BMS through Triphase Accelerator to develop a first-in-class therapy for leukaemia (bit.ly/37aP4b5).

Other excellent examples of successful chemical probe development using open science, in a collaborative manner, include the identification and prediction of the mechanism of action of IRE1α endoribonuclease inhibitor of hydroxyarylaldehyde class (IC_{50} = 2.3 μM, PDB codes 4PL4 and 4PL5),[145] *de novo* design of the first human boron-containing peptidomimetic ClpP/ClpX inhibitor (IC_{50} = 0.8 μM, PDB code 3STA)[146] and discovery of an MLLT1/3 YEATS domain chemical probe (IC_{50} = 0.26 μM, PDB code 6HT1).[147] A novel live-cell platform for high-throughput identification of small molecules targeting functional protein–protein interactions of receptor tyrosine kinases, called mammalian membrane two-hybrid drug screening (MaMTH-DS), was able to identify known drugs midostaurin (IC_{50} = 30 nM), gilteritinib (IC_{50} = 500 nM) and EMI1 (ChemBridge 5213777, IC_{50} = 24 nM) as potent EGFR triple mutant (EGFR L858R/T790M/C797S) inhibitors.[148] It is important to note that all three compounds inhibited the EGFR triple mutant (including the T790M gatekeeper residue mutation), known to be notoriously difficult to inhibit. In addition, EMI1 is a novel allosteric inhibitor of EGFR. Its exact mechanism of action is yet to be determined.

3.10 Database of Known and Potential Probes

The Chemical Probes portal (https://www.chemicalprobes.org) is a freely available Web-based resource. Users can propose small molecules as chemical probes or use the portal before they begin a research programme or write a grant application and search for the most appropriate chemical probe for their experiment. The portal provides accessible recommendations about chemical probes based on expert input from their Scientific Advisory Board (SAB), a diverse group of scientific thought leaders from the fields of chemical biology, medicinal chemistry, pharmacology and related disciplines.

A similar database called Probe Miner (https://probeminer.icr.ac.uk) was created as a chemical probes objective assessment resource to allow the use of public medicinal chemistry data to empower the quantitative, objective, data-driven evaluation of chemical probes. Over 1.8 million compounds were assessed for their suitability as chemical tools against 2220 human targets and the biases and limitations encountered were dissected. Probe Miner represents a valuable resource to aid the identification of potential chemical probes.[149]

Another useful database that contains structures and data for >4000 probes and >12 000 drugs can be accessed using the Probes & Drugs (P&D) portal at https://www.probes-drugs.org. It contains information on a large number of commercially available focused libraries [*e.g.* EU-OS Bioactive Compound Library, Novartis Chemogenetic Library (NIBR MoA Box), opnMe Portal, Open Science Probes, Kinase Chemogenomics Set (KCGS) and many others] that could be used for screening to identify initial hits for probe development. More information on probe databases is available in Section 1.5 in Chapter 1.

3.11 The Future: A Probe for Every Protein

By one estimate, small-molecule probes are currently available for only 4% of the human proteome. On a summer's day in 2018, 54 industry and academic researchers gathered at Bayer's headquarters in Germany to discuss an old and bold idea: what would it take to develop a small-molecule or antibody probe for every protein in the human proteome? Despite a path that is fraught with pitfalls, the meeting's organizers saw a way out of this darkness. With the ambitious Target 2035 plan, they are calling for a global federation of linked but independent research groups to work together to develop a comprehensive set of small-molecule and antibody probes by 2035 that will cover the entire druggable genome. This is an enormous and ambitious plan that will definitely require new approaches, including computational methods.[150]

Abbreviations

AFMoC	Adaptation of the fields for molecular comparison
AI	Artificial intelligence
ANN	Artificial neural network
BICLAIM	Boehringer Ingelheim's Comprehensive Library of Accessible and Innovative Molecules
BLAST	Basic local alignment search tool
BPMC	Biased probability Monte Carlo
CNN	Convolution neural network
CoMFA	Comparative molecular field analysis
CoRIA	Comparative residue interaction analysis
CPU	Central processing unit
CSAR	Community Structure–Activity Resource
DL	Deep learning
DNA	Deoxyribonucleic acid
DNN	Deep neural network
EBI	European Bioinformatics Institute
GA	Genetic algorithm
GOLD	Genetic optimization for ligand docking
GPU	Graphics processing unit
KSS	Kinase selectivity screen
LOMETS	Local meta-threading server
MaMTH-DS	Mammalian membrane two-hybrid drug screening
MCM	Monte Carlo search algorithms with Metropolis criterion
MD	Molecular dynamics
MLL1	Mixed lineage leukaemia 1
MLLT1/3	Mixed lineage leukaemia translocated to chromosome 1/3 protein
MLRA	Multiple linear regression analysis
MM	Molecular mechanics
MOAD	Mother of all databases
MSA	Multiple sequence alignment
NMR	Nuclear magnetic resonance
OICR	Ontario Institute for Cancer Research
OS	Operating system
PDB	Protein Data Bank
PDSP	Psychoactive Drug Screening Program
PIR	Protein Information Resources
PLS	Partial least squares
PTM	Post-translational modifications
QM	Quantum mechanics
QSAR	Quantitative structure–activity relationship
RF	Random forest
RL	Reinforcement learning
RMSE	Root mean square error

RNN	Recurrent neural network
ROCS	Rapid overlay of chemical structure
RTK	Receptor tyrosine kinase
SAB	Scientific Advisory Board
SAR	Structure–activity relationship
SCORPIO	Structure–Calorimetry of Reported Protein Interactions Online
SGC	Structural Genomics Consortium
SMILES	Simplified molecular-input line-entry system
S-SAR	Spectral structure–activity relationship
WDR5	WD-repeat domain 5 protein

Acknowledgements

G.P. acknowledges the support of the Ontario Institute for Cancer Research and its funding from the Government of Ontario.

References

1. T. Lengauer and M. Rarey, *Curr. Opin. Struct. Biol.*, 1996, **6**, 402.
2. I. Kuntz, J. Blaney and S. Oatley, *et al.*, *J. Mol. Biol.*, 1982, **161**, 269.
3. T. Ewing, S. Makino and A. Skillman, *et al.*, *J. Comput.-Aided Mol. Des.*, 2001, **15**, 411.
4. G. Morris, D. Goodsell and R. Halliday, *et al.*, *J. Comput. Chem.*, 1998, **19**, 1639.
5. G. Jones, P. Willett and R. Glen, *et al.*, *J. Mol. Biol.*, 1997, **267**, 727.
6. R. Friesner, J. Banks and R. Murphy, *et al.*, *J. Med. Chem.*, 2004, **47**(7), 1739.
7. T. Halgren, R. Murphy and R. Friesner, *et al.*, *J. Med. Chem.*, 2004, **47**, 1750.
8. R. Abagyan, M. Totrov and D. Kuznetsov, *J. Comput. Chem.*, 1994, **15**, 488.
9. R. Abagyan and M. Totrov, *J. Mol. Biol.*, 1994, **235**, 983.
10. A. Jain, *J. Med. Chem.*, 2003, **46**, 499.
11. A. Jain, *J. Comput.-Aided Mol. Des.*, 2007, **21**, 281.
12. O. Trott and A. Olson, *J. Comput. Chem.*, 2009, **31**, 455.
13. C. Corbeil, C. Williams and P. Labute, *J. Comput.-Aided Mol. Des.*, 2012, **26**(6), 775.
14. H. Gohlke, M. Hendlich and G. Klebe, *J. Mol. Biol.*, 2000, **295**, 337.
15. H. Velec, H. Gohlke and G. Klebe, *J. Med. Chem.*, 2005, **48**, 6296.
16. K. Chaudhary and N. Mishra, *JSM Chem.*, 2016, **4**, 1.
17. J. Li, A. Fu and L. Zhang, *Interdiscip. Sci.*, 2019, **11**(2), 320.
18. A. Cho, V. Guallar, B. Berne and R. Friesner, *J. Comput. Chem.*, 2005, **26**, 915.
19. Y. Chung, J. Hah and A. Cho, *J. Chem. Inf. Model.*, 2009, **49**, 2382.

20. A. Cho and D. Rinaldo, *J. Comput. Chem.*, 2009, **30**, 2609.
21. M. Olsson and U. Ryde, *J. Chem. Theory Comput.*, 2017, **13**(5), 2245.
22. I. Halperin, B. Ma and H. Wolfson, *et al.*, *Proteins*, 2000, **47**, 409.
23. J. Changeux and S. Edelstein, *F1000 Biol. Rep.*, 2011, **3**, 19.
24. V. Tanchuk, V. Tanin and A. Vovk, *Chem. Biol. Drug Des.*, 2012, **80**, 121.
25. M. Allen and D. Tildesley, *Computer Simulations of Liquids*, Clarendon Press, Oxford, 1989.
26. K. Binder and D. Heermann, *Monte Carlo Simulation in Statistical Physics*, Springer, New York 2002.
27. U. Yadava, *Endocrinol. Metabol. Int. J.*, 2018, **6**(6), 359.
28. R. Friesner, J. Banks and R. Murphy, *et al.*, *J. Med. Chem.*, 2004, **47**, 1739.
29. Z. Li and H. Scheraga, *Proc. Natl. Acad. Sci. U. S. A.*, 1987, **84**(19), 6611.
30. J. Degen, C. Wegscheid-Gerlach and A. Zaliani, *et al.*, *ChemMedChem*, 2008, **3**, 1503.
31. M. Rarey, B. Kramer and T. Lengauer, *et al.*, *J. Mol. Biol.*, 1996, **261**, 470.
32. M. Miller, S. Kearsley and D. Underwood, *et al.*, *J. Comput.-Aided Mol. Des.*, 1994, **8**, 153.
33. A. Jain, *J. Med. Chem.*, 2003, **46**, 499.
34. N. Majeux, M. Scarsi and A. Caflisch, *Proteins*, 2001, **42**, 256.
35. R. DesJarlais, R. Sheridan and J. Dixon, *et al.*, *J. Med. Chem.*, 1986, **29**, 2149.
36. E. Kellenberger, J. Rodrigo and P. Muller, *et al.*, *Proteins*, 2004, **57**, 225.
37. X. Li, Y. Li and T. Cheng, *et al.*, *J. Comput. Chem.*, 2010, **31**, 2109.
38. M. Su, Q. Yang and Y. Du, *et al.*, *J. Chem. Inf. Model.*, 2019, **5**, 895.
39. K. Lexa and H. Carlson, *Q. Rev. Biophys.*, 2012, **45**(3), 301.
40. A. Lavecchia and C. Di Giovanni, *Curr. Med. Chem.*, 2013, **20**, 2839.
41. J. Liu and R. Wang, *J. Chem. Inf. Model.*, 2015, **55**, 475.
42. D. Kitchen, H. Decornez and J. Furr, *et al.*, *Nat. Rev. Drug Discovery*, 2004, **3**(11), 935.
43. X. Meng, H. Zhang and M. Mezei, *et al.*, *Curr. Comput.-Aided Drug*, 2011, **7**(2), 146.
44. Z. Wang, H. Sun and X. Yao, *et al.*, *Phys. Chem. Chem. Phys.*, 2016, **18**, 12964.
45. G. Warren, C. Andrews and A. Capelli, *et al.*, *J. Med. Chem.*, 2006, **49**, 5912.
46. W. Xu, A. Lucke and D. Fairlie, *J. Mol. Graphics Modell.*, 2015, **57**, 76.
47. D. Houston and M. Walkinshaw, *J. Chem. Inf. Model.*, 2013, **53**, 384.
48. S. Zhong, S. Zhong and Y. Zhang, *et al.*, *Curr. Opin. Drug Discovery Dev.*, 2010, **13**, 326.
49. I. Kuntz, J. Blaney and S. Oatley, *et al.*, *J. Mol. Biol.*, 1982, **161**, 269.
50. R. Smith, J. Dunbar Jr and P. Ung, *et al.*, *J. Chem. Inf. Model.*, 2011, **51**, 2115.
51. H. Carlson, R. Smith and K. Damm-Ganamet, *J. Chem. Inf. Model.*, 2016, **56**, 1063.
52. Z. Liu, J. Li and L. Han, *et al.*, *J. Chem. Inf. Model.*, 2014, **54**, 1700.
53. Z. Liu, J. Li and L. Han, *et al.*, *J. Chem. Inf. Model.*, 2014, **54**, 1717.

54. I. Muegge and Y. Martin, *J. Med. Chem.*, 1999, **42**, 791.
55. L. Breiman, *Mach. Learn.*, 2001, **45**, 5.
56. P. Ballester and J. Mitchell, *Bioinformatics*, 2010, **26**, 1169.
57. H. Li, K. Leung and P. Ballester, *et al.*, *PLoS One*, 2014, **9**(1), e85678.
58. R. Wang, X. Fang and Y. Lu, *et al.*, *J. Med. Chem.*, 2004, **47**, 2977.
59. V. Tanchuk, V. Tanin and A. Vovk, *et al.*, *Chem. Biol. Drug Des.*, 2015, **87**(4), 618.
60. R. Abagyan and M. Totrov, *Proceedings of the Third Annual International Conference on Computational Molecular Biology*, 1999, p. 312.
61. P. Lam, R. Abagyan and M. Totrov, *J. Comput.-Aided Mol. Des.*, 2018, **33**, 35.
62. C. Corbeil, C. Williams and P. Labute, *J. Comput.-Aided Mol. Des.*, 2012, **26**, 775.
63. C. M. Venkatachalam, X. Jiang, T. Oldfield and M. Waldman, *J. Mol. Graphics Modell.*, 2003, **21**, 289.
64. C. McMartin and R. S. Bohacek, *J. Comput.-Aided Mol. Des.*, 1997, **11**, 333.
65. K. Arnold, L. Bordoli and J. Kopp, *et al.*, *Bioinformatics*, 2006, **22**, 195.
66. J. Fraser, H. van den Bedem and A. Samelson, *et al.*, *Proc. Natl. Acad. Sci. U. S. A.*, 2011, **108**(39), 16247.
67. M. Fischer, B. Shoichet and J. Fraser, *ChemBioChem*, 2015, **16**(11), 1560.
68. L. Verlet, *Phys. Rev.*, 1967, **159**(1), 98.
69. J. McCammon, B. Gelin and M. Karplus, *Nature*, 1977, **267**, 585.
70. A. MacKerell, D. Bashford and M. Bellott, *et al.*, *J. Phys. Chem. B*, 1998, **102**, 3586.
71. W. Cornell, P. Cieplak and C. Bayly, *et al.*, *J. Am. Chem. Soc.*, 1995, **117**, 5179.
72. W. Jorgensen and J. Tirado-Rives, *J. Am. Chem. Soc.*, 1988, **110**, 1657.
73. C. Oostenbrink, A. Villa and A. Mark, *et al.*, *J. Comput. Chem.*, 2004, **25**, 1656.
74. S. Kirkpatrick, S. Gelatt and C. Vecchi, *Science*, 1983, **220**(4598), 671.
75. V. Salmaso and S. Moro, *Front. Pharmacol.*, 2018, **9**, 923.
76. D. Sabbadin and S. Moro, *J. Chem. Inf. Model.*, 2014, **54**, 372.
77. A. Cuzzolin, M. Sturlese and G. Deganutti, *et al.*, *J. Chem. Inf. Model.*, 2016, **56**, 687.
78. F. Melo and A. Sali, *Protein Sci.*, 2007, **16**, 2412.
79. R. Todeschini and V. Consonni, *Handbook of Molecular Descriptors*, Wiley-VCH, 2000.
80. https://chm.kode-solutions.net/products_dragon.php.
81. http://www.yapcwsoft.com/dd/padeldescriptor/.
82. R. Cramer, D. Patterson and J. Bunce, *J. Am. Chem. Soc.*, 1988, **110**, 5959.
83. R. Cramer, *J. Med. Chem.*, 2003, **46**, 374.
84. M. Putz and A. Lacrama, *Int. J. Mol. Sci.*, 2007, **8**, 363.
85. H. Gohlke and G. Klebe, *J. Med. Chem.*, 2002, **45**(19), 4153.

86. P. Datar, S. Khedkar and A. Malde, *et al.*, *J. Comput.-Aided Mol. Des.*, 2006, **20**, 343.
87. M. Damale, S. Harke and F. Kalam Khan, *et al.*, *Mini-Rev. Med. Chem.*, 2014, **14**, 35.
88. J. Cartmell, S. Enoch and D. Krstajic, *et al.*, *J. Comput.-Aided Mol. Des.*, 2005, **19**, 821.
89. A. M. Davis and D. J. Wood, *Mol. Pharmaceutics*, 2010, **10**, 1183.
90. K. Roy, S. Kar and P. Ambure, *Chemom. Intell. Lab. Syst.*, 2015, **145**, 22.
91. A. Golbraikh and A. Tropsha, *J. Mol. Graphics Modell.*, 2002, **20**(4), 269.
92. P. Kirkpatrick and C. Ellis, *Nature*, 2004, **432**, 823.
93. C. Wermuth, C. Ganellin and P. Lindberg, *et al.*, *Pure Appl. Chem.*, 1998, **70**, 1129.
94. A. Vuorinen and D. Schuster, *Methods*, 2015, **71**, 113.
95. C. Melo-Filho, R. Braga and C. Andrade, *Curr. Comput.-Aided Drug Des.*, 2014, **10**(2), 148.
96. S. Dixon, A. Smondyrev and E. Knoll, *et al.*, *J. Comput.-Aided Mol. Des.*, 2006, **20**, 647.
97. G. Wolber and T. Langer, *J. Chem. Inf. Model.*, 2005, **45**, 160.
98. H. Böhm, *J. Comput.-Aided Mol. Des.*, 1992, **6**, 593.
99. H. Böhm, *J. Comput.-Aided Mol. Des.*, 1994, **8**, 243.
100. H. Böhm, *J. Comput.-Aided Mol. Des.*, 1994, **8**, 623.
101. I. Muegge, A. Bergner and J. Kriegl, *J. Comput. Aided. Mol. Des.*, 2017, **31**, 275.
102. U. Lessel, B. Wellenzohn and M. Lilienthal, *et al.*, *J. Chem. Inf. Model.*, 2009, **49**, 270.
103. M. Rarey and J. S. Dixon, *J. Comput.-Aided Mol. Des.*, 1998, **12**, 471.
104. ROCS 3.4.0.4: OpenEye Scientific Software, Santa Fe, NM. http://www.eyesopen.com.
105. P. Hawkins, A. Skillman and A. Nicholls, *J. Med. Chem.*, 2007, **50**, 74.
106. J. Venhorst, S. Nunez and J. W. Terpstra, *et al.*, *J. Med. Chem.*, 2008, **51**, 3222.
107. R. Sheridan, G. McGaughey and W. Cornell, *J. Comput.-Aided Mol. Des.*, 2008, **22**, 257.
108. J. Grant, M. Gallardo and B. Pickup, *J. Comput. Chem.*, 1996, **17**, 1653.
109. T. Rush, J. Grant and L. Mosyak, *et al.*, *J. Med. Chem.*, 2005, **48**, 1489.
110. B. Wellenzohn, U. Lessel and A. Beller, *et al.*, *J. Med. Chem.*, 2012, **55**, 11031.
111. U. Lessel, B. Wellenzohn and J. R. Fischer, *et al.*, *J. Chem. Inf. Model.*, 2012, **52**, 373.
112. T. Tuccinardi, G. Ortore and M. Santos, *et al.*, *J. Chem. Inf. Model.*, 2007, **47**, 2293.
113. J. Sutherland, R. Nandigam and J. Erickson, *et al.*, *J. Chem. Inf. Model.*, 2007, **49**, 1715.
114. F. Gentile, V. Agrawal and M. Hsing, *et al.*, *ACS Central Sci.*, 2020, **6**(6), 939.
115. Y. Gil, M. Greaves and J. Hendler, *et al.*, *Science*, 2014, **346**, 171.
116. C. Krittanawong, H. Zhang and Z. Wang, *et al.*, *J. Am. Coll. Cardiol.*, 2017, **69**, 2657.

117. K. Chockley and E. Emanuel, *J. Am. Coll. Radiol.*, 2016, **13**, 1415.
118. H. Altae-Tran, B. Ramsundar and A. S. Pappu, *et al.*, *ACS Cent. Sci.*, 2017, **3**, 283.
119. E. Gawehn, J. Hiss and G. Schneider, *Mol. Inf.*, 2016, **35**, 3.
120. M. Ragoza, J. Hochuli and E. Idrobo, *et al.*, *J. Chem. Inf. Model.*, 2017, **57**, 942.
121. M. Segler and M. Waller, *Chemistry*, 2017, **23**, 6118.
122. J. Smith, O. Isayev and A. E. Roitberg, *Chem. Sci.*, 2017, **8**, 3192.
123. K. Schütt, F. Arbabzadah and S. Chmiela, *et al.*, *Nat. Commun.*, 2017, **8**, 13890.
124. D. Ciresan, U. Meier, J. Masci *et al.*, *Proceeding of the 22nd International Joint Conference on Artificial Intelligence*, Barcelona, Catalonia, Spain, July, 2011.
125. https://www.kaggle.com/c/MerckActivity.
126. V. Schnecke and J. Boström, *Drug Discovery Today*, 2006, **11**, 43.
127. R. Macarron, *Drug Discovery Today*, 2006, **11**, 277.
128. B. Sanchez-Lengeling, C. Outeiral, G. L. Guimaraes, *et al.*, *Chemistry (ORGANIC)*, 2017.
129. P. Polishchuk, T. Madzhidov and A. Varnek, *J. Comput.-Aided Mol. Des.*, 2013, **27**, 675.
130. D. Reker, P. Schneider and G. Schneider, *Chem. Sci.*, 2016, **7**, 3919.
131. D. Reker and G. Schneider, *Drug Discovery Today*, 2015, **20**, 458.
132. R. Gómez-Bombarelli, J. Wei and D. Duvenaud, *et al.*, *ACS Cent. Sci.*, 2018, **4**, 268.
133. M. Segler, T. Kogej and C. Tyrchan, *et al.*, *ACS Cent. Sci.*, 2018, **4**, 120.
134. A. Kadurin, A. Aliper and A. Kazennov, *et al.*, *Oncotarget*, 2017, **8**, 10883.
135. K. De Asis, J. Hernandez-Garcia, G. Z. Holland, et al., arXiv, 2017, 1703.01327.
136. M. Krakovsky, *Commun. ACM*, 2016, **59**(8), 12.
137. D. Silver, A. Huang and C. Maddison, *et al.*, *Nature*, 2016, **529**, 484.
138. H. van den Herik, J. Uiterwijk and J. van Rijswijck, *Artif. Intell.*, 2002, **134**, 277.
139. M. Popova, O. Isayev and A. Tropsha, *Sci. Adv.*, 2018, **4**, 7885.
140. G. Senisterra, H. Wu and A. Allali-Hassani, *et al.*, *Biochem. J.*, 2013, **449**, 151–159.
141. Y. Bolshan, M. Getlik and E. Kuznetsova, *et al.*, *ACS Med. Chem. Lett.*, 2013, **4**, 353.
142. F. Grebien, M. Vedadi and M. Getlik, *et al.*, *Nat. Chem. Biol.*, 2015, **11**, 571.
143. M. Getlik, D. Smil and C. Zepeda-Velazquez, *et al.*, *J. Med. Chem.*, 2016, **59**, 2478.
144. J. Zhu, M. Sammons and G. Donahue, *et al.*, *Nature*, 2015, **525**, 206.
145. M. Sanches, N. Duffy and M. Talukdar, *et al.*, *Nat. Commun.*, 2014, **5**, 4202.
146. J. Tan, J. Grouleff and Y. Jitkova, *et al.*, *J. Med. Chem.*, 2019, **62**, 6377.

147. M. Moustakim, T. Christott and O. Monteiro, *et al.*, *Angew. Chem., Int. Ed.*, 2018, **57**(50), 16302.
148. P. Saraon, J. Snider and Y. Kalaidzidis, *et al.*, *Nat. Chem. Biol.*, 2020, **16**, 577.
149. A. Antolin, J. Tym and A. Komianou, *et al.*, *Cell Chem. Biol.*, 2018, **25**(2), 194.
150. A. Mullard, *Nat. Rev. Drug Discovery*, 2019, **18**(10), 733.

CHAPTER 4

Best Practices for Design and Characterization of Covalent Chemical Probes

RONEN GABIZON[†a], EFRAT RESNICK[†a] AND NIR LONDON*[a]

[a]Department of Organic Chemistry, Weizmann Institute of Science, Rehovot 7610001, Israel
*E-mail: nir.london@weizmann.ac.il

4.1 Introduction

Small molecules that are able to form a covalent bond with a specific residue at their target binding site offer several unique advantages in the pursuit of useful chemical probes. Such covalent binders often show superior potency compared with their non-covalent counterparts[1,2] and, as a corollary, the ability to target proteins or binding sites that are considered challenging, such as protein–protein interactions[3–5] and allosteric sites.[6,7] The non-equilibrium binding kinetics also allow irreversible binders to compete with endogenous substrates such as ATP that are found in very high concentrations in the cell. Covalent inhibitors display increased and tuneable residence times, which are often determined by the protein turnover.[8] Lastly, covalent binders have the potential for exquisite selectivity when targeting

[†]These authors contributed equally.

Chemical Biology No. 16
The Discovery and Utility of Chemical Probes in Target Discovery
Edited by Paul Brennan and Saleta Vazquez Rodriguez

Published by the Royal Society of Chemistry, www.rsc.org

non-conserved residues[9,10] and, moreover, in combination with activity-based probes and proteomic approaches, their proteomic selectivity can be directly assessed,[11,12] in addition to target engagement in cells and *in vivo*.

These advantages are not limited to research or tool compounds and in recent years several rationally designed covalent drugs have been approved for various indications. Afatinib, neratinib and osimertinib covalently and irreversibly target cysteine 797 in the kinase domain of EGFR (epidermal growth factor receptor) and are approved drugs for non-small cell lung cancer and breast cancer indications. Ibrutinib and acalabrutinib irreversibly bind to cysteine 481 in BTK (Bruton's tyrosine kinase) and are approved for the treatment of chronic lymphocytic leukaemia, Waldenström's macroglobulinaemia and mantle cell lymphoma. Bortezamib and carfilzomib are earlier drugs that both inhibit the proteasome covalently and are used to treat multiple myeloma.

Despite all of these advantages and their proven track record, covalent chemical probes also present a unique set of challenges, including the necessity to account for reactivity, proteomic selectivity and non-traditional evaluation of potency. Here we set out to review the best practices in the design and evaluation of covalent chemical probes.

4.2 Discovery of Covalent Chemical Probes

When contemplating the design of a covalent probe against a target protein, several considerations may guide the design and discovery process. Is there an available structure for the target? Are there known inhibitors? Are there known substrates? What residues in the target binding site may lend themselves to covalent binding? Are they conserved? Below we outline common strategies for covalent hit discovery that might be suitable for various scenarios (Figure 4.1).

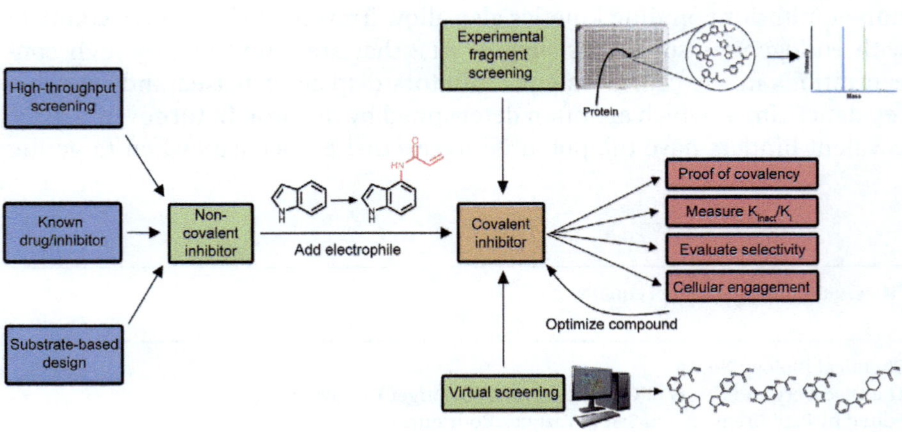

Figure 4.1 Covalent probe discovery methods and attributes for optimization.

4.2.1 Target Residues for Covalent Probes

Cysteine is the most popular amino acid to target in the design of covalent binders.[13] It is one of the rarest amino acids that instils an additional level of selectivity to cysteine-targeting small molecules. Concomitantly, it is the most nucleophilic amino acid, which allows for lower intrinsic reactivity in cysteine-targeting compounds, further reducing the number of possible off-targets. Many electrophilic moieties have been used over the years to target cysteine residues. Acrylamides are of special note, since they are present in almost all rationally designed approved covalent drugs, such as afatinib, osimertinib, ibrutinib and neratinib. Other common electrophiles that target cysteines include additional α,β-unsaturated carbonyls, chloroacetamides, halomethyl ketones, vinyl sulfones/sulfonamides, propionamides and epoxides. More recently, cyanoacrylamides[8,14–16] and α-chlorofluoroacetamide[17] were demonstrated to be effective reversible covalent cysteine-targeting agents.

Although cysteine is by far the most popular amino acid for covalent targeting, there are many examples of electrophiles targeting other amino acids. Examples include aziridines,[18] stabilized diazo compounds[19] and isoxazolium salts[20] to target aspartates or glutamates. Certain spiroepoxides[21] and cyanoenones[22] were able to target histidine residues. Michael acceptors such as vinyl sulfones and acrylamides can target lysine residues, in addition to sulfonyl fluorides[23] and aldehydes.[24] Sulfonyl fluorides[25] and fluorosulfates[26] have also been used to target hydroxyls in serine, threonine and tyrosine residues. Even methionine, which traditionally is not considered reactive, was selectively targeted recently by oxaziridines,[27] based on its redox potential. These and additional "warheads" have been widely reviewed recently;[13,28,29] therefore, we shall not focus on the choice of the exact electrophile, but rather on its general properties such as intrinsic reactivity.

4.2.2 Development of Covalent Binders Starting from Non-covalent Binders

Some of the earliest reported rationally designed covalent inhibitors, targeting enzymes, were mechanism-based inhibitors (also known as "suicide inhibitors").[30] These are derived from known substrates of the target enzyme. This approach has been widely used to design covalent protease inhibitors, as previously reviewed.[31–34] Examples of other enzymes that proved tractable for this approach are aromatase[35] and β-lactamase.[36] More recent examples include protein arginine deiminases (PADs), whose substrate is a guanidine group on an arginine side chain. By changing this guanidine to a fluoroacetamidine[37] or chloroacetamidine,[38] covalent inhibitors and activity-based probes could be designed for these enzymes. Nicotinamide *N*-methyltransferase (NNMT) was similarly targeted by adding a chlorine atom to its substrate, nicotinamide, rendering it electrophilic following the enzymatic reaction.[39]

A similar simple approach for the development of a covalent binder is to start from a known non-covalent binder or inhibitor (not necessarily a substrate) and modify it with a suitable covalent warhead. This approach is particularly useful if the structure of the target–inhibitor complex is known. In such a case, if the target contains a nucleophilic residue in a nearby position, a covalent binder can be obtained by installing an appropriately located electrophile on the known inhibitor. Typically, several positions, geometries and electrophile chemistries are tested. Since frequently the reversible binding has already been optimized previously, this method can generate potent covalent inhibitors while synthesizing and testing a relatively small number of compounds. The structure of the reversible complex or a known structure–activity relationship of the reversible inhibitor can serve to focus the efforts by ruling out positions that would compromise the interaction with the target or create unfavourable geometries for covalent bond formation.

A few recent examples of this strategy include the design of selective KDM5 (histone lysine demethylase 5) inhibitors *via* structure-based derivatization of non-covalent inhibitors, by targeting a cysteine residue that is not found in KDM4.[40] Similarly, a cysteine residue unique to PRMT5 (protein arginine methyltransferase 5) was used to derivatize a known non-covalent binder into an irreversible binder.[41] Mons *et al.* derivatized odanacatib, a potent cathepsin K inhibitor, with a very mild terminal alkyne to achieve an irreversible inhibitor.[42]

Derivatization of potent reversible binders has proven especially prolific for the design of covalent kinase inhibitors. Several FDA-approved covalent inhibitors were developed using this method. In the development of ibrutinib, a high-throughput screening campaign led to the discovery of a nanomolar reversible inhibitor of BTK. A series of derivatives harbouring an electrophile in the vicinity of cysteine 481 eventually led to the highly potent and selective ibrutinib.[43] Other covalent BTK inhibitors have been developed in a similar manner by starting from other potent reversible inhibitors.[44,45] Selective covalent inhibitors for JAK3 (Janus kinase 3) were developed in a similar fashion,[46,47] by starting from tofacitinib, an approved non-covalent JAK3 inhibitor and using structural modelling to incorporate electrophiles in various positions. This effort led to several potent JAK3 inhibitors, which were also highly selective over other JAK proteins as they targeted a cysteine residue unique to JAK3. Such covalent kinase inhibitors are often effective inhibitors of typical resistance mutations, a concept which was extensively explored in the context of EGFR.[48–50] In some cases, covalent inhibitors of one kinase can serve as starting points for the development of a covalent binder for another kinase, with a cysteine at a corresponding position, as in a recent example of the development of an MKK7 (mitogen-activated protein kinase kinase 7) inhibitor[51] inspired by a previous EGFR inhibitor. Covalent kinase inhibitor discovery has been very widely reviewed and many more examples can be found therein.[52–55]

An interesting approach to instil selectivity for covalent kinase inhibitors is to introduce a specific unique cysteine near the kinase active site and design a corresponding covalent probe to bind it selectively.[56] This can form a useful system for investigating the biology of the target kinase[57,58] and can also be applied to other protein families.[59]

It should be noted that if an existing inhibitor is not available, the development can be divided into two stages: an initial campaign for the discovery of non-covalent inhibitors, followed by the addition of the electrophile. For example, a non-covalent inhibitor of the interaction between BCL6 (B-cell lymphoma 6) and its corepressor (BCOR) was found by enzyme-linked immunosorbent assay (ELISA)-based high-throughput screening. Using the crystal structure of the complex as a guide, a chloroacetamide group was added to form a potent covalent inhibitor.[4]

4.2.3 Covalent Docking and Virtual Screening

The past few years have seen the advent of several covalent docking and virtual screening software packages[60–68] that have also been described in Chapter 3. Most of these methods are geared towards docking or pose prediction, that is, modelling of the covalent adduct form of a single compound in the target binding site. A recent benchmark study[69] compared the performances of six covalent docking tools (AutoDock4, CovDock, FITTED, GOLD, ICM-Pro and MOE) across 207 covalent ligand–protein complexes. The top-predicted pose showed RMSD <2 Å in 40–60% of the cases. Choosing the best pose out of the top 10 predictions improved the performance to 50–90% of the set. The results tended to be better for smaller and less flexible ligands. Similar results were reported by a second benchmark[70] that evaluated the performance of MOE, GOLD, CovDock and ICM-Pro on a set of 330 covalent complexes and assessed the best-scoring poses and the best-sampled poses. The top-scoring pose showed RMSD <2 Å in 43–56% of the cases and the best-sampled pose reached this level of accuracy in 61–80% of the cases.

Application of such methods to large libraries of putative covalent ligands can lead to effective virtual screening for the identification of novel covalent binders. We have developed DOCKovalent,[60] which was prospectively applied to several protein targets including RSK2 (p90 ribosomal protein S6 kinase-2), JAK3, AmpC, MKK7,[71] K-Ras^{G12C},[72] and αE7CBE (α7 carboxylic ester hydrolase).[73] Hit rates from these campaigns ranged from 6.5% to 100% with typical potencies (either IC_{50} or K_i) in the low nanomolar to even the picomolar range. Moreover, co-crystal structures with newly discovered hits almost always showed RMSD <2 Å to the docking prediction. Additional successful applications of various covalent virtual screening include the identification of covalent prolyl oligopeptidase inhibitors with cellular activity using the Fitted software;[68] application of DUckCov[67] to discover new covalent inhibitors of JAK3 and binders for K-Ras^{G12C}; the use of GOLD to identify covalent cathepsin K inhibitors;[66] the use of CovDock to screen 1648 fragments and discover a cathepsin L covalent inhibitor;[74] and the use of the"SCAR" method[75] to discover a new covalent inhibitor for *S*-adenosylmethionine decarboxylase. As these methods are just starting to become established, we believe that the near future will see an increase in the use of covalent virtual screening as a discovery technology.

4.2.4 Covalent Fragment Screening and Disulfide Tethering

Fragments are small molecules, typically defined with a molecular weight (MW) of <300 Da, moderately lipophilic (highly soluble), with few rotatable bonds.[76] As chemical space grows exponentially with increase in molecular size, the small size of fragments, as opposed to larger compounds, allows better coverage of fragment chemical space, even with relatively small libraries of only a few thousand compounds.[77,78] Moreover, their low complexity increases their probability of binding target proteins[79] and, once a hit is found, it offers more opportunities to grow and develop fragment hits while still remaining within Lipinski's "rule of five".[80] These properties have made fragment-based drug discovery a very popular and successful technique.[81–83] A major limitation of fragment screening, however, is the low affinity of initial hits, which requires sensitive detection methods such as NMR spectroscopy, X-ray crystallography, native mass spectrometry or surface plasmon resonance. The typical weak affinity of hits also necessitates a long and elaborate process of optimization to improve potency. By screening *electrophile* fragments, these issues can be addressed: the covalent bond created between the compound and the protein allows hits to be easily detected *via* intact-protein liquid chromatography–mass spectrometry (LC-MS) and dramatically increases the initial potency of the fragment hits.

A common and established method using covalent fragments for probe development is disulfide tethering. In this approach, fragments containing a disulfide are screened against a protein of interest containing a native or an introduced cysteine residue.[84] Because disulfides are not suitable as cellular probes, following hit identification and probe optimization the disulfide element is removed or replaced with an electrophile. Conversion from a disulfide to an electrophile can be complicated as the exact geometry of the electrophile needs to be carefully optimized (similarly to the introduction of an electrophile to a non-covalent compound). Disulfide tethering has been used to develop stabilizers for protein–protein interactions[85] and for finding the first inhibitors for K-Ras^{G12C}.[7]

An alternative to disulfide tethering is the direct screening of compounds with mild electrophilic warheads that are suitable as cellular probes. For such molecules, probe development will include optimization of both the compound and the warhead. Pioneering screens of around 10–100 covalent fragments were performed against recombinant proteins[86,87] and an inhibitor for the cysteine protease papain,[88] a selective probe for CDK2[89] and a probe for the PasB domain of HIF-1a[90] were discovered. We recently implemented a wider scope screen with a library of ~1000 chloroacetamides and acrylamides against a variety of targets and developed selective probes for OTUB2 (ovarian tumour domain deubiquitinase) and NUDT7,[91] neither of which had previously reported inhibitors. Other covalent fragment screens were also performed in a cellular context, simultaneously identifying protein targets for covalent fragments and optimizing them towards chemical probes.[11,29,92,93] Cellular screens typically rely on proteomic methods that compare samples treated with and without the query compounds either by mass spectrometry or by visualizing targets in gel, as discussed later in this chapter. These can be useful starting points for phenotypic screening.

Finally, it should be noted that whereas high-throughput screening approaches typically avoid covalent inhibitors for fear of non-selective inhibition, there are several examples of DNA-[94] or PNA-encoded[95,96] libraries specifically designed to identify such binders (see Chapter 2).

4.3 Characterization of Covalent Chemical Probes

Following the discovery of a covalent hit, several unique features need to be addressed and assessed in the process of its optimization into a covalent chemical probe. These include validation of the formation of a covalent bond with the protein target, evaluation of its potency, its intrinsic reactivity and the potency–reactivity trade-off, cellular target engagement and evaluation of proteomic selectivity. Below we describe the best-practices in each of these categories.

4.3.1 Proof of Covalency or Irreversibility

There are a few possible approaches to prove that a compound binds its target protein in a covalent manner (Figure 4.2). A co-crystal structure, if one can be obtained, can provide information on the binding mode and possible directions for the optimization of a covalent probe. In addition, it can show the formation of the covalent bond between the compound and the protein, as continuous electron density between the electrophilic centre and target side chain (Figure 4.2A).[7,71,91,97,98] Other methods for validation of covalency rely on eliminating the possibility of the creation of the covalent bond either by mutating the target amino acid that takes part in the covalent bond[99] (Figure 4.2B) or by synthesizing a version of the compound lacking the electrophilic warhead.[71,100,101] These approaches, although typically very informative, still leave some room for doubt, as a mutation might distort the protein and not allow binding, or alternatively a compound might still bind potently in the absence of the electrophile.

Other methods allow more direct proof of irreversibility, but do not provide structural information about the location of the covalent bond. One such method is intact-protein LC-MS.[88,91] Since the protein is denatured and unfolded during the LC, in case the correct molecular adduct mass is detected for the protein, this would indicate that the probe is covalently and irreversibly bound (Figure 4.2C). Further combination with LC-MS/MS can provide evidence as to the location of the covalent modification.

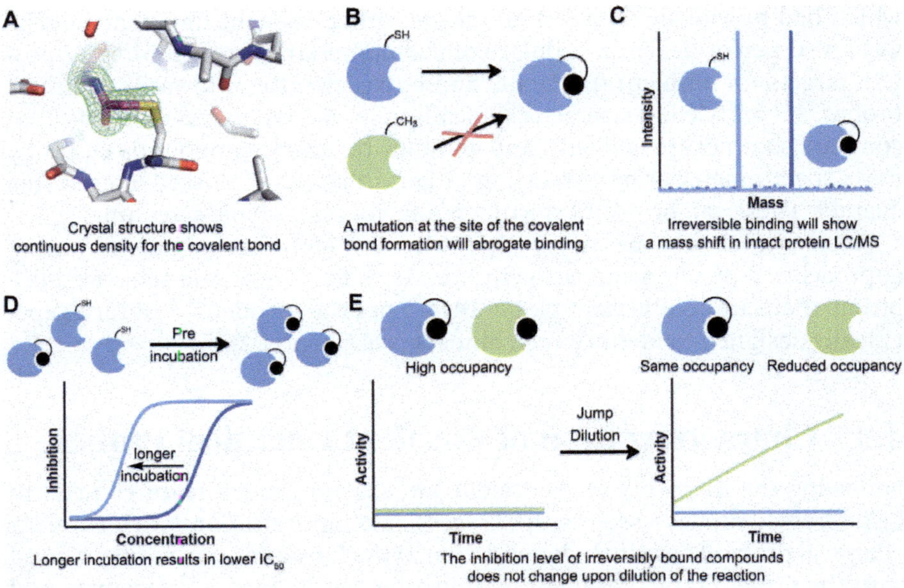

Figure 4.2 Methods for proof of covalency or irreversibility. (A) Continuous density in X-ray crystallography for the covalent bond between the probe and the protein. (B) A protein with a mutation of the target amino acid that mediates the covalent bond cannot form such a bond with the probe. (C) Mass shift corresponding to the mass of the irreversible adduct in intact-protein LC-MS. (D) Inhibition of an irreversible probe will increase with increase in incubation time. (E) Inhibition of a protein by an irreversible probe would not decrease due to jump dilution of the reaction, as opposed to a reversible probe.

An additional option is to measure the time dependence of the inhibition (Figure 4.2D). For a reversible binder, there is an equilibrium between the bound and unbound states of the protein. For an irreversible covalent binder, however, the off-rate is insignificant, hence over time more and more protein will be labelled and the measured inhibition would be stronger as a function of pre-incubation time.[102] Another option that relies on enzymatic activity is a jump dilution experiment (Figure 4.2E). In this experiment, the inhibition of a candidate compound is measured at a certain protein and compound concentration. Following an initial inhibition assay, the sample is diluted and the inhibition is reassessed. A reversible inhibitor would equilibrate following the dilution and the bound fraction of the protein will be reduced, as will the measured inhibition. An irreversible inhibitor, however, will not dissociate upon dilution, so the inhibition will not be changed.[60,91] Another method to prove irreversibility is by buffer exchange (*e.g.* by gel filtration) for the target protein following incubation with the inhibitor.[43] Irreversible inhibitors will retain their inhibition after this procedure, whereas reversible inhibitors will be separated from the protein and would no longer inhibit the target. Chowdhury *et al.* used this method to show the irreversible nature of vinyl sulfone cathepsin L inhibitors.[74] These methods are valid for the measurement of both reversible and irreversible covalent binders and attest only to the irreversibility and not to the formation of the covalent bond.

4.3.2 Characterization of the Potency of Irreversible Covalent Probes

Non-covalent enzyme inhibitors are often characterized by their IC_{50} – the concentration at which they inhibit half of the enzymatic activity under the assay conditions. However, for irreversible inhibitors, which are out of equilibrium, such measurements are very sensitive to the assay conditions and in particular to protein concentration and time of incubation.[103] Kinetic evaluation of irreversible binding is typically a more informative approach.

The binding of irreversible probes to their target can be described as a two-step process. In the first step, the probe reversibly binds to the protein, forming an encounter complex, also referred to as a Michaelis complex. Following binding, the probe can either dissociate or form the covalent bond (Figure 4.3). The equilibrium of the first reversible binding step is approximated

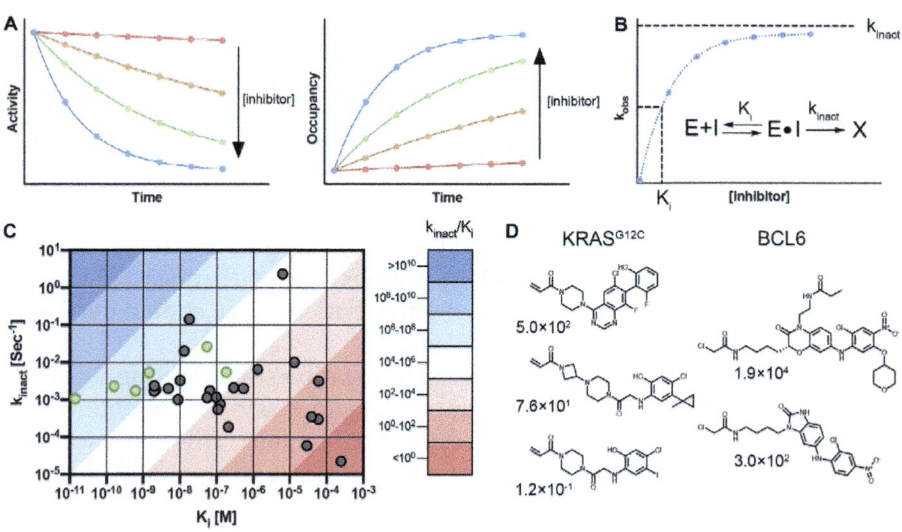

Figure 4.3 k_{inact} and K_i determination and typical values. (A) Inhibition or occupancy *versus* time at different inhibitor concentrations is used to determine the observed first-order rate constant (k_{obs}). (B) Non-linear fit of k_{obs} *versus* inhibitor concentration using $Y = k_{inact}X/(K_i + X)$. (C) Typical k_{inact} and K_i values for covalent compounds at different optimization stages for various targets (see Table 4.1). k_{inact}/K_i range is indicated by the diagonals of the plot. FDA-approved drugs are in green. (D) Examples of progress in k_{inact}/K_i values (in M^{-1} s^{-1}) with optimization of probes for two proteins.

by K_i and the rate of the formation of the irreversible covalent bond in the encounter complex is given by k_{inact}. From these, one can calculate k_{inact}/K_i, also referred to as the specificity constant or the bimolecular rate constant, describing the overall rate of inactivation.

4.3.2.1 *Determination of* k_{inact} *and* K_i

In a seminal paper published in 1962, Kitz and Wilson[104] derived linear equations that under a few assumptions allowed the determination of k_{inact} and K_i from enzyme activity measurements under various concentrations of irreversible inhibitor for various pre-incubation times. Modern fitting methods allow more accurate fitting of the raw data (Figure 4.3A and B) without resorting to linearization. However, they are still dependent on the basic assumptions: (1) the inhibitor concentration in the assay is much higher than the enzyme concentration; (2) the enzyme activity without inhibition is linear in time, *i.e.* no substrate depletion; and (3) the reversible bound complex is in rapid equilibrium with the free enzyme, *i.e.* the binding and dissociation are much faster than the covalent inactivation step. Although these assumptions may hold for many covalent chemical probes, they typically do not hold for very potent covalent inhibitors in which the inhibitor concentration in the assay might be very low and the off-rate from the reversible encounter complex might be very slow.

Several alternative methods were developed to estimate these rate constants more easily. Krippendorff *et al.* reported a method to determine k_{inact} and K_i directly from time-dependent IC_{50} values.[102] Miyahisa *et al.* developed an alternative method to determine k_{inact}/K_i for a series of compounds by using a competition experiment with an irreversible probe.[105] To use it, one has to determine k_{inact}/K_i for the probe kinetically, but can then merely determine IC_{50}s for subsequent assayed compounds, greatly simplifying the experimental setup. With regard to IC_{50}, Thorarensen *et al.* suggested that in certain cases measuring IC_{50} under long incubation times should correlate with k_{inact}/K_i, while being easier to measure.[46]

Finally, differential equation numerical integration solvers, using software such as DynaFit,[106] can be used to extrapolate both K_i and k_{inact}, without relying on the assumptions taken by the traditional models. This was used, for example, to characterize several potent covalent EGFR inhibitors.[107]

k_{inact}/K_i values can also be calculated for irreversible binding to non-enzymatic targets. In that scenario, the observed rate of the percentage occupancy of the protein should be measured instead of enzymatic activity (Figure 4.3A). This is often achieved *via* intact-protein mass spectrometry using LC-MS. For example, Johansson *et al.*[98] determined k_{inact}/K_i for irreversible fragment binders to HOIP E3 ligase by quantification of the percentage labelling at different time points and concentrations. A similar technique was used to determine k_{inact}/K_i for covalent K-Ras^{G12C} binders.[108] For some inhibitors, the formation of the covalent bond can be directly detected spectroscopically[109] or by following a cofactor.[110] Further reading on the kinetic evaluation of irreversible inhibitors can be found in extensive reviews on the subject.[111,112]

4.3.2.2 *Typical* k_{inact}/K_i *Values*

We collected examples of recently reported covalent binders for which the kinetic parameters were measured (Table 4.1, Figure 4.3C). In cases where multiple SAR (structure–activity relationship) series have been reported we selected representatives. Several interesting trends arise from examination of these data. First, FDA-approved covalent kinase inhibitors have among the highest reported k_{inact}/K_i values. This may stem from the fact that covalent kinase inhibitors as a class show very high k_{inact}/K_i values, owing to the typical very high-affinity reversible binding of the initial non-covalent scaffold. Binders of more challenging target classes such as deubiquitinases and E3 ligases show more modest k_{inact}/K_i values. This does not mean, however, that they cannot be further optimized. These binders are typically primary hits from fragment or virtual screens and their low values may reflect lack of sufficient optimization rather than intrinsic properties of the protein target.

In several cases, k_{inact}/K_i values are reported along the optimization route, which can provide a perspective for the expected magnitude of improvements (Figure 4.3D). For instance, one of the original irreversible binders of K-Ras^{G12C}, derived from a tethering screen hit and itself heavily optimized,[7] had a reported k_{inact}/K_i value of only 1.20×10^{-1} M^{-1} s^{-1},[113] representing the challenging nature of the allosteric switch II pocket. A subsequent analogue ARS-853 showed an improvement of >600-fold with $k_{inact}/K_i = 7.60 \times 10^1$ M^{-1} s^{-1}.[113] Finally, scaffold hopping and further optimization led to a binder with $k_{inact}/K_i = 5.01 \times 10^2$ M^{-1} s^{-1}.[108] In another example, an irreversible BCL6 binder developed by derivatizing a reversible fragment hit showed $k_{inact}/K_i = 3.00 \times 10^1$ M^{-1} s^{-1}; following optimization, binders with improvements of up to 630-fold could be found with $k_{inact}/K_i = 1.90 \times 10^4$ M^{-1} s^{-1}.

An important point to make is that k_{inact} is governed by several contributions, including the reactivity of the electrophile, the reactivity of the protein nucleophile and the acceleration of the reaction by optimal positioning of the two. A striking example of this is a recent acrylamide-based covalent inhibitor of JAK3,[46] with an extremely high k_{inact} of 2.32 s^{-1} (see Table 4.1; Figure 4.3C). Such a high k_{inact} value could be mistaken to represent a high reactivity of the electrophile. However, Thorarensen *et al.* showed that for other kinases such as ITK (interleukin-2-inducible T-cell kinase), for which the compound still has an nM range of K_i, its k_{inact} is almost four orders of magnitude lower (1.44×10^{-4} s^{-1}).[46] Another proof that the potency is not driven by reactivity was provided by improved GST (glutathione *S*-transferase) clearance. Hence optimization of k_{inact} while controlling for reactivity can be an efficient approach to introduce additional selectivity for covalent binders.

Table 4.1 Examples of recently reported k_{inact} and K_i values of various covalent binders.

Target	k_{inact}/s^{-1}	K_i/M	k_{inact}/K_i/M^{-1} s^{-1}	Electrophile	Discovery	Reference
AKT	1.85×10^{-3}	2.20×10^{-9}	8.41×10^{5}	Acrylamide	Optimization	175
BCL6			1.90×10^{4}	Chloroacetamide	Optimization	4
BCL6			3.00×10^{1}	Chloroacetamide	Derivatization	4
BTK	2.33×10^{-3}	2.00×10^{-9}	1.17×10^{6}	Acrylamide	Derivatization	45
BTK	5.60×10^{-3}	1.81×10^{-7}	3.09×10^{4}	Methylpropionamide	Acalabrutinib	176
BTK	2.70×10^{-2}	5.42×10^{-8}	4.98×10^{5}	Acrylamide	Ibrutinib	176
BTK	1.17×10^{-3}	9.70×10^{-8}	1.00×10^{4}	Acrylamide	Optimization	45
BTK	1.00×10^{-3}	8.90×10^{-9}	1.10×10^{5}	Acrylamide	Optimization	44
Cathepsin K	1.83×10^{-4}	2.10×10^{-7}	8.73×10^{2}	Alkyne	Derivatization	42
Cathepsin L	3.00×10^{-4}	5.90×10^{-5}	5.08×10^{0}	Vinyl sulfone	Virtual screening	74
Cathepsin L	1.40×10^{-1}	1.80×10^{-8}	7.78×10^{6}	Vinyl sulfone	Optimization	32
$EGFR^{L858R/T790M}$	1.80×10^{-3}	6.30×10^{-10}	2.86×10^{6}	Acrylamide	Dacomitinib	107
$EGFR^{L858R/T790M}$	2.40×10^{-3}	1.60×10^{-10}	1.50×10^{7}	Acrylamide	Afatinib	107
$EGFR^{L858R/T790M}$	1.10×10^{-3}	1.40×10^{-11}	7.86×10^{7}	Acrylamide	Neratinib	107
$EGFR^{L858R/T790M}$	5.50×10^{-3}	1.46×10^{-9}	3.77×10^{6}	Acrylamide	Osimertinib	177
$EGFR^{L858R/T790M}$	2.00×10^{-2}	1.30×10^{-8}	1.53×10^{6}	Acrylamide	Optimization	49
FAAH			4.03×10^{4}	Piperazine urea	Optimization	176
FAAH			1.26×10^{4}	Piperazine urea	Optimization	178
FLT3	3.30×10^{-3}	1.01×10^{-8}	3.25×10^{5}	Chloroacetamide	Derivatization	179
FLT3	1.70×10^{-3}	1.98×10^{-9}	8.59×10^{5}	Vinyl sulfone	Derivatization	179
GBA1	2.00×10^{-3}	4.80×10^{-9}	4.17^{5}	Aziridine	Optimization	180
Glutamate racemase	3.50×10^{-4}	3.86×10^{-5}	9.07×10^{0}	Rhodanine	Virtual screening	181
HOIP-RBR			9.70×10^{-1}	Methyl acrylate	Fragment screening	98
JAK3			3.00×10^{5}	Acrylamide	Derivatization	182
JAK3	2.32×10^{0}	6.31×10^{-6}	3.68×10^{5}	Acrylamide	Optimization	46
KDM5	2.00×10^{-3}	5.36×10^{-7}	3.73×10^{3}	Chloroacetamide	Derivatization	40

KDM5	2.10×10^{-3}	2.89×10^{-7}	7.27×10^{3}	Acrylamide	Derivatization	40
KDM5	7.60×10^{-4}	1.24×10^{-7}	6.13×10^{3}	Acrylamide	Derivatization	40
$KRAS^{G12C}$			5.01×10^{2}	Acrylamide	Optimization	108
$KRAS^{G12C}$			7.60×10^{1}	Acrylamide	Optimization	113
$KRAS^{G12C}$			1.20×10^{-1}	Acrylamide	Tethering screening	113
MAO B	1.73×10^{-3}	6.50×10^{-8}	2.67×10^{4}	Cyclopropylamine	Mechanism-based inhibitor	110
NEDD4	5.80×10^{-5}	2.93×10^{-5}	1.98×10^{0}	Methyl acrylate	Optimization	97
NEDD4	2.20×10^{-5}	2.46×10^{-4}	8.95×10^{-2}	Methyl acrylate	Fragment screening	97
NUDT7	1.00×10^{-2}	1.32×10^{-5}	7.57×10^{2}	Chloroacetamide	Optimization	91
NUDT7			2.33×10^{2}	Chloroacetamide	Fragment screening	91
OTUB2			3.75×10^{0}	Chloroacetamide	Fragment screening	91
PAD3			2.60×10^{2}	Chloroacetamidine	Mechanism-based inhibitor	38
PDK1	5.47×10^{-4}	1.09×10^{-7}	5.02×10^{3}	Disulfide	Optimization	183
PfGAPDH			1.07×10^{1}	3-Bromoisoxazoline	Optimization	184
Pp 4-OT	3.17×10^{-3}	6.00×10^{-5}	5.30×10^{1}	Allyl chloride	Mechanism-based inhibitor	185
PRMT5	1.13×10^{-3}	5.50×10^{-8}	2.06×10^{4}	Aldehyde	Derivatization	41
Transglutaminase 2	6.45×10^{-3}	1.31×10^{-6}	4.92×10^{3}	Acrylamide	Optimization	186

4.3.3 Evaluation of Reactivity

As discussed, the reactivity of covalent chemical probes has to be tightly controlled. Cysteines themselves have a wide range of reactivity,[114–117] hence the reactivity of the covalent probe has to be matched, such that it will allow efficient labelling on the one hand, without being too reactive, which will result in non-selective labelling of too many off-targets on the other hand. Reactivity evaluation of covalent binders is important from the very first stages of the discovery process (as early as library design or electrophile selection) in order to find meaningful starting points and it is important still to keep watch throughout the process of probe development so that potency gains will be mediated through increased specificity rather than *via* development of more reactive compounds.

There are several methods to assess the intrinsic reactivity of covalent-acting compounds. The most commonly used method is measurement of the glutathione (GSH) modification rate. This method is very widely used for the characterization of various electrophiles[118,119] and also for SAR studies during probe development.[49] The wide use of this method provides a good dataset for the comparative evaluation of new covalent compounds with respect to many other compounds in different stages of development. GSH serves in such assays as a model thiol and by determining the rate of the reaction between the electrophilic compounds and GSH it is possible to make an estimation of the intrinsic reactivity of the compounds. GSH is a good model for several reasons. First, it is cheap and readily available, and also soluble up to high concentrations. More importantly, however, cells contain high levels of GSH (1–10 mM)[120] that could potentially mitigate the cellular activity of covalent compounds. For these reasons, it might be preferable to model other thiols such as cysteine[121] or *N*-acetylcysteine methyl ester.[88]

Reaction rates in such assays are typically followed by LC-MS or NMR spectroscopy and the reactivity is reported as GSH $t_{½}$. Measuring GSH modification allows a wide dynamic range for reactivity measurements by varying the experimental conditions, *i.e.* increasing the temperature for slow-acting compounds or the use of low concentrations or short incubations for more reactive compounds, as was done in the evaluation of reactivity for compounds with different electrophilic warheads.[118,122] One of the major limitations in the use of GSH modification measurements is the low throughput; most measurements are performed one compound at a time, using time-consuming and specialized instruments (LC-MS or NMR), which might also have limited accessibility. Cee *et al.* addressed this issue to some extent by pooling 10–15 compounds for a single LC-MS measurement.[123]

A few methods were developed that allow GSH reactivity measurements using a simple plate reader. Epps and Taylor used a competition with a maleimide-modified fluorophore that fluoresces upon reaction with GSH, in order to assess the reaction rate.[124] However, because the fluorophore reacts rapidly with GSH, it does not allow measurements of weakly reactive compounds. Craven *et al.* used 7-diethylamino-3-(4′-maleimidylphenyl)-4-methylcoumarin (CPM) as a reporter for free thiols.[89] The reaction between electrophiles and GSH is sampled at different time points by adding CPM, which fluoresces in the presence of free thiols. This allows one to collect all of the different time points and read them together. Both of these methods allow the use of a simple plate reader for the readout and have high throughput, but do not consider the effect that the query compounds might have on the fluorophore.

An alternative is to use DTNB [Ellman's reagent: 5,5′-dithiobis (2-nitrobenzoic acid)] instead of GSH. Similarly to GSH, TNB^{2-} (reduced DTNB) has a free thiol and it also shows absorbance at 412 nm. DTNB has been used mostly to profile various enzymes such as catechol oxidase and lactase[125] and myeloperoxidase.[126] We introduced the opportunity to use DTNB for reactivity measurements.[91] When an electrophile alkylates TNB^{2-} it loses its 412 nm absorbance, and it is possible to follow this absorbance loss in order to calculate directly a second-order reaction rate. Because the procedure only requires absorbance readout, it is possible to measure easily a large number of compounds simultaneously on a simple plate reader. The limitations of this assay include possible interference by compounds that have intrinsic absorbance at a similar wavelength (although background subtractions mitigate most of these) that might interfere with the measurements. Also, the dynamic range of this method might not be suitable for compounds with extremely low reactivity.

Methods using a model thiol are useful as they allow a robust and simple way to assess the general reactivity of electrophilic compounds; however, model thiols may not fully mimic protein thiols. ALARM NMR (a La assay to detect reactive molecules by nuclear magnetic resonance)[127] is a method that relies on a surface-exposed reactive cysteine in the La antigen protein to evaluate adduct formation with query compounds. The La antigen protein is stable to oxidation, does not rely on a secondary reporter for GSH concentrations and this cysteine better represents thiols in a protein context. However, this method does not directly measure the rate of the reaction. Another method relying on NMR spectroscopy was recently reported[128] in which the reaction rates between various model compounds and isolated amino acids were directly measured.

A complementary approach for reactivity assessment that is gaining traction relies on computational predictions. Preliminary substructure-based algorithms predict the probability of various substructures to be thiol reactive, based on an extensive experimental dataset.[129] More recently, density functional theory (DFT)-based calculations were used to correlate calculated kinetic reaction barriers with the experimental reactivity of various acrylamides.[118] The electrophilicity index,[130] based on calculation of the highest occupied and lowest unoccupied molecular orbitals (HOMO/LUMO) near the electrophile, also correlated well with the experimental reactivity of acrylamides.[131] Lonsdale *et al.* performed an extensive evaluation of various computational and experimental methods and proposed a decision tree based on molecular properties to select the most suitable computational protocol to predict a given electrophile reactivity.[119]

Since non-specific, reactivity-driven protein binding is the most common confounding factor for covalent-acting compounds, it is extremely important to take it into consideration at the beginning, and also during the optimization process of a covalent chemical probe.

4.3.4 Proof of Cellular Engagement

A molecule that binds covalently and potently to the target may still not efficiently label the target in the cellular environment due to factors such as poor cell penetration, lack of stability or competition with cellular nucleophiles. Therefore, it is imperative to confirm that the compound engages the target in the cell. Since there is usually no direct way to observe the binding of the compound to the target in the cell, several methods have been developed to test indirectly for cellular engagement.

A method frequently used to test the cellular engagement of a compound is by probing the function of the target, such as phosphorylation of substrates for kinases.[71,132,133] Another method utilizes the fact that the binding of a compound to the target frequently increases the stability of the protein at high temperatures and reduces the extent of unfolding and degradation. This is used as a basis for cellular thermal shift assays (CETSAs),[134] which provide a quick and simple method for determining cellular engagement that do not require any advanced apparatus. Both of these approaches can be applied for testing cellular engagement for both covalent and non-covalent inhibitors.[91,135]

Activity-based probes (ABPs) can serve as tools for determining cellular engagement. ABPs can bind various protein families and give direct readouts such as fluorescence signals[136,137] or serve to identify the target proteins using LC-MS/MS.[138] Pre-incubation of the samples with the covalent probe followed by a test with the ABP can give a quantitative estimation of cellular engagement, and also estimates of the selectivity of the probe. Finally, competitive proteomics methods or pull-downs from cells can also validate cellular engagement, as elaborated below.

4.3.5 Evaluation of Selectivity

In addition to the ability to bind and inhibit its target, a crucial parameter for a chemical probe is its selectivity profile. The more selective the probe is, the better is our ability to use it to draw correct biological insights regarding its target. Therefore, estimation of a probe's selectivity is an inherent part of its development and should be closely monitored as part of its design and optimization. A historical concern regarding the use of covalent inhibitors is their potential for non-selective irreversible binding to off-targets.[139]

Although this concern must be addressed, covalent inhibitors have two significant advantages over non-covalent inhibitors with regard to selectivity: first, the ability to tune the reactivity of the electrophile gives an additional variable that can be modified to optimize the selectivity and potency of the molecule (as discussed earlier); and second, the fact that the inhibitor binds covalently to its targets permits detailed characterization of the off-targets to a level not possible for non-covalent inhibitors. Covalent inhibitors can be used as a tag for isolating their bound targets and for identifying the targets and the labelling sites.[12,108,140] Below we review some of the approaches for the evaluation of the selectivity of covalent chemical probes.

4.3.5.1 *Non-proteomic Methods*

Estimating the selectivity of covalent inhibitors can be performed on several levels. At the most basic level, candidate molecules are tested in functional assays against a panel of enzymes with similar activity or homology to the main target (Figure 4.4). The selectivity is estimated by the relative activity of the compound on the target *versus* the other proteins in the panel. The Structural Genomics Consortium (SGC) (https://www.thesgc.org/) recommended as a threshold criterion that a chemical probe should possess >30-fold selectivity relative to other sequence-related proteins of the same target family. A similar criterion should apply to covalent chemical probes. This method is particularly simple to apply to families that have many well-characterized members, such as kinases, for which screening of such panels is available through commercial services. It is very common to use commercial kinase panels to evaluate the selectivity of covalent kinase inhibitors.[44,51,71,100,101] Screening against other enzymes such as E3 ubiquitin ligases[97,98] and protein arginine deimidases[38,141] is typically performed on small panels of proteins made in-house.

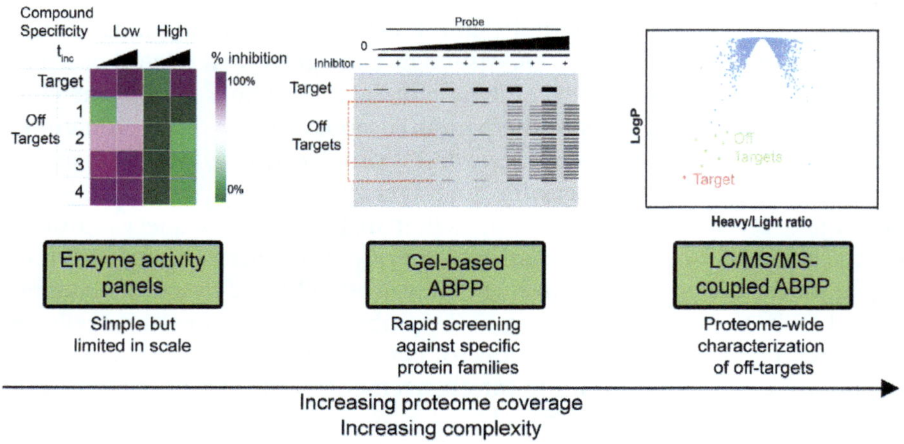

Figure 4.4 Methods for determination of the selectivity of covalent probes.

Although high selectivity in such enzyme panel screens will not guarantee high selectivity across the proteome, it can serve as an initial step to filter out low-selectivity candidates. However, beyond the obvious limitation that it can sample only a small fraction of the proteome, these methods are typically expensive, require the preparation or purchase of multiple purified proteins and require large amounts of compound and protein. It is also limited to proteins that can be purified easily and tested *in vitro* for binding or for enzymatic activity.[142]

4.3.5.2 Chemical Proteomics Approaches

Covalent inhibitors are particularly amenable to characterization using chemical proteomics approaches owing to their irreversible binding to their targets. These approaches use the query molecules or specific probes as tools to bind and isolate proteins across the entire proteome (Figure 4.4). The studied molecules are incubated with lysates, cells or alternative relevant physiological systems and the bound proteins are isolated, quantified and characterized either using gel-based methods or *via* trypsinization and LC-MS/MS, using instrumentation that is now widely available.[138,140,143] Using chemical proteomics methods, the engagement of the tested compounds with the entire proteome or with a wide array of proteins can be tested simultaneously and the concentration and time dependences of the engagement can be measured. In particular, activity-based protein profiling (ABPP)[140,144–146] has become a powerful tool for the characterization of the targets of covalent inhibitors[11,92,93,147] and natural products.[148–150]

In order to isolate or detect the proteins that react irreversibly with the inhibitor, it needs to possess an additional chemical moiety that will function as a reporter tag, such as biotin, a fluorophore[151] or a reactive group such as an alkyne or azide that can be modified downstream.[152,153] Although some molecules do contain such groups and can be used directly for this purpose, such as ethinylestradiol[154] and LAS17,[155] most compounds cannot be used directly as ABPP probes. Although analogues with alkyne or azide groups can be effective ABPP probes,[12,156] they frequently have somewhat different selectivity profiles to the parent compounds.[12,157] This problem is addressed by the use of competitive ABPP protocols, in which a separate, general probe is used to react with the target and as many possible relevant off-targets. The identification of targets that are bound by the query inhibitor is then achieved by pre-incubation with the query, resulting in a reduction of signal for the general probe binding to any protein bound by the query compound.[158–160] Because the signal from the probed targets is measured relative to a reference signal (in the absence of competing inhibitor), it allows the quantitative estimation of inhibitor engagement to the various targets in a dose-dependent manner.[12,159,160]

Another advantage of using competitive ABPP is the ability to tailor the ABPP probe for testing the inhibitor's selectivity *versus* defined subgroups within the proteome. On one side of the scale, the probe can be designed as a close analogue of the inhibitor itself, with an additional functional group (such as an alkyne or fluorophore) placed strategically to prevent it from interfering with binding to the target.[12,156,157] Since the probe is similar to the inhibitor, it is expected to bind the desired target tightly and to bind a significant fraction of the inhibitor's off-targets, in addition to other proteins not bound by the inhibitor. Since the number of proteins targeted by such a probe will be small, such probes can be used in gel-based competitive ABPP for the rapid, quantitative assessment of inhibitor engagement to the target and off-targets. The weakness of this method lies in the requirement to synthesize a specialized probe for each individual inhibitor and with the chance of missing potential off-targets not identified by the probe. Larger subgroups of the proteome can be analysed using probes that target specific classes of proteins, particularly enzymes. These include, for example, serine hydrolases,[161–163] glycosidases,[160,164,165] lipases,[159] kinases[23,166] and other enzyme families. Finally, ABPP can be used to detect and quantify targets of covalent inhibitors at a proteome-wide level, by using highly reactive, non-specific probes and analysing the labelled proteins using LC-MS/MS.[167–171] New probe chemistries are constantly under development to increase the fraction of the proteome that can be analysed using this method, both for drug discovery for new targets and for characterizing off-targets.[27,172,173]

Although competitive ABPP is a powerful method to identify and characterize protein targets and off-targets of covalent inhibitors, it is limited in its ability to determine the exact sites of modifications at the amino acid level.[147] This most likely results from the complex fragmentation patterns of the bound covalent inhibitors, which makes identification of the modified peptides difficult. This information is important to be able to characterize fully the mechanism of binding of an inhibitor to off-targets, which may prove useful for further optimization. While the labelling sites can be inferred indirectly, recent advances in analysing the fragmentation patterns of bound inhibitors along with enrichment protocols for inhibitor-bound proteins permit the reliable determination of amino acid-level modification sites.[138,174]

In conclusion, characterizing the selectivity of covalent inhibitors should be performed as an inherent part of their design and optimization. At the most basic level, screening the inhibitor against panels of structurally and functionally similar proteins can provide a preliminary assessment of the inhibitor's selectivity. More thorough characterization should be performed using chemical proteomics methods that can identify and quantify unknown off-targets, and also identify the binding sites of the inhibitor on various targets.

4.4 Summary and Suggested Covalent Chemical Probe Criteria

Covalent chemical probes need to combine potency, selectivity and covalent binding to their targets in cells. However, since the potency of covalent probes increases with time (as do their off-target labelling), it is not straightforward to define optimization criteria for their development. Covalent probes should efficiently label their target at non-toxic concentrations on a time scale that is short compared with the target's half-life in the cell. These parameters are highly variable and therefore it is difficult to define clear threshold values that would define a general covalent probe. However, inspection of the values reported for various compounds are still insightful. Initial hits from fragment screenings frequently have k_{inact}/K_i values on the order of 1 M^{-1} s^{-1} and require concentrations >10 μM combined with long incubation times for significant labelling of the target. Such hits are not sufficiently potent to be covalent probes and require further optimization. At the other end of the spectrum lie clinically approved inhibitors (Table 4.1; Figure 4.3C) with k_{inact}/K_i values higher than 1×10^5 M^{-1} s^{-1} and fully label the target within minutes at nanomolar concentrations. Many effective covalent probes have intermediate values of k_{inact}/K_i.

Concerning selectivity, although typical enzymatic activity panels can and should still be applied to assess selectivity within the target's protein family, it is also important to assess the proteomic selectivity of covalent probes. Such methods that are available for covalent binders typically give a much broader view and importantly can assess the selectivity within the cellular environment. An optimal covalent probe would exhibit minimal off-target binding, particularly within its target protein family.

Selectivity for covalent probes is closely related to intrinsic reactivity. Since there are many ways to measure intrinsic reactivity, specifying target values according to one approach might not be productive. Still, using TNB^{2-} as a probe for reactivity, we saw that covalent fragments with TNB^{2-} reaction rates faster than 10^{-7} M^{-1} s^{-1} tend to be toxic and suggested this value as a guiding principal in triaging hits.

For non-covalent compounds, the SGC suggested criteria for designation as a *bona fide* "chemical probe": *in vitro* potency of <100 nM, >30-fold selectivity *versus* other family members and significant on-target cell activity at 1 μM. We propose to expand this definition for covalent chemical probes as follows:

- Covalent binding to the target should be confirmed by at least two independent methods.
- *In vitro* k_{inact}/K_i >100 M^{-1} s^{-1}.
- Cellular engagement at <1 μM.
- Proteomic selectivity assessment should show no subfamily off-targets at 1 μM.

We hope that these criteria will help guide the development of useful and selective covalent chemical probes for the myriad of protein targets that still lack chemical probes.

References

1. I. D. Kuntz, K. Chen, K. A. Sharp and P. A. Kollman, *Proc. Natl. Acad. Sci. U. S. A.*, 1999, **96**, 9997–10002.
2. A. J. T. Smith, X. Zhang, A. G. Leach and K. N. Houk, *J. Med. Chem.*, 2009, **52**, 225–233.
3. J. C. Way, *Curr. Opin. Chem. Biol.*, 2000, **4**, 40–46.
4. T. Sameshima, T. Yamamoto, O. Sano, S. Sogabe, S. Igaki, K. Sakamoto, K. Ida, M. Gotou, Y. Imaeda, J. Sakamoto and I. Miyahisa, *Biochemistry*, 2018, **57**, 1369–1379.
5. M.-C. Lu, J.-A. Ji, Z.-Y. Jiang and Q.-D. You, *Med. Res. Rev.*, 2016, **36**, 924–963.
6. J. Weisner, R. Gontla, L. van der Westhuizen, S. Oeck, J. Ketzer, P. Janning, A. Richters, T. Mühlenberg, Z. Fang, A. Taher, V. Jendrossek, S. C. Pelly, S. Bauer, W. A. L. van Otterlo and D. Rauh, *Angew. Chem., Int. Ed.*, 2015, **54**, 10313–10316.
7. J. M. Ostrem, U. Peters, M. L. Sos, J. A. Wells and K. M. Shokat, *Nature*, 2013, **503**, 548–551.
8. J. M. Bradshaw, J. M. McFarland, V. O. Paavilainen, A. Bisconte, D. Tam, V. T. Phan, S. Romanov, D. Finkle, J. Shu, V. Patel, T. Ton, X. Li, D. G. Loughhead, P. A. Nunn, D. E. Karr, M. E. Gerritsen, J. O. Funk, T. D. Owens, E. Verner, K. A. Brameld, R. J. Hill, D. M. Goldstein and J. Taunton, *Nat. Chem. Biol.*, 2015, **11**, 525–531.
9. M. S. Cohen, C. Zhang, K. M. Shokat and J. Taunton, *Science*, 2005, **308**, 1318–1321.
10. R. Lagoutte, R. Patouret and N. Winssinger, *Curr. Opin. Chem. Biol.*, 2017, **39**, 54–63.
11. K. M. Backus, B. E. Correia, K. M. Lum, S. Forli, B. D. Horning, G. E. González-Páez, S. Chatterjee, B. R. Lanning, J. R. Teijaro, A. J. Olson, D. W. Wolan and B. F. Cravatt, *Nature*, 2016, **534**, 570–574.
12. B. R. Lanning, L. R. Whitby, M. M. Dix, J. Douhan, A. M. Gilbert, E. C. Hett, T. O. Johnson, C. Joslyn, J. C. Kath, S. Niessen, L. R. Roberts, M. E. Schnute, C. Wang, J. J. Hulce, B. Wei, L. O. Whiteley, M. M. Hayward and B. F. Cravatt, *Nat. Chem. Biol.*, 2014, **10**, 760–767.
13. D. A. Shannon and E. Weerapana, *Curr. Opin. Chem. Biol.*, 2015, **24**, 18–26.
14. I. M. Serafimova, M. A. Pufall, S. Krishnan, K. Duda, M. S. Cohen, R. L. Maglathlin, J. M. McFarland, R. M. Miller, M. Frödin and J. Taunton, *Nat. Chem. Biol.*, 2012, **8**, 471–476.
15. R. M. Miller, V. O. Paavilainen, S. Krishnan, I. M. Serafimova and J. Taunton, *J. Am. Chem. Soc.*, 2013, **135**, 5298–5301.
16. S. Krishnan, R. M. Miller, B. Tian, R. D. Mullins, M. P. Jacobson and J. Taunton, *J. Am. Chem. Soc.*, 2014, **136**, 12624–12630.
17. N. Shindo, H. Fuchida, M. Sato, K. Watari, T. Shibata, K. Kuwata, C. Miura, K. Okamoto, Y. Hatsuyama, K. Tokunaga, S. Sakamoto,

S. Morimoto, Y. Abe, M. Shiroishi, J. M. M. Caaveiro, T. Ueda, T. Tamura, N. Matsunaga, T. Nakao, S. Koyanagi, S. Ohdo, Y. Yamaguchi, I. Hamachi, M. Ono and A. Ojida, *Nat. Chem. Biol.*, 2019, **15**, 250–258.
18. L. M. McGregor, M. L. Jenkins, C. Kerwin, J. E. Burke and K. M. Shokat, *Biochemistry*, 2017, **56**, 3178–3183.
19. K. A. Mix and R. T. Raines, *Org. Lett.*, 2015, **17**, 2358–2361.
20. P. Martín-Gago, E. K. Fansa, M. Winzker, S. Murarka, P. Janning, C. Schultz-Fademrecht, M. Baumann, A. Wittinghofer and H. Waldmann, *Cell Chem. Biol.*, 2017, **24**, 589–597.e5.
21. M. Morgen, C. Jöst, M. Malz, R. Janowski, D. Niessing, C. D. Klein, N. Gunkel and A. K. Miller, *ACS Chem. Biol.*, 2016, **11**, 1001–1011.
22. C. G. Jakob, A. K. Upadhyay, P. L. Donner, E. Nicholl, S. N. Addo, W. Qiu, C. Ling, S. M. Gopalakrishnan, M. Torrent, S. P. Cepa, J. Shanley, A. R. Shoemaker, C. C. Sun, A. Vasudevan, K. R. Woller, J. B. Shotwell, B. Shaw, Z. Bian and J. E. Hutti, *J. Med. Chem.*, 2018, **61**, 6647–6657.
23. Q. Zhao, X. Ouyang, X. Wan, K. S. Gajiwala, J. C. Kath, L. H. Jones, A. L. Burlingame and J. Taunton, *J. Am. Chem. Soc.*, 2017, **139**, 680–685.
24. A. Dal Corso, M. Catalano, A. Schmid, J. Scheuermann and D. Neri, *Angew. Chem., Int. Ed.*, 2018, **130**, 17424–17428.
25. A. Narayanan and L. H. Jones, *Chem. Sci.*, 2015, **6**, 2650–2659.
26. O. O. Fadeyi, L. R. Hoth, C. Choi, X. Feng, A. Gopalsamy, E. C. Hett, R. E. Kyne Jr, R. P. Robinson and L. H. Jones, *ACS Chem. Biol.*, 2017, **12**, 2015–2020.
27. S. Lin, X. Yang, S. Jia, A. M. Weeks, M. Hornsby, P. S. Lee, R. V. Nichiporuk, A. T. Iavarone, J. A. Wells, F. D. Toste and C. J. Chang, *Science*, 2017, **355**, 597–602.
28. S. Ray and A. S. Murkin, *Biochemistry*, 2019, **58**, 5234–5244.
29. K. M. Backus, *Curr. Top. Microbiol. Immunol.*, 2019, **420**, 375–417.
30. C. T. Walsh, *Annu. Rev. Biochem.*, 1984, **53**, 493–535.
31. M. Garcia-Calvo, E. P. Peterson, B. Leiting, R. Ruel, D. W. Nicholson and N. A. Thornberry, *J. Biol. Chem.*, 1998, **273**, 32608–32613.
32. J. T. Palmer, D. Rasnick, J. L. Klaus and D. Brömme, *J. Med. Chem.*, 1995, **38**, 3193–3196.
33. R. H. Abeles and A. L. Maycock, *Acc. Chem. Res.*, 1976, **9**, 313–319.
34. J. C. Powers, J. L. Asgian, O. D. Ekici and K. E. James, *Chem. Rev.*, 2002, **102**, 4639–4750.
35. P. J. Bednarski, D. J. Porubek and S. D. Nelson, *J. Med. Chem.*, 1985, **28**, 775–779.
36. S. M. Drawz and R. A. Bonomo, *Clin. Microbiol. Rev.*, 2010, **23**, 160–201.
37. Y. Luo, B. Knuckley, Y.-H. Lee, M. R. Stallcup and P. R. Thompson, *J. Am. Chem. Soc.*, 2006, **128**, 1092–1093.
38. H. Jamali, H. A. Khan, J. R. Stringer, S. Chowdhury and J. A. Ellman, *J. Am. Chem. Soc.*, 2015, **137**, 3616–3621.
39. S. Sen, S. Mondal, L. Zheng, A. J. Salinger, W. Fast, E. Weerapana and P. R. Thompson, *ACS Chem. Biol.*, 2019, **14**, 613–618.

40. S. Vazquez-Rodriguez, M. Wright, C. M. Rogers, A. P. Cribbs, S. Velupillai, M. Philpott, H. Lee, J. E. Dunford, K. V. M. Huber, M. B. Robers, J. D. Vasta, M. Thezenas, S. Bonham, B. Kessler, J. Bennett, O. Fedorov, F. Raynaud, A. Donovan, J. Blagg, V. Bavetsias, U. Oppermann, C. Bountra, A. Kawamura and P. E. Brennan, *Angew. Chem., Int. Ed.*, 2019, **58**, 515–519.
41. H. Lin, M. Wang, Y. W. Zhang, S. Tong, R. A. Leal, R. Shetty, K. Vaddi and J. I. Luengo, *ACS Med. Chem. Lett.*, 2019, **10**, 1033–1038.
42. E. Mons, I. D. C. Jansen, J. Loboda, B. R. van Doodewaerd, J. Hermans, M. Verdoes, C. A. A. van Boeckel, P. A. van Veelen, B. Turk, D. Turk and H. Ovaa, *J. Am. Chem. Soc.*, 2019, **141**, 3507–3514.
43. Z. Pan, H. Scheerens, S.-J. Li, B. E. Schultz, P. A. Sprengeler, L. C. Burrill, R. V. Mendonca, M. D. Sweeney, K. C. K. Scott, P. G. Grothaus, D. A. Jeffery, J. M. Spoerke, L. A. Honigberg, P. R. Young, S. A. Dalrymple and J. T. Palmer, *ChemMedChem*, 2007, **2**, 58–61.
44. Q. Liang, Y. Chen, K. Yu, C. Chen, S. Zhang, A. Wang, W. Wang, H. Wu, X. Liu, B. Wang, L. Wang, Z. Hu, W. Wang, T. Ren, S. Zhang, Q. Liu, C.-H. Yun and J. Liu, *Eur. J. Med. Chem.*, 2017, **131**, 107–125.
45. B. Wang, Y. Deng, Y. Chen, K. Yu, A. Wang, Q. Liang, W. Wang, C. Chen, H. Wu, C. Hu, W. Miao, W. Hur, W. Wang, Z. Hu, E. L. Weisberg, J. Wang, T. Ren, Y. Wang, N. S. Gray, Q. Liu and J. Liu, *Eur. J. Med. Chem.*, 2017, **137**, 545–557.
46. A. Thorarensen, M. E. Dowty, M. E. Banker, B. Juba, J. Jussif, T. Lin, F. Vincent, R. M. Czerwinski, A. Casimiro-Garcia, R. Unwalla, J. I. Trujillo, S. Liang, P. Balbo, Y. Che, A. M. Gilbert, M. F. Brown, M. Hayward, J. Montgomery, L. Leung, X. Yang, S. Soucy, M. Hegen, J. Coe, J. Langille, F. Vajdos, J. Chrencik and J.-B. Telliez, *J. Med. Chem.*, 2017, **60**, 1971–1993.
47. M. Forster, A. Chaikuad, S. M. Bauer, J. Holstein, M. B. Robers, C. R. Corona, M. Gehringer, E. Pfaffenrot, K. Ghoreschi, S. Knapp and S. A. Laufer, *Cell Chem. Biol.*, 2016, **23**, 1335–1340.
48. J. Engel, C. Becker, J. Lategahn, M. Keul, J. Ketzer, T. Mühlenberg, L. Kollipara, C. Schultz-Fademrecht, R. P. Zahedi, S. Bauer and D. Rauh, *Angew. Chem., Int. Ed.*, 2016, **55**, 10909–10912.
49. H. Cheng, S. K. Nair, B. W. Murray, C. Almaden, S. Bailey, S. Baxi, D. Behenna, S. Cho-Schultz, D. Dalvie, D. M. Dinh, M. P. Edwards, J. L. Feng, R. A. Ferre, K. S. Gajiwala, M. D. Hemkens, A. Jackson-Fisher, M. Jalaie, T. O. Johnson, R. S. Kania, S. Kephart, J. Lafontaine, B. Lunney, K. K.-C. Liu, Z. Liu, J. Matthews, A. Nagata, S. Niessen, M. A. Ornelas, S. T. M. Orr, M. Pairish, S. Planken, S. Ren, D. Richter, K. Ryan, N. Sach, H. Shen, T. Smeal, J. Solowiej, S. Sutton, K. Tran, E. Tseng, W. Vernier, M. Walls, S. Wang, S. L. Weinrich, S. Xin, H. Xu, M.-J. Yin, M. Zientek, R. Zhou and J. C. Kath, *J. Med. Chem.*, 2016, **59**, 2005–2024.
50. S. Tomassi, J. Lategahn, J. Engel, M. Keul, H. L. Tumbrink, J. Ketzer, T. Mühlenberg, M. Baumann, C. Schultz-Fademrecht, S. Bauer and D. Rauh, *J. Med. Chem.*, 2017, **60**, 2361–2372.

51. P. Wolle, J. Hardick, S. J. F. Cronin, J. Engel, M. Baumann, J. Lategahn, J. M. Penninger and D. Rauh, *J. Med. Chem.*, 2019, **62**, 2843–2848.
52. A. Chaikuad, P. Koch, S. A. Laufer and S. Knapp, *Angew. Chem., Int. Ed.*, 2018, **57**, 4372–4385.
53. Q. Liu, Y. Sabnis, Z. Zhao, T. Zhang, S. J. Buhrlage, L. H. Jones and N. S. Gray, *Chem. Biol.*, 2013, **20**, 146–159.
54. Z. Zhao and P. E. Bourne, *Drug Discovery Today*, 2018, **23**, 727–735.
55. Z. Zhao, Q. Liu, S. Bliven, L. Xie and P. E. Bourne, *J. Med. Chem.*, 2017, **60**, 2879–2889.
56. A. L. Garske, U. Peters, A. T. Cortesi, J. L. Perez and K. M. Shokat, *Proc. Natl. Acad. Sci. U. S. A.*, 2011, **108**, 15046–15052.
57. C. A. Ocasio, A. A. Warkentin, P. J. McIntyre, K. J. Barkovich, C. Vesely, J. Spencer, K. M. Shokat and R. Bayliss, *ACS Chem. Biol.*, 2018, **13**, 2956–2965.
58. A. Kung, M. Schimpl, A. Ekanayake, Y.-C. Chen, R. Overman and C. Zhang, *ACS Chem. Biol.*, 2017, **12**, 1499–1503.
59. K. J. Barkovich, M. K. Moore, Q. Hu and K. M. Shokat, *Nucleic Acids Res.*, 2018, **46**, 8689–8699.
60. N. London, R. M. Miller, S. Krishnan, K. Uchida, J. J. Irwin, O. Eidam, L. Gibold, P. Cimermančič, R. Bonnet, B. K. Shoichet and J. Taunton, *Nat. Chem. Biol.*, 2014, **10**, 1066–1072.
61. X. Ouyang, S. Zhou, C. T. T. Su, Z. Ge, R. Li and C. K. Kwoh, *J. Comput. Chem.*, 2013, **34**, 326–336.
62. D. Toledo Warshaviak, G. Golan, K. W. Borrelli, K. Zhu and O. Kalid, *J. Chem. Inf. Model.*, 2014, **54**, 1941–1950.
63. C. Scholz, S. Knorr, K. Hamacher and B. Schmidt, *J. Chem. Inf. Model.*, 2015, **55**, 398–406.
64. G. Bianco, S. Forli, D. S. Goodsell and A. J. Olson, *Protein Sci.*, 2016, **25**, 295–301.
65. K. Zhu, K. W. Borrelli, J. R. Greenwood, T. Day, R. Abel, R. S. Farid and E. Harder, *J. Chem. Inf. Model.*, 2014, **54**, 1932–1940.
66. J. Schröder, A. Klinger, F. Oellien, R. J. Marhöfer, M. Duszenko and P. M. Selzer, *J. Med. Chem.*, 2013, **56**, 1478–1490.
67. M. Rachman, A. Scarpino, D. Bajusz, G. Pálfy, I. Vida, A. Perczel, X. Barril and G. M. Keserű, *ChemMedChem*, 2019, **14**, 1011–1021.
68. S. De Cesco, S. Deslandes, E. Therrien, D. Levan, M. Cueto, R. Schmidt, L.-D. Cantin, A. Mittermaier, L. Juillerat-Jeanneret and N. Moitessier, *J. Med. Chem.*, 2012, **55**, 6306–6315.
69. A. Scarpino, G. G. Ferenczy and G. M. Keserű, *J. Chem. Inf. Model.*, 2018, **58**, 1441–1458.
70. C. Wen, X. Yan, Q. Gu, J. Du, D. Wu, Y. Lu, H. Zhou and J. Xu, *Molecules*, 2019, **24**, 2183.
71. A. Shraga, E. Olshvang, N. Davidzohn, P. Khoshkenar, N. Germain, K. Shurrush, S. Carvalho, L. Avram, S. Albeck, T. Unger, B. Lefker, C. Subramanyam, R. L. Hudkins, A. Mitchell, Z. Shulman, T. Kinoshita and N. London, *Cell Chem. Biol.*, 2019, **26**, 98–108.e5.

72. C. I. Nnadi, M. L. Jenkins, D. R. Gentile, L. A. Bateman, D. Zaidman, T. E. Balius, D. K. Nomura, J. E. Burke, K. M. Shokat and N. London, *J. Chem. Inf. Model.*, 2018, **58**, 464–471.
73. G. J. Correy, D. Zaidman, S. Carvalho, P. D. Mabbitt, P. J. James, A. C. Kotze, N. London and C. J. Jackson, *bioRxiv*, 2017, 161430.
74. S. R. Chowdhury, S. Kennedy, K. Zhu, R. Mishra, P. Chuong, A.-U. Nguyen, S. G. Kathman and A. V. Statsyuk, *Bioorg. Med. Chem. Lett.*, 2019, **29**, 36–39.
75. Y. Ai, L. Yu, X. Tan, X. Chai and S. Liu, *J. Chem. Inf. Model.*, 2016, **56**, 1563–1575.
76. M. Congreve, R. Carr, C. Murray and H. Jhoti, *Drug Discovery Today*, 2003, **8**, 876–877.
77. T. Hesterkamp and M. Whittaker, *Curr. Opin. Chem. Biol.*, 2008, **12**, 260–268.
78. C. Lipinski and A. Hopkins, *Nature*, 2004, **432**, 855–861.
79. M. M. Hann, A. R. Leach and G. Harper, *J. Chem. Inf. Comput. Sci.*, 2001, **41**, 856–864.
80. C. A. Lipinski, F. Lombardo, B. W. Dominy and P. J. Feeney, *Adv. Drug Delivery Rev.*, 2001, **46**, 3–26.
81. D. A. Erlanson, S. W. Fesik, R. E. Hubbard, W. Jahnke and H. Jhoti, *Nat. Rev. Drug Discovery*, 2016, **15**, 605–619.
82. D. E. Scott, A. G. Coyne, S. A. Hudson and C. Abell, *Biochemistry*, 2012, **51**, 4990–5003.
83. C. W. Murray and D. C. Rees, *Nat. Chem.*, 2009, **1**, 187–192.
84. D. A. Erlanson, A. C. Braisted, D. R. Raphael, M. Randal, R. M. Stroud, E. M. Gordon and J. A. Wells, *Proc. Natl. Acad. Sci. U. S. A.*, 2000, **97**, 9367–9372.
85. E. Sijbesma, K. K. Hallenbeck, S. Leysen, P. J. de Vink, L. Skóra, W. Jahnke, L. Brunsveld, M. R. Arkin and C. Ottmann, *J. Am. Chem. Soc.*, 2019, **141**, 3524–3531.
86. R. H. Nonoo, A. Armstrong and D. J. Mann, *ChemMedChem*, 2012, **7**, 2082–2086.
87. C. Jöst, C. Nitsche, T. Scholz, L. Roux and C. D. Klein, *J. Med. Chem.*, 2014, **57**, 7590–7599.
88. S. G. Kathman, Z. Xu and A. V. Statsyuk, *J. Med. Chem.*, 2014, **57**, 4969–4974.
89. G. B. Craven, D. P. Affron, C. E. Allen, S. Matthies, J. G. Greener, R. M. L. Morgan, E. W. Tate, A. Armstrong and D. J. Mann, *Angew. Chem., Int. Ed.*, 2018, **57**, 5257–5261.
90. R. Cardoso, R. Love, C. L. Nilsson, S. Bergqvist, D. Nowlin, J. Yan, K. K.-C. Liu, J. Zhu, P. Chen, Y.-L. Deng, H. J. Dyson, M. J. Greig and A. Brooun, *Protein Sci.*, 2012, **21**, 1885–1896.
91. E. Resnick, A. Bradley, J. Gan, A. Douangamath, T. Krojer, R. Sethi, P. P. Geurink, A. Aimon, G. Amitai, D. Bellini, J. Bennett, M. Fairhead, O. Fedorov, R. Gabizon, J. Gan, J. Guo, A. Plotnikov, N. Reznik, G. F. Ruda, L. Díaz-Sáez, V. M. Straub, T. Szommer, S. Velupillai, D. Zaidman,

Y. Zhang, A. R. Coker, C. G. Dowson, H. M. Barr, C. Wang, K. V. M. Huber, P. E. Brennan, H. Ovaa, F. von Delft and N. London, *J. Am. Chem. Soc.*, 2019, **141**, 8951–8968.
92. A. M. Roberts, D. K. Miyamoto, T. R. Huffman, L. A. Bateman, A. N. Ives, D. Akopian, M. J. Heslin, C. M. Contreras, M. Rape, C. F. Skibola and D. K. Nomura, *ACS Chem. Biol.*, 2017, **12**, 899–904.
93. L. Bar-Peled, E. K. Kemper, R. M. Suciu, E. V. Vinogradova, K. M. Backus, E. D. Horning, T. A. Paul, T.-A. Ichu, R. U. Svensson, J. Olucha, M. W. Chang, B. P. Kok, Z. Zhu, N. T. Ihle, M. M. Dix, P. Jiang, M. M. Hayward, E. Saez, R. J. Shaw and B. F. Cravatt, *Cell*, 2017, **171**, 696–709.e23.
94. G. Zimmermann, U. Rieder, D. Bajic, S. Vanetti, A. Chaikuad, S. Knapp, J. Scheuermann, M. Mattarella and D. Neri, *Chemistry*, 2017, **23**, 8152–8155.
95. J.-P. Daguer, C. Zambaldo, D. Abegg, S. Barluenga, C. Tallant, S. Müller, A. Adibekian and N. Winssinger, *Angew. Chem., Int. Ed.*, 2015, **54**, 6057–6061.
96. C. Zambaldo, J.-P. Daguer, J. Saarbach, S. Barluenga and N. Winssinger, *Medchemcomm*, 2016, **7**, 1340–1351.
97. S. G. Kathman, I. Span, A. T. Smith, Z. Xu, J. Zhan, A. C. Rosenzweig and A. V. Statsyuk, *J. Am. Chem. Soc.*, 2015, **137**, 12442–12445.
98. H. Johansson, Y.-C. Isabella Tsai, K. Fantom, C.-W. Chung, S. Kümper, L. Martino, D. A. Thomas, H. C. Eberl, M. Muelbaier, D. House and K. Rittinger, *J. Am. Chem. Soc.*, 2019, **141**, 2703–2712.
99. W. Zhou, D. Ercan, P. A. Jänne and N. S. Gray, *Bioorg. Med. Chem. Lett.*, 2011, **21**, 638–643.
100. N. Kwiatkowski, T. Zhang, P. B. Rahl, B. J. Abraham, J. Reddy, S. B. Ficarro, A. Dastur, A. Amzallag, S. Ramaswamy, B. Tesar, C. E. Jenkins, N. M. Hannett, D. McMillin, T. Sanda, T. Sim, N. D. Kim, T. Look, C. S. Mitsiades, A. P. Weng, J. R. Brown, C. H. Benes, J. A. Marto, R. A. Young and N. S. Gray, *Nature*, 2014, **511**, 616–620.
101. T. Zhang, F. Inesta-Vaquera, M. Niepel, J. Zhang, S. B. Ficarro, T. Machleidt, T. Xie, J. A. Marto, N. Kim, T. Sim, J. D. Laughlin, H. Park, P. V. LoGrasso, M. Patricelli, T. K. Nomanbhoy, P. K. Sorger, D. R. Alessi and N. S. Gray, *Chem. Biol.*, 2012, **19**, 140–154.
102. B.-F. Krippendorff, R. Neuhaus, P. Lienau, A. Reichel and W. Huisinga, *J. Biomol. Screening*, 2009, **14**, 913–923.
103. H. J. Burt, A. Galetin and J. B. Houston, *Xenobiotica*, 2010, **40**, 331–343.
104. R. Kitz and I. B. Wilson, *J. Biol. Chem.*, 1962, **237**, 3245–3249.
105. I. Miyahisa, T. Sameshima and M. S. Hixon, *Angew. Chem., Int. Ed.*, 2015, **127**, 14305–14308.
106. P. Kuzmic, *Anal. Biochem.*, 1996, **237**, 260–273.
107. P. A. Schwartz, P. Kuzmic, J. Solowiej, S. Bergqvist, B. Bolanos, C. Almaden, A. Nagata, K. Ryan, J. Feng, D. Dalvie, J. C. Kath, M. Xu, R. Wani and B. W. Murray, *Proc. Natl. Acad. Sci. U. S. A.*, 2014, **111**, 173–178.

108. A. Wijeratne, J. Xiao, C. Reutter, K. W. Furness, R. Leon, M. Zia-Ebrahimi, R. N. Cavitt, J. M. Strelow, R. D. Van Horn, S.-B. Peng, D. A. Barda, T. A. Engler and M. J. Chalmers, *ACS Med. Chem. Lett.*, 2018, **9**, 557–562.
109. S. Klüter, J. R. Simard, H. B. Rode, C. Grütter, V. Pawar, H. C. A. Raaijmakers, T. A. Barf, M. Rabiller, W. A. L. van Otterlo and D. Rauh, *Chembiochem*, 2010, **11**, 2557–2566.
110. T. Malcomson, K. Yelekci, M. T. Borrello, A. Ganesan, E. Semina, N. De Kimpe, S. Mangelinckx and R. R. Ramsay, *FEBS J.*, 2015, **282**, 3190–3198.
111. J. M. Strelow, *J. Biomol. Screening*, 2017, **22**, 3–20.
112. M. Mohutsky and S. D. Hall, in *Enzyme Kinetics in Drug Metabolism: Fundamentals and Applications*, ed. S. Nagar, U. A. Argikar and D. J. Tweedie, Humana Press, Totowa, NJ, 2014, pp. 57–91.
113. M. P. Patricelli, M. R. Janes, L.-S. Li, R. Hansen, U. Peters, L. V. Kessler, Y. Chen, J. M. Kucharski, J. Feng, T. Ely, J. H. Chen, S. J. Firdaus, A. Babbar, P. Ren and Y. Liu, *Cancer Discovery*, 2016, **6**, 316–329.
114. G. Roos, N. Foloppe and J. Messens, *Antioxid. Redox Signaling*, 2013, **18**, 94–127.
115. E. Awoonor-Williams and C. N. Rowley, *J. Chem. Theory Comput.*, 2016, **12**, 4662–4673.
116. E. Awoonor-Williams and C. N. Rowley, *J. Chem. Inf. Model.*, 2018, **58**, 1935–1946.
117. R. Liu, Z. Yue, C.-C. Tsai and J. Shen, *J. Am. Chem. Soc.*, 2019, **141**, 6553–6560.
118. M. E. Flanagan, J. A. Abramite, D. P. Anderson, A. Aulabaugh, U. P. Dahal, A. M. Gilbert, C. Li, J. Montgomery, S. R. Oppenheimer and T. Ryder, *et al.*, *J. Med. Chem.*, 2014, **57**, 10072–10079.
119. R. Lonsdale, J. Burgess, N. Colclough, N. L. Davies, E. M. Lenz, A. L. Orton and R. A. Ward, *J. Chem. Inf. Model.*, 2017, **57**, 3124–3137.
120. P. Baudouin-Cornu, G. Lagniel, C. Kumar, M.-E. Huang and J. Labarre, *J. Biol. Chem.*, 2012, **287**, 4552–4561.
121. P. A. MacFaul, A. D. Morley and J. J. Crawford, *Bioorg. Med. Chem. Lett.*, 2009, **19**, 1136–1138.
122. P. Ábrányi-Balogh, L. Petri, T. Imre, P. Szijj, A. Scarpino, M. Hrast, A. Mitrović, U. P. Fonovič, K. Németh, H. Barreteau, D. I. Roper, K. Horváti, G. G. Ferenczy, J. Kos, J. Ilaš, S. Gobec and G. M. Keserű, *Eur. J. Med. Chem.*, 2018, **160**, 94–107.
123. V. J. Cee, L. P. Volak, Y. Chen, M. D. Bartberger, C. Tegley, T. Arvedson, J. McCarter, A. S. Tasker and C. Fotsch, *J. Med. Chem.*, 2015, **58**, 9171–9178.
124. D. E. Epps and B. M. Taylor, *Anal. Biochem.*, 2001, **295**, 101–106.
125. H. Esterbauer, E. Schwarzl and M. Hayn, *Anal. Biochem.*, 1977, **77**, 486–494.
126. J. M. Dypbukt, C. Bishop, W. M. Brooks, B. Thong, H. Eriksson and A. J. Kettle, *Free Radical Biol. Med.*, 2005, **39**, 1468–1477.
127. J. R. Huth, R. Mendoza, E. T. Olejniczak, R. W. Johnson, D. A. Cothron, Y. Liu, C. G. Lerner, J. Chen and P. J. Hajduk, *J. Am. Chem. Soc.*, 2005, **127**, 217–224.

128. J. S. Martin, C. J. MacKenzie, D. Fletcher and I. H. Gilbert, *Bioorg. Med. Chem.*, 2019, **27**, 2066–2074.
129. J. T. Metz, J. R. Huth and P. J. Hajduk, *J. Comput.-Aided Mol. Des.*, 2007, **21**, 139–144.
130. R. G. Parr, L. v. Szentpály and S. Liu, *J. Am. Chem. Soc.*, 1999, **121**, 1922–1924.
131. F. Palazzesi, M. A. Grundl, A. Pautsch, A. Weber and C. S. Tautermann, *J. Chem. Inf. Model.*, 2019, **59**, 3565–3571.
132. R. D. Caldwell, H. Qiu, B. C. Askew, A. T. Bender, N. Brugger, M. Camps, M. Dhanabal, V. Dutt, T. Eichhorn, A. S. Gardberg, A. Goutopoulos, R. Grenningloh, J. Head, B. Healey, B. L. Hodous, B. R. Huck, T. L. Johnson, C. Jones, R. C. Jones, I. Mochalkin, F. Morandi, N. Nguyen, M. Meyring, J. R. Potnick, D. C. Santos, R. Schmidt, B. Sherer, A. Shutes, K. Urbahns, A. V. Follis, A. A. Wegener, S. C. Zimmerli and L. Liu-Bujalski, *J. Med. Chem.*, 2019, **62**, 7643–7655.
133. H. Xu, M. I. Jesson, U. I. Seneviratne, T. H. Lin, M. N. Sharif, L. Xue, C. Nguyen, R. A. Everley, J. I. Trujillo, D. S. Johnson, G. R. Point, A. Thorarensen, I. Kilty and J.-B. Telliez, *ACS Chem. Biol.*, 2019, **14**, 1235–1242.
134. D. Martinez Molina, R. Jafari, M. Ignatushchenko, T. Seki, E. A. Larsson, C. Dan, L. Sreekumar, Y. Cao and P. Nordlund, *Science*, 2013, **341**, 84–87.
135. H. Xu, A. Gopalsamy, E. C. Hett, S. Salter, A. Aulabaugh, R. E. Kyne, B. Pierce and L. H. Jones, *Org. Biomol. Chem.*, 2016, **14**, 6179–6183.
136. M. B. C. Dillon, D. A. Bachovchin, S. J. Brown, M. G. Finn, H. Rosen, B. F. Cravatt and K. A. Mowen, *ACS Chem. Biol.*, 2012, **7**, 1198–1204.
137. A. Borodovsky, H. Ovaa, N. Kolli, T. Gan-Erdene, K. D. Wilkinson, H. L. Ploegh and B. M. Kessler, *Chem. Biol.*, 2002, **9**, 1149–1159.
138. C. M. Browne, B. Jiang, S. B. Ficarro, Z. M. Doctor, J. L. Johnson, J. D. Card, S. C. Sivakumaren, W. M. Alexander, T. M. Yaron, C. J. Murphy, N. P. Kwiatkowski, T. Zhang, L. C. Cantley, N. S. Gray and J. A. Marto, *J. Am. Chem. Soc.*, 2019, **141**, 191–203.
139. A. K. Ghosh, I. Samanta, A. Mondal and W. R. Liu, *ChemMedChem*, 2019, **14**, 889–906.
140. K. T. Barglow and B. F. Cravatt, *Nat. Methods*, 2007, **4**, 822–827.
141. B. Knuckley, J. E. Jones, D. A. Bachovchin, J. Slack, C. P. Causey, S. J. Brown, H. Rosen, B. F. Cravatt and P. R. Thompson, *Chem. Commun.*, 2010, **46**, 7175–7177.
142. D. Leung, C. Hardouin, D. L. Boger and B. F. Cravatt, *Nat. Biotechnol.*, 2003, **21**, 687–691.
143. U. Rix and G. Superti-Furga, *Nat. Chem. Biol.*, 2009, **5**, 616–624.
144. N. Jessani and B. F. Cravatt, *Curr. Opin. Chem. Biol.*, 2004, **8**, 54–59.
145. B. F. Cravatt, A. T. Wright and J. W. Kozarich, *Annu. Rev. Biochem.*, 2008, **77**, 383–414.
146. A. B. Berger, P. M. Vitorino and M. Bogyo, *Am. J. PharmacoGenomics*, 2004, **4**, 371–381.

147. E. Weerapana, A. E. Speers and B. F. Cravatt, *Nat. Protoc.*, 2007, **2**, 1414–1425.
148. M. H. Wright and S. A. Sieber, *Nat. Prod. Rep.*, 2016, **33**, 681–708.
149. C. Drahl, B. F. Cravatt and E. J. Sorensen, *Angew. Chem., Int. Ed.*, 2005, **44**, 5788–5809.
150. D. K. Nomura and T. J. Maimone, *Curr. Top. Microbiol. Immunol.*, 2019, **420**, 351–374.
151. K. T. Barglow and B. F. Cravatt, *Chem. Biol.*, 2004, **11**, 1523–1531.
152. A. E. Speers, G. C. Adam and B. F. Cravatt, *J. Am. Chem. Soc.*, 2003, **125**, 4686–4687.
153. A. E. Speers and B. F. Cravatt, *Chem. Biol.*, 2004, **11**, 535–546.
154. H.-C. Liang, Y.-C. Liu, H. Chen, M.-C. Ku, Q.-T. Do, C.-Y. Wang, S.-F. Tzeng and S.-H. Chen, *J. Proteome Res.*, 2018, **17**, 2590–2599.
155. L. A. Crawford and E. Weerapana, *Mol. BioSyst.*, 2016, **12**, 1768–1771.
156. N. Liu, S. Hoogendoorn, B. van de Kar, A. Kaptein, T. Barf, C. Driessen, D. V. Filippov, G. A. van der Marel, M. van der Stelt and H. S. Overkleeft, *Org. Biomol. Chem.*, 2015, **13**, 5147–5157.
157. H. Lebraud, D. J. Wright, C. E. East, F. P. Holding, M. O'Reilly and T. D. Heightman, *Mol. BioSyst.*, 2016, **12**, 2867–2874.
158. A. T. Wright, J. D. Song and B. F. Cravatt, *J. Am. Chem. Soc.*, 2009, **131**, 10692–10700.
159. J. Zhou, E. D. Mock, A. Martella, V. Kantae, X. Di, L. Burggraaff, M. P. Baggelaar, K. Al-Ayed, A. Bakker, B. I. Florea, S. H. Grimm, H. den Dulk, C. T. Li, L. Mulder, H. S. Overkleeft, T. Hankemeier, G. J. P. van Westen and M. van der Stelt, *ACS Chem. Biol.*, 2019, **14**, 164–169.
160. C.-L. Kuo, W. W. Kallemeijn, L. T. Lelieveld, M. Mirzaian, I. Zoutendijk, A. Vardi, A. H. Futerman, A. H. Meijer, H. P. Spaink, H. S. Overkleeft, J. M. F. G. Aerts and M. Artola, *FEBS J.*, 2019, **286**, 584–600.
161. K. Otrubova, S. Chatterjee, S. Ghimire, B. F. Cravatt and D. L. Boger, *Bioorg. Med. Chem.*, 2019, **27**, 1693–1703.
162. A. F. Kornahrens, A. B. Cognetta 3rd, D. M. Brody, M. L. Matthews, B. F. Cravatt and D. L. Boger, *J. Am. Chem. Soc.*, 2017, **139**, 7052–7061.
163. H. Xu, J. D. Majmudar, D. Davda, P. Ghanakota, K. H. Kim, H. A. Carlson, H. D. Showalter, B. R. Martin and G. L. Amidon, *Mol. Pharm.*, 2015, **12**, 3399–3407.
164. W. W. Kallemeijn, K.-Y. Li, M. D. Witte, A. R. A. Marques, J. Aten, S. Scheij, J. Jiang, L. I. Willems, T. M. Voorn-Brouwer and C. P. van Roomen, *et al.*, *Angew. Chem., Int. Ed.*, 2012, **51**, 12529–12533.
165. L. I. Willems, J. Jiang, K.-Y. Li, M. D. Witte, W. W. Kallemeijn, T. J. N. Beenakker, S. P. Schröder, J. M. Aerts, G. A. van der Marel and J. D. C. Codée, *et al.*, *Chem. - Eur. J.*, 2014, **20**, 10864–10872.
166. M.-C. Yee, S. C. Fas, M. M. Stohlmeyer, T. J. Wandless and K. A. Cimprich, *J. Biol. Chem.*, 2005, **280**, 29053–29059.
167. A. J. Maurais and E. Weerapana, *Curr. Opin. Chem. Biol.*, 2019, **50**, 29–36.

168. M. Abo, C. Li and E. Weerapana, *Mol. Pharm.*, 2018, **15**, 743–749.
169. S. M. Hacker, K. M. Backus, M. R. Lazear, S. Forli, B. E. Correia and B. F. Cravatt, *Nat. Chem.*, 2017, **9**, 1181–1190.
170. C. C. Ward, J. I. Kleinman and D. K. Nomura, *ACS Chem. Biol.*, 2017, **12**, 1478–1483.
171. E. Weerapana, C. Wang, G. M. Simon, F. Richter, S. Khare, M. B. D. Dillon, D. A. Bachovchin, K. Mowen, D. Baker and B. F. Cravatt, *Nature*, 2010, **468**, 790–795.
172. M. Abo and E. Weerapana, *J. Am. Chem. Soc.*, 2015, **137**, 7087–7090.
173. F. Yang, J. Gao, J. Che, G. Jia and C. Wang, *Anal. Chem.*, 2018, **90**, 9576–9582.
174. S. B. Ficarro, C. M. Browne, J. D. Card, W. M. Alexander, T. Zhang, E. Park, R. McNally, S. Dhe-Paganon, H.-S. Seo, I. Lamberto, M. J. Eck, S. J. Buhrlage, N. S. Gray and J. A. Marto, *Anal. Chem.*, 2016, **88**, 12248–12254.
175. N. Uhlenbrock, S. Smith, J. Weisner, I. Landel, M. Lindemann, T. A. Le, J. Hardick, R. Gontla, R. Scheinpflug, P. Czodrowski, P. Janning, L. Depta, L. Quambusch, M. P. Müller, B. Engels and D. Rauh, *Chem. Sci.*, 2019, **10**, 3573–3585.
176. T. Barf, T. Covey, R. Izumi, B. van de Kar, M. Gulrajani, B. van Lith, M. van Hoek, E. de Zwart, D. Mittag, D. Demont, S. Verkaik, F. Krantz, P. G. Pearson, R. Ulrich and A. Kaptein, *J. Pharmacol. Exp. Ther.*, 2017, **363**, 240–252.
177. S. Smith, M. Keul, J. Engel, D. Basu, S. Eppmann and D. Rauh, *ACS Omega*, 2017, **2**, 1563–1575.
178. D. S. Johnson, C. Stiff, S. E. Lazerwith, S. R. Kesten, L. K. Fay, M. Morris, D. Beidler, M. B. Liimatta, S. E. Smith, D. T. Dudley, N. Sadagopan, S. N. Bhattachar, S. J. Kesten, T. K. Nomanbhoy, B. F. Cravatt and K. Ahn, *ACS Med. Chem. Lett.*, 2011, **2**, 91–96.
179. D. Bensinger, D. Stubba, A. Cremer, V. Kohl, T. Waßmer, J. Stuckert, V. Engemann, K. Stegmaier, K. Schmitz and B. Schmidt, *J. Med. Chem.*, 2019, **62**, 2428–2446.
180. B. T. Adams, S. Niccoli, M. A. Chowdhury, A. N. K. Esarik, S. J. Lees, B. P. Rempel and C. P. Phenix, *Chem. Commun.*, 2015, **51**, 11390–11393.
181. N. R. Vance, K. R. Witkin, P. W. Rooney, Y. Li, M. Pope and M. A. Spies, *ChemMedChem*, 2018, **13**, 2514–2521.
182. J. Kempson, D. Ovalle, J. Guo, S. T. Wrobleski, S. Lin, S. H. Spergel, J. J.-W. Duan, B. Jiang, Z. Lu, J. Das, B. V. Yang, J. Hynes Jr, H. Wu, J. Tokarski, J. S. Sack, J. Khan, G. Schieven, Y. Blatt, C. Chaudhry, L. M. Salter-Cid, A. Fura, J. C. Barrish, P. H. Carter and W. J. Pitts, *Bioorg. Med. Chem. Lett.*, 2017, **27**, 4622–4625.
183. Y. Liu, Z. Xie, D. Zhao, J. Zhu, F. Mao, S. Tang, H. Xu, C. Luo, M. Geng, M. Huang and J. Li, *J. Med. Chem.*, 2017, **60**, 2227–2244.

184. G. Cullia, S. Bruno, S. Parapini, M. Margiotta, L. Tamborini, A. Pinto, A. Galbiati, A. Mozzarelli, M. Persico, A. Paladino, C. Fattorusso, D. Taramelli and P. Conti, *ACS Med. Chem. Lett.*, 2019, **10**, 590–595.
185. T. M. M. Stack, W. Li, W. H. Johnson Jr, Y. J. Zhang and C. P. Whitman, *Biochemistry*, 2018, **57**, 1012–1021.
186. E. Badarau, Z. Wang, D. L. Rathbone, A. Costanzi, T. Thibault, C. E. Murdoch, S. El Alaoui, M. Bartkeviciute and M. Griffin, *Chem. Biol.*, 2015, **22**, 1347–1361.

CHAPTER 5

Cyclic Peptides as Chemical Probes

L. SERRANO[†a], G. ROPER[†a] AND A. KAWAMURA*[a,b]

[a]University of Oxford, Chemistry Research Laboratory, 12 Mansfield Road, Oxford OX1 3TA, UK; [b]Newcastle University, School of Natural and Environmental Sciences, Chemistry, Bedson Building, Newcastle upon Tyne NE1 7RU, UK
*E-mail: akane.kawamura@newcastle.ac.uk

5.1 Introduction

Peptides have historically been used both as therapeutics and as chemical probes. Initially, this field was inspired by Nature and, since the isolation of the 51-mer insulin peptide in 1921, many natural biologically active peptides have been characterized and extensively studied, synthesized/derivatized and used as therapeutics. Other relevant examples are oxytocin, which is a natural hormone used for inducing labour, vancomycin and cyclosporine, which are cyclic peptides with non-canonical amino acids (nCAAs) produced by bacteria and fungi and used as antibiotics, and leuprolide, which is a derivative of a natural 10-mer hormone gonadotropin-releasing hormone used in *in vitro* fertilization (IVF) in addition to treating prostate cancer, endometriosis and uterine fibroids.[1]

[†]These authors contributed equally.

Chemical Biology No. 16
The Discovery and Utility of Chemical Probes in Target Discovery
Edited by Paul Brennan and Saleta Vazquez Rodriguez

Published by the Royal Society of Chemistry, www.rsc.org

The development of solid-phase peptide synthesis (SPPS) methods, started in 1963 by Robert Bruce Merrifield[2] (with the subsequent award of the Nobel Prize in Chemistry in 1984), have made peptide synthesis accessible to most laboratory settings and, more recently, the availability of automated peptide synthesizers has reduced synthesis times from weeks to hours. This has allowed unprecedented access to peptide libraries and peptide analogues, which has led to an increased understanding of peptide properties such as membrane permeability, 3D structure, stability and solubility, aiding in the design of peptides with the required features to be used as chemical probes and therapeutics.

The next big step in this area of research was the ability to identify *de novo* binders for targets. Initially, the study of peptides was limited to analogues of natural product peptides or protein segments identified as protein–protein interaction (PPI) hotspots. This approach, although useful, cannot be used to develop probes for targets that have not been extensively studied. Taking inspiration from the vast small-molecule libraries employed for high-throughput screening, peptide libraries of random peptide sequences have been developed that can be screened against any target to identify binders with completely novel sequences. The generation of these libraries was largely facilitated by the development of biochemical methods where the cellular machinery can be highjacked. An example of this is the groundbreaking peptide phage display technology, for which George P. Smith and Sir Gregory P. Winter were awarded one half of the Nobel Prize in Chemistry in 2018. Other methods have since been developed, based on similar display principles but applied in various innovative ways.

In this chapter, we discuss the use of *de novo* cyclic peptides as probes. Cyclic peptides have attracted increasing attention in recent years, owing to their advantageous properties in comparison with their linear counterparts. The chemistry of cyclic peptides is discussed, followed by an overview of selection methods available for identifying *de novo* peptide probes. Finally, key examples of cyclic peptide probes that have recently been developed and their applications are presented.

5.2 Advantages and Disadvantages of Peptides as Probes

Chemical probes are traditionally associated with small molecules[3] but can also span the chemical space beyond small molecules, including peptides and biologics. There are unique advantages and challenges associated with each modality (Table 5.1), with peptides often bridging the gap between small molecules and biologics.[4,5] Peptides occupy the chemical space between small molecules ($<\sim$500 Da) and biologics ($>\sim$5000 Da)[6] (Figure 5.1). The larger size and the 3D chemical diversity of peptides compared with small molecules increase the interaction surface area, improving the affinity and selectivity.[7] Peptides are also able to target effectively larger, flatter protein sites than small molecules,

Table 5.1 Summary of the advantages and disadvantages of using peptides as probes.

Advantages	Disadvantages
Good size match for the flat and shallow sites associated with PPIs	Intracellular targets are challenging owing to poor cellular permeability
Flexibility of peptides allows them to "fit" into target cavities or bind across large surfaces	Many hydrogen bond donors/acceptors - poor drug-like properties
High structural diversity facilitated by modularity	Cleavage by proteolytic enzymes
High potency and selectivity facilitated by high structural diversity	Potential immunogenicity
High-diversity screening techniques	Low blood plasma stability
Biocompatible, so tend to have low toxicity	High hepatic and renal clearance
Ease/speed of synthesis and SAR analysis; both active and inactive control probes can readily be generated	More difficult to model computationally than small molecules
Ease of modification for improvement of probe	

- Cyclic peptides display increased stability relative to linear peptides.
- Cyclic peptides have fewer degrees of conformational freedom, reducing the entropic penalty of binding their target. Bicyclic peptides are even more rigid.

owing to their size and more flexible structures.[7,8] This makes peptides a good choice for targeting PPIs, which tend to be ~1500–3000 Å^2, whereas small molecules are more suited for targeting smaller, deeper pockets of ~300–1000 Å^2 associated with protein–small molecule interactions.[6] Antibodies and other biological probes can have even greater affinity and specificity for their targets than peptides, but are generally mainly suitable for extracellular targets because they are seldom membrane permeable.[5,9] However, engineering cell permeability for peptides is possible, hence peptides can give access to a wider range of targets than antibodies.[5,10] The advent of innovative technologies for the rapid screening and identification of *de novo* peptides[11] has increased their popularity as probes in recent years.

Peptides have vast structural diversity as they can be varied in length and be made up of any combination of the 20 canonical amino acids (CAAs), in addition to countless nCAAs.[7] The diversity in peptide libraries increases the chances of identifying selective and highly potent hits against the target of interest[11] and the efficiency of library production can be enhanced by using ribosomal processes.[12] Automated Fmoc SPPS allows hits from these libraries to be quickly and easily synthesized and facilitates rapid progress of structure–activity relationship (SAR) studies to optimize the properties of the resulting probe. In addition, many Fmoc-protected nCAA building blocks are commercially available, allowing efficient automated synthesis of peptides containing nCAAs.[13]

Computational modelling of peptide–protein binding is more challenging than for predicting small molecule–protein binding,[14] owing to the increased complexity of the interactions between peptides and proteins.[8] Models are continually being improved (see Chapter 3), but there is still progress to be made to match the quality of modelling available for small molecules and proteins.[15,16] Although there are many advantages of peptides over small molecules, there are challenges that need to be overcome. Generating cell-permeable peptide probes can be very difficult and time consuming; whereas this is not problematic for targeting cell-surface proteins, it is a major barrier for intracellular targets. This is partly because peptides tend to be beyond the traditional Lipinski's "rule of five" in terms of both molecular weight and number of hydrogen bonds.[17] Most peptides also have low blood plasma stability as they are cleaved by proteases, and further issues include high clearance and some illicit immunogenicity.[8] Optimization of these properties is necessary before peptides can be used in *in vivo* models.

In some cases, Nature has addressed the issues associated with peptides *via* evolution.[18] Some bioactive natural peptides are cell permeable and proteolytically stable and have desirable pharmacokinetic properties. A classic example is cyclosporine A, a cyclic peptide that contains nCAAs, including D-amino acids and backbone N^{α}-methylated amino acids[19,20] (Figure 5.1). Such bioactive peptides have inspired the design of *de novo* peptides. For instance, cyclization reduces the degree of conformational freedom of a peptide, thus lowering the entropic penalty upon binding to a protein target and potentially increasing the resulting affinity and selectivity.[12] In addition, intramolecular backbone hydrogen bonds are more likely in cyclic peptides, reducing the number of solvent-exposed hydrogen bond donors and acceptors and thus increasing cell permeability (this also increases the rigidity of the peptide, increasing potency and selectivity).[21,22] Finally, many methods of peptide cyclization shield the N- and/or C-terminus of the peptides, thus reducing their susceptibility to proteolysis, as many proteases bind the N-/C-termini of proteins.[4] In designing peptide probes, cyclization has also been used to stabilize α-helices, usually by "stapling" the i and $(i + 3/4/7)$ amino acids with a linker, which resulted in peptide inhibitors of PPIs involving α-helices.[23] In general, stapled peptides originate from proteins[24] and *de novo* linear peptides[25] that are active in an α-helical conformation. Beyond cyclization, incorporation of nCAAs (such as D-amino acids, N^{α}-methylation and side-chain cross-linking), which are often found in naturally occurring bioactive peptides, can help to improve physicochemical properties. Proline can also be installed to encourage the formation of β-hairpin turns, positioning the backbone favourably for intramolecular hydrogen bonding.[21] A combination of these strategies is often implemented to improve the properties of peptide probes.[8]

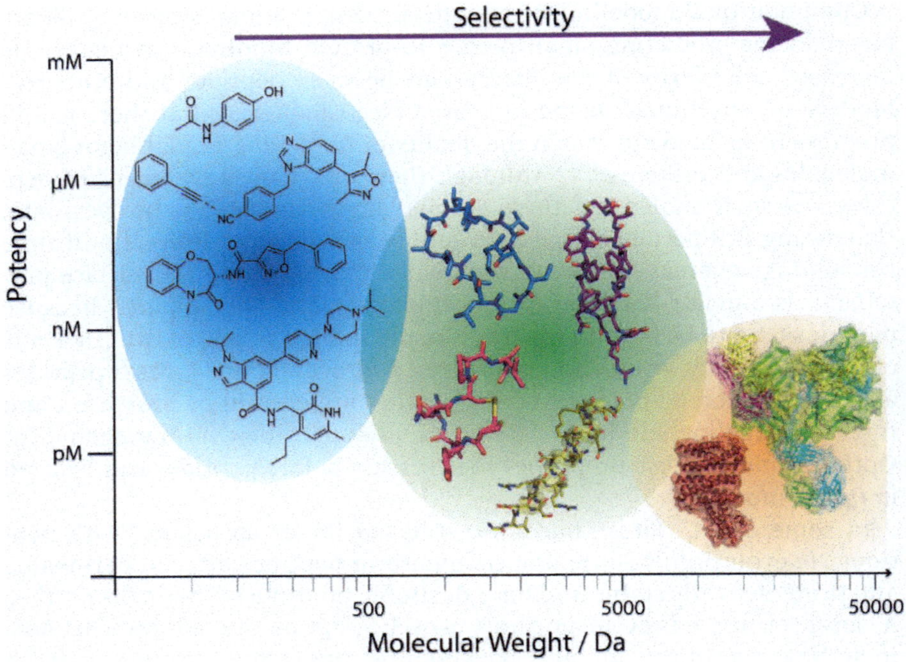

Figure 5.1 Peptides fall into the molecular weight range between small molecules and proteins. Generalized illustration of size, potency and selectivity properties of small molecules (blue shading), peptides (green shading) and biologics/antibodies (orange shading). Examples of each class are given as representation only; small molecules clockwise from the top include paracetamol, compound 28a,[82] a poised fragment,[83] GSK481 and UNC1999; peptides (green shading) clockwise from the top left include cyclosporine (PDB code 1cwa), CP2 (PDB code 5LY1), Mcl-1–SAHBD (PDB code 3MK8) and oxytocin (PDB code 1XY2); biologics (orange shading), the top protein is immunoglobulin antibody (PDB code 1IGT) and the bottom protein is granulocyte (PDB code 1CWA).

5.3 Selection Methods for Cyclic Peptides

Earlier research on cyclic peptides was primarily based on natural products and (bio)synthesis of their derivatives or the design of peptide fragments derived from protein sequences of known PPIs – with the most common strategy being to chemically staple or cyclize the peptide fragment to retain the 3D structure of their proteins. The application of these approaches has thus been limited to well-characterized protein targets/PPIs. Recent advances in peptide synthesis, library generation and screening technologies make the identification of high-affinity *de novo* cyclic peptides for a variety of targets possible. The most efficient way of developing *de novo* peptides is through library screening against a target of interest, involving multiple rounds of selection/enrichment for binding peptides, followed by hit generation and functional validation.[26]

Peptide library generation can be achieved through chemical methods (*e.g.* automated peptide synthesis) or biological methods (*e.g.* using cellular or synthetic ribosome apparatus). Both methods can be used to achieve rapid synthesis of a very large and diverse pool of peptides (see Table 5.2). In chemically synthesized libraries, the diversity can be achieved by split and mix randomization. In comparison, sequence variety in libraries made with biological methods can be generated through varying the length and composition of the DNA library encoding the peptides. This is achieved through saturation mutagenesis in which codons can be randomized to represent all the available CAAs at each position (*e.g.* NNS, NNK) or reprogrammed to incorporate additional nCAAs.[27] For example, a 5-mer NNK peptide library using 20 CAAs can generate a theoretical library diversity of 20^5 (3.2 million).

In display libraries, the peptides are linked to their encoding genetic tags to allow for rapid identification of hits binding to the target of interest. In biological methods, this is achieved either by tethering the peptide produced to its encoding RNA/DNA (*e.g.* using puromycin) or by compartmentalization, which ensures that every template is in a separate compartment (*e.g.* a cell or vesicle) and that the peptides produced in the compartment are all generated from this template.[26] The development of DNA next-generation sequencing (NGS) has revolutionized the encoded library screening technologies. Previously, hit deconvolution was a time- and labour-intensive process with the DNA from each hit prepared individually for Sanger sequencing. With the development of NGS, a quicker and more in-depth analysis of the hits is now possible, giving access to datasets that can be used for SAR analysis, to determine trends and to inform further study of the targets.[28] One bead, one compound (OBOC) libraries are an exception as the peptides synthesized are not tagged for identification but instead binding peptides are identified through chemical methods. The hit deconvolution is therefore difficult and more time consuming than DNA sequencing.[29]

5.3.1 Phage Display

Phage display was initially developed for displaying antibody fragments.[30] It is a biological method and has since been used as a mainstream method for screening linear and cyclic peptide libraries (kits are commercially available). Phage display involves fusing a peptide sequence to the coat protein of a phage (M13, fd or f1 filamentous phages have been used for cyclic peptide selections). The variable peptide sequences can be inserted at the solvent-exposed N-terminus of either the pVIII protein which coats the sides of the phage particles (~2700 proteins per phage) or the pIII protein of which five copies are displayed at one end of the phage particle.[31] The use of the pVIII protein is advantageous as avidity effects (several hundred peptides displayed per phage) can help identify low-affinity binders in the first rounds of screening against a new target. Using the pIII protein in which 1–5 copies of the peptide can be displayed is more appropriate during later rounds of selection when trying to enrich for high-affinity binders. Various systems are available for inserting a peptide sequence into the phage proteins, the most common method being the use of phagemids.[32] A phagemid vector is designed from a filamentous phage that encodes the pIII/pVIII protein, with a variable region corresponding to a peptide for the library, and contains the required information for the production of this modified protein. As the phagemid encodes only the modified protein, a helper phage containing all the genetic information is required for propagation of the phagemid virion.[33]

Initially, a phagemid vector library containing variable inserts is transformed into *Escherichia coli* to produce the phage starting library (Figure 5.2A). The phagemid vector contains an antibiotic resistance gene that selects for transformed cells. The cells are then infected with a helper phage to propagate the phagemid virion. The phagemid virions can then be screened by affinity capture against immobilized targets. Unbound phages are removed by successive washes and bound phages subsequently recovered. These are used to infect *E. coli* and infected colonies are cultured to produce more copies of the phage displaying hit peptides. The process can be repeated through several rounds of enrichment, then the phagemid DNA can be isolated and sequenced to determine the target-binding peptide sequences. In phage display, the size of the library is limited by the transformation efficiency of *E. coli* (~10^9, depending on vector size). The use of phagemid systems presents a number of advantages compared with unmodified filamentous phage systems. The number of copies of peptide displayed can be varied to as low as one peptide per phage on the pIII protein. This can be helpful as it removes any avidity effects from the selection and ensures that a greater proportion of the binding peptides are recovered as fewer peptides are lost during the infection of *E. coli* (modifying the pIII proteins decreases the ability of the virion to infect cells).[34] In addition, as the quantity of genetic material in phagemid vectors is half that of filamentous phages, the transfection efficiency is higher, increasing the possible size of the library.[35]

The phage display system has been modified to produce cyclic peptide libraries (Figure 5.2B). Cysteine residues can be inserted for disulfide bond formation[36] or to react with small organic molecules (*e.g.* di- or tribromomethylbenzene) to form cyclic or bicyclic peptides.[37] As both of these methods have limitations (lability of disulfide bonds and off-target reaction with cysteines outside the peptide sequence), a method has been developed to insert N^{ε}-acryloyllysine, which reacts under physiological conditions with a cysteine in the peptide sequence irreversibly to form a cyclic peptide.[38]

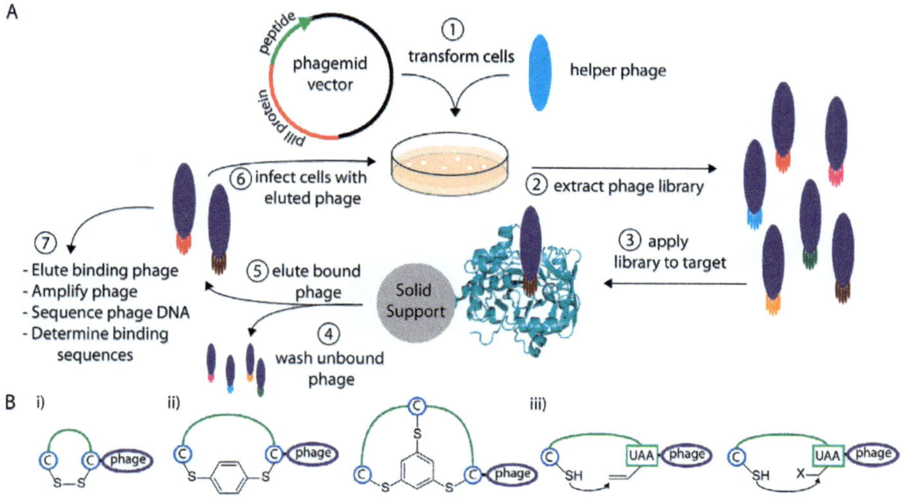

Figure 5.2 Phage display selection and peptide cyclization methods. (A) Selection flow for phage display. Step 1: phagemid vectors are prepared, each containing a specific variable peptide sequence. These vectors are transformed into cells infected with helper phages, allowing the cells to produce new phagemids displaying the variable peptide sequences. Step 2: colonies are picked, grown in media on a larger scale and the phagemids are recovered from the growth media. Step 3: the phagemids are incubated with the immobilized target. Step 4: the target is washed to remove unbound phages. Step 5: bound phages are eluted and recovered. These can either be used to (Step 6) infect cells in a new round of selection for further enrichment or (Step 7) be sent for sequencing to identify hits. (B) Cyclization methods used in cyclic peptide phage libraries under physiological conditions: (i) disulfide bond formation; (ii) use of dibromomethylbenzene for cyclic and tribromomethylbenzene for bicyclic non-reversible bond formation; analogues of di- and tribromomethylbenzene can also be used; (iii) cyclization between a cysteine residue and an electrophilic nCAA.

5.3.2 mRNA Display

mRNA display uses a cell-free *in vitro* selection method, in which peptides are fused to encoding mRNA from which they are translated by a linker (*e.g.* puromycin)[39] (Figure 5.3A). A starting DNA library containing a variable region of determined length, flanked by two defined amino acids for cyclization purposes, is designed. The starting DNA library is transcribed to mRNA to which the puromycin linker is ligated. The peptides are then synthesized by translation components either purified from cells or recombinantly generated, in the *in vitro* translation (IVT) step.[40] The reverse transcription step provides each peptide with a stable hybrid DNA/RNA tag. An affinity capture screen

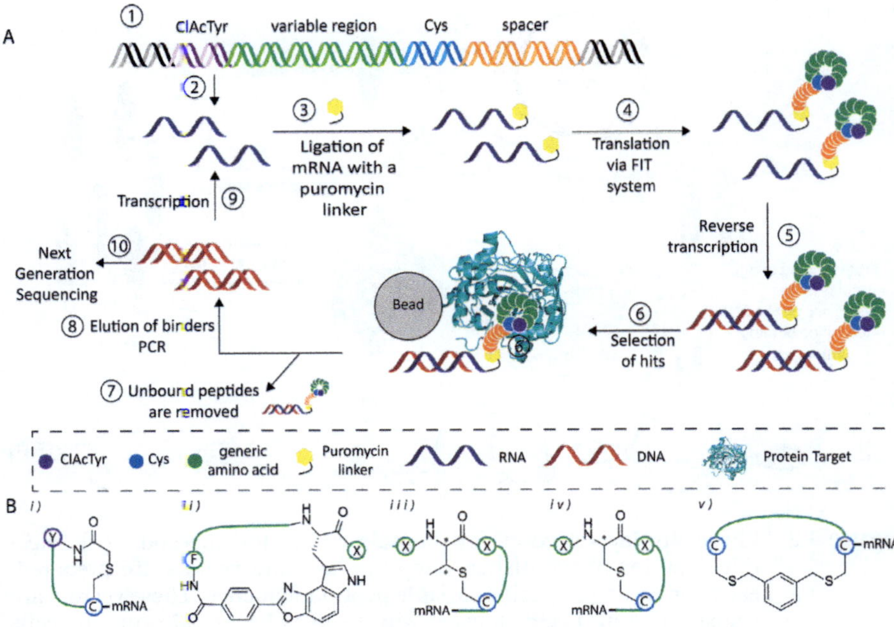

Figure 5.3 mRNA display-based cyclic peptide library selection and cyclization chemistries used. (A) An example of a selection flow for mRNA display (based on RaPID technology). Step 1: DNA template is produced containing a T7 promoter site for RNA polymerase recognition, a ribosome binding site, a start codon for translation, a variable peptide sequence flanked by two selected amino acids for cyclization, a spacer and a sequence for puromycin ligation. Step 2: transcription of DNA template to mRNA. Step 3: ligation of DNA-linked puromycin to mRNA strands. Step 4: peptide translation using an *in vitro* translation mixture to reprogram the Met starting codon with nCAAs [*e.g.* chloroacetyltyrosine (ClAcTyr)] for cyclization. The final amino acid in the sequence is also selected for cyclization, often a nucleophilic cysteine residue. Ligation of peptide to mRNA strand occurs through the puromycin linker and spontaneous cyclization takes place. Step 5: reverse transcription to give an RNA–DNA duplex for each peptide. Step 6: incubation of peptide library with immobilized target. Step 7: unbound targets washed away. Step 8: single-stranded DNA recovered by heat elution, qPCR quantification of recovered DNA followed by PCR amplification. Step 9: DNA used in transcription reaction to produce mRNA used in further selection rounds for sequence enrichment. Step 10: if sufficient enrichment is achieved, DNA is analysed by NGS for hit identification. (B) Cyclization methods for mRNA display: (i) cyclization between a chloroacetamide on an N-terminal residue (e.g. tyrosine, Y, and a cysteine sulfhydryl group, C; (ii) oxidative coupling of 5-hydroxytryptophan and benzylamine on a phenylalanine residue, F, (X) indicates any AA can be used in the position; (iii) Michael addition of cysteine, C, to dehydrobutyrine giving two racemic stereocentres; (iv) 4-selenalanine is converted to dehydroalanine by oxidative elimination, which reacts with a cysteine *via* Michael addition to give one racemic stereocentre; (v) α,α′-dibromo-*m*-xylene used to bridge two cysteine residues.[45]

against an immobilized target is carried out with this library, in which non-binders are removed by successive washes and DNA from binders is recovered by heat elution and polymerase chain reaction (PCR). The recovered DNA can be fed back into the screen for multiple rounds of enrichment. When satisfactory enrichment has been achieved (monitored by qPCR), the DNA can be analysed by NGS to deconvolute hit sequences. With this method, linear and cyclic peptide libraries of 10^{12}–10^{13} peptides can be screened. The advantage of the cell-free system is that a variety of nCAAs can be incorporated to diversify and expand the library. By reassigning codons, nCAAs can be incorporated into peptides in a site-specific manner. This is limited by the ability of aminoacyl-tRNA synthetases to acylate unrecognized nCAAs onto tRNA. To insert chemically diverse nCAAs, natural tRNAs and aminoacyl-tRNA synthetases can be replaced with a tRNA preloaded with an nCAA.

Engineered nCAA–tRNA can be achieved by chemically modifying the amino acid on a pre-acylated tRNA,[41] acylating chemically synthesized nCAAs using natural or modified aminoacyl-tRNA synthetases[42] or by using synthetic ribozymes [*e.g.* flexizymes used in Random non-standard Peptide Integrated Discovery (RaPID) libraries].[43,44] These last two methods allow for the greatest diversity of nCAA incorporation into mRNA selections and are the basis of most cyclization methods reported to date. The diversity and physicochemical properties of the libraries synthesized can also be improved through these strategies.[40] A variety of cyclization methods are compatible with the technique, including head-to-tail or side-chain cyclization (Figure 5.3B), with some limitations on the chemistry used due to aqueous compatibility.[45]

5.3.3 Split-intein Circular Ligation of Peptides and Proteins (SICLOPPS)

Split-intein circular ligation of peptides and proteins (SICLOPPS) is a selection method in which the peptides are produced and tested for activity against their target within a single cell (Figure 5.4A). In this technique, the variable peptide sequence is inserted into a plasmid flanked by split-intein protein sequences which, when translated, are recognized and linked by biological splicing mechanisms, releasing the variable peptide sequence with head-to-tail cyclization by a peptide bond (Figure 5.4B).[46] The designed plasmid is transformed/transfected into cells (*E. coli*, *Saccharomyces cerevisiae* and human cells have been used[46–48]), which limits the library size to the transformation efficiency of the cell type used (10^9 for *E. coli*, lower for eukaryotic cells).[49] The method requires a cysteine or serine residue at the peptide N-terminus for cyclization, but otherwise can accommodate a random amino acid region of variable length. Incorporation of nCAAs has been achieved,[50,51] but it remains limited to one nCAA per peptide, unlike mRNA display in which multiple nCAAs can be inserted. The intein splicing is coupled with a cellular assay with an identifiable readout that allows hits to be isolated by collecting positive cells, recovering the plasmid and amplifying it for sequencing.

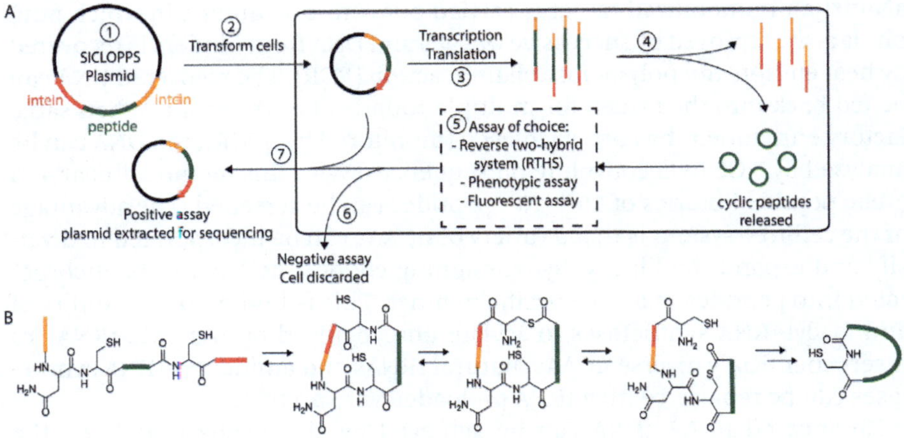

Figure 5.4 SICLOPPS cyclic peptide library selection and intein-based cyclization chemistry. (A) Selection flow for SICLOPPS. Step 1: SICLOPPS plasmids are prepared, each containing a specific variable peptide sequence. Step 2: these vectors are transformed into cells designed to carry out the assay of choice. Step 3: the plasmids are transcribed and translated through cellular machinery. Step 4: cyclic peptide is spliced from the protein sequence. Step 5: sufficient cyclic peptide is produced and cells containing hit peptides are identified through a selected assay. Step 6: cells are sorted and negative cells are discarded. Step 7: positive cells are recovered and the plasmid is extracted and sequenced for hit deconvolution. (B) Peptide splicing mechanism for SICLOPPS.

One such assay developed for use with *E. coli* is a bacterial reverse two-hybrid system (RTHS) that can identify peptides which disrupt PPIs.[52] The main advantage of this method is that functional assays are used to identify active peptides. In contrast, most other methods are affinity capture methods that deliver tight binding peptides but do not guarantee activity. In this method, the peptides are produced in the cell, which allows them to be tested directly in *in vivo* functional assays. This is in contrast to other *in vitro* methods where testing of intracellular functional effect of peptides may be limiting due to cell permeability issues. However, peptides produced by SICLOPPS are not guaranteed cell permeability when synthesized chemically.

5.3.4 Split and Mix Chemical Synthesis

Split and mix synthesis of peptides is an efficient way of making a large and diverse library of peptides chemically.[53] The peptides are covalently linked onto a surface, which can be a bead, a chemical scaffold or a membrane. The starting scaffold is split equally between multiple reaction vessels, each of which contains reagents to add a specific amino acid to the scaffold. After washing to remove reagents, the products are mixed, then split further for addition of a second amino acid. This process is repeated to assemble the peptide, increasing the library diversity exponentially with each successive reaction (Figure 5.5A). This method permits the incorporation of nCAAs (limited

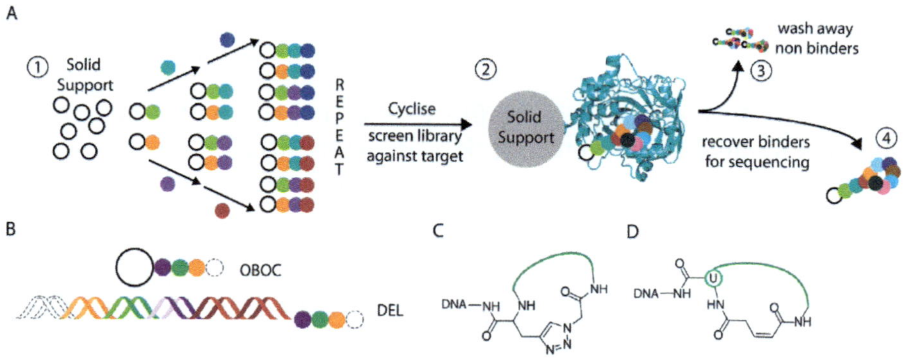

Figure 5.5 Chemically synthesized cyclic peptide libraries. (A) Selection flow for split and mix peptide synthesis. Step 1: the scaffold onto which peptides are grown is split into sections, different peptides are added to each of the sections and the sections are then mixed, then split randomly for addition of a new amino acid. Step 2: once the library has been synthesized to the wanted peptide length and cyclized, an affinity screen can be performed against an immobilized target. Step 3: non-binders are washed away. Step 4: binding peptides are recovered and identified. (B) The scaffold onto which peptides are grown can vary. In one bead, one compound (OBOC) libraries, TentaGel beads are usually used. In DNA-encoded libraries (DELs), a DNA scaffold is used with one end for peptide synthesis and one end for ligating-encoding DNA sequences. (C) A DEL cyclization strategy using Huisgen cycloaddition between two nCAAs, one containing an alkyne and the other an azide. (D) A DEL cyclization strategy using a Wittig reaction between two nCAAs, one containing a Wittig reagent and the other an aldehyde.

only by coupling efficiency) and allows various modes of cyclization with more freedom than enzymatic conditions allow. The libraries made in this way can be used in affinity-based selections against immobilized or labelled targets. Two examples in which this method is used will be discussed: one bead, one compound (OBOC) libraries and DNA-encoded libraries (DELs) (Figure 5.5B).

OBOC libraries are made synthetically using SPPS and split and mix methods.[53,54] The peptides are synthesized on beads, most commonly TentaGel. OBOC libraries can comprise 10^6–10^8 library members and can yield linear, cyclic, bicyclic or branched peptides. Modifications such as glycosylation can also be added to the peptide post-synthesis.[55] Post-screen against a selected target with a fluorescent tag, positive beads can be identified by fluorescence microscopy and picked out manually (low throughput) or automatically using a Complex Object Parametric Analyzer and Sorter (COPAS, high throughput).[56] The positive hits can then be identified by either Edman degradation microsequencing,[57] tandem mass spectrometry [in which a cleavable hydrophilic linker to cleave the peptide from the bead and a bromine tag for better mass spectrometric (MS) recognition are used][58] or partial Edman degradation-mass spectrometry (PED-MS).[59] This last method was designed specifically to sequence cyclic peptides in OBOC libraries which are difficult to deconvolute using traditional methods. Although MS remains the favoured method for sequencing owing to its low cost and rapid throughput, the identification of long sequences (15+ amino acids) and cyclic/bicyclic/branched peptides remains challenging.

DELs use DNA segments to identify each amino acid added by split and mix. The scaffold on which the peptides are synthesized is a piece of double-stranded DNA with an nCAA at one end onto which the peptide sequence can be synthesized. For each amino acid added, an encoding segment of DNA is ligated onto the DNA scaffold. All reactions must be water compatible for the DNA to remain in solution and for the DNA ligation to occur. There are currently two DEL methods for cyclic peptides. In the first, macrocyclization is achieved using a copper-catalysed Huisgen cycloaddition;[60] at the end of each peptide addition, an ether precipitation is used to recover the library members, washing away reagents before the next step (Figure 5.5C). The second method uses one-pot synthesis rather than split and mix and uses the Wittig reaction for cyclization[61] (Figure 5.5D). DNA-encoded libraries have an advantage over OBOC libraries in that DNA tags can easily be decoded by NGS, whereas hit identification remains complicated for cyclic and nCAA-containing peptides in OBOC libraries.[29]

5.3.5 Comparison of Methods

There are advantages and disadvantages to each of the methods described (Table 5.2). mRNA display can generate the largest library of peptides ($>10^{12}$) as it is unconstrained by the transformation efficiency of cells or by issues regarding peptide length or limited by chemical synthesis/capacity. OBOC and DELs have the most access to nCAAs; however, they have some limitations: in the case of OBOC, deconvoluting hit sequences that are cyclic and/or contain nCAAs can be challenging, and for DELs few cyclization strategies exist owing to DNA compatibility issues. Although mRNA display offers the greatest diversity in nCAA incorporation amongst biological methods, successful efforts have been made to incorporate nCAAs into both phage display and SICLOPPS. In terms of peptide function, SICLOPPS presents clear advantages, as cyclic peptides are naturally generated *in vivo* and hits are selected based on function rather than affinity for a target. This is particularly useful for identifying inhibitors of PPIs. However, prioritization of hits due to function means that peptide optimization is likely required to achieve the desired binding potencies for probe development. Phage display is a well-established technique, with fewer *in vitro* manipulations than mRNA display. It has benefited from advances such as phagemid development and cyclization strategies but has limitations due to the difficulty of nCAA incorporation and *in vivo* steps limiting chemical cyclization approaches. The application of phages/OBOC for certain targets may also be limited by sterics, since the phages/beads are very large compared with the target. Steric issues are minimized in the case of mRNA display and DELs, which have small encoding tags, and SICLOPPS, which uses untagged peptides.

Table 5.2 Summary of selection methods for cyclic peptides.

Selection method	Size of library	Cyclization	Type of screening	nCAA integration	Notes
Phage display	Limited by cell transformation step; $<10^9$ for *E. coli*	Linear, cyclic and bicyclic possible. Most examples rely on cysteine residue placement in sequence	Cyclic peptides produced *in vivo* Target screening *in vitro*	Limited to one nCAA	Hit deconvolution by DNA sequencing (NGS possible)
mRNA display	10^{12}–10^{13}	Many cyclization strategies available	Peptides produced *in vitro* with ribosomes Target screening *in vitro*	Multiple nCAAs can be inserted	Hit deconvolution by sequencing (NGS possible) Multiple cyclization methods available
SICLOPPS	$>10^8$ Limited by cell transformation step	Intein splicing	Peptides produced *in vivo* Target screening *in vivo*	Limited to one nCAA	Hit deconvolution by sequencing (NGS possible) Function-based screening
OBOC (one bead, one compound)	$\sim10^7$ Theoretically can be higher but physically limited by synthesizer capacity	Any chemical method compatible with SPPS	Peptides chemically produced *in vitro* Target screening *in vitro*	Multiple nCAAs can be inserted	Chemistry for nCAAs and cyclization not restricted Hit identification by MS or Edman degradation challenging for cyclic peptides and nCAAs
DEL (DNA-encoded library)	10^{12} for cycloaddition 256 000 for Wittig cyclization	Huisgen “click” cycloaddition or Wittig cyclization	Peptides chemically produced *in vitro* Target screening *in vitro*	Multiple nCAAs can be inserted	Hit deconvolution by sequencing (NGS possible) Chemistry limited by aqueous conditions necessary for DNA solubility

5.4 Examples of Cyclic Peptides

There are now abundant and diverse examples of *de novo* cyclic peptides that have been developed as chemical tools/probes or as starting points for therapeutic development. In this section, we highlight some recent examples of different types of libraries and screening methodologies used to develop cyclic peptides against different targets of biomedical and therapeutic interest.

5.4.1 Stapled Peptides Targeting p53:Mdm2/Mdm4

Mdm2/Mdm4 are oncoproteins upregulated in many cancer cells and that bind to the N-terminus of tumour suppressor protein p53 and inhibit regulation of the cell cycle.[62] The disruption of the Mdm–p53 complex is thus critical in p53 activation and this PPI has been extensively explored as a therapeutic strategy for cancer. As the Mdm-p53 PPI involves an α-helix of p53, earlier work in this area was focused on the generation of α-helical peptide inhibitors or "stapled peptides" to stabilize the α-helix conformation by cyclization, to mimic this interaction better. Brown *et al.*[62] discovered linear peptide binders of Mdm2/Mdm4 using phage display, which were subsequently stabilized *via* cyclization to generate stapled peptides.[25] The original peptide was modified by removing the C-terminal Pro12 residue before a hydrocarbon *i*, *i* + 7 staple was added; subsequent SAR studies resulted in a stapled peptide Asn8Ala analogue that displayed a very strong p53 transcriptional response in cells. The peptide had greater potency than SAH-8,[24] a stapled-peptide based on the α-helix of p53 involved in the interaction and with activity comparable to that of nutlin, a small-molecule probe targeting Mdm2, without the cytotoxicity observed at higher concentrations of nutlin.[62] Another stapled peptide derived from a different linear peptide from phage display has made it to phase I/II clinical trials.[63,64]

Stapled peptide SAH-8 (Figure 5.6G) was functionalized by Lau *et al.*[65] using a two-component stapling approach, the *i* and *i* + 7 positions were modified to azide-containing nCAAs that could be joined to a separate linker using click chemistry. This two-component stapling approach allowed functionalization of the linker; one peptide included a cell-penetrating tri-Arg peptide in the linker, resulting in a cell-permeable stapled peptide from the cell-impermeable parent.[65]

5.4.2 Investigating the Biology of HIF-1 in Response to Hypoxia–Inhibiting Dimerization

Cyclic peptides can also be used to disrupt protein dimerization. Hypoxia-inducible factor (HIF) is a transcription factor stabilized under hypoxic conditions and is important for the upregulation of genes necessary for cell survival under low-oxygen conditions; HIF is a promising cancer drug target

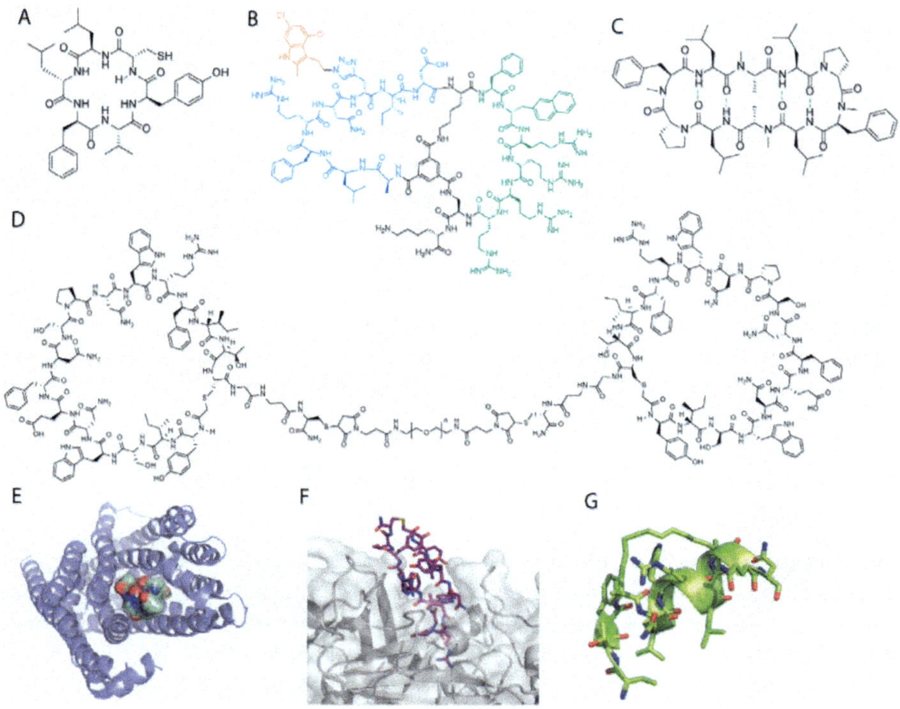

Figure 5.6 Examples of cyclic peptides. (A) Cyclic peptide inhibiting HIF heterodimerization, *cyclo*-CLLFVY, discovered using SICLOPPs (Section 5.4.2). (B) A cell-penetrating bicyclic peptide that inhibits K-Ras, discovered from a library of peptides on beads (Section 5.4.6); the peptide library consists of a cell-penetrating ring (green), a five amino acid variable region (blue) and peptides containing L-propargylglycine were coupled to 4,6-dichloro-2-methyl-3-aminoethylindole (a small-molecule inhibitor of K-Ras) (orange). (C) A peptide produced using a decapeptide scaffold, which was designed based on bioavailable peptides (Section 5.4.8). The scaffold has a β-hairpin structure that positions the backbone to allow four intramolecular hydrogen bonds; backbone amide nitrogens not involved in hydrogen bonding are methylated. (D) A MET-binding cyclic peptide was developed into a dimerization probe by connecting two peptides together using a poly(ethylene glycol) linker (Section 5.4.3). Dimerization of MET receptors using the probe led to artificial activation. (E) Macrocyclic peptides that stabilize PfMATE were used as co-crystallization ligands, later enabling the optimization of crystallization of the apo form (Section 5.4.4) (PDB code 3VVR). (F) KDM4A–C-selective cyclic peptide inhibitor (CP2) in complex with KDM4A; the crystal structure was used to inform modifications to the parent peptide, resulting in a cell-active peptide (Section 5.4.5) (PDB code 5LY1). (G) Crystal structure of a natural product-derived peptide, stabilized using peptide stapling, SAH-8, which inhibits Mdm–p53 complex formation (Section 5.4.1) (PDB code 3V3B).

because tumours are often hypoxic.[66] SICLOPPS was used to generate a cyclic peptide, *cyclo*-CLLFVY, that inhibits the heterodimerization of HIF-1α and HIF-1β (Figure 5.6A); the peptide identified was the first inhibitor to disrupt HIF-1 dimerization but not HIF-2.[67] *Cyclo*-CLLFVY ligated to a Tat-tag cell-penetrating peptide (CPP) sequence inhibited HIF-1 dimerization under hypoxic conditions (when HIF is expressed) in a luciferase reporter assay in two different cell lines.[67] The Tat-tagged peptide was later used as a tool to investigate the presence of an epigenetically controlled HIF-1α transactivation loop on HIF-1α mRNA and protein levels under hypoxic conditions in cancer cells.[68] This probe was useful as siRNA and genetic knockouts affect the quantity of HIF-1α in cells. Engineered cells that chromosomally encode the inhibitory peptide were generated and used to study the hypoxic response,[69] allowing the cyclic peptide probes to be used as heterologously expressed, inducible heterodimerization inhibitors in cells.

5.4.3 Activation of the MET Receptor – Induction of Dimerization

Cyclic peptides have also been used as tools to induce PPIs to probe the target biology. The MET receptor is activated by hepatocyte growth factor (HGF) binding and the response plays a role in wound healing, regeneration, anti-fibrosis and development.[70] Monomeric MET-binding macrocycles identified using RaPID screening were dimerized with a poly(ethylene glycol) linker, creating a probe that induced dimerization of MET receptors (Figure 5.6D) that led to artificial receptor activation.[71] The dimerizing probe has a similar effect to HGF and facilitates cell migration, induces phosphorylation of MET and other downstream kinase proteins also phosphorylated by HGF and elicits the same signalling cascade/gene expression profile as HGF.[70] This example elegantly demonstrates the ability of cyclic peptide probes to investigate oligomerization/complex formation of the target protein in biology.

5.4.4 Macrocyclic Peptides as Co-crystallization Ligands

Cyclic peptides can act to stabilize the target protein, often by stabilizing a flexible region of the protein or "locking" in a protein conformation to aid protein crystallization. Tanaka *et al.* generated macrocyclic peptides (Figure 5.6E) that stabilize and inhibit *Pyrococcus furiosus* multidrug and toxic compound extrusion (PfMATE) transporter protein using RaPID screening.[72] Structures of PfMATE obtained in complexes with macrocyclic peptides revealed subtle pH control over the conformation of a transmembrane helix, enabling the authors later to generate high-resolution apo structures,[73] providing a promising strategy for crystallization.

5.4.5 Substrate-competitive Selective KDM4A–C Inhibitor

Peptide screening techniques have proved useful for finding binders/inhibitors of proteins for which there are no selective small-molecule alternatives. Kawamura *et al.* used RaPID screening to develop a selective cyclic peptide inhibitor of histone demethylases 4A–C (KDM4A–C),[74] CP2 (Figure 5.6F). Generating chemical probes that target KDMs has been challenged, owing to the high active-site homology between many of the KDMs and other groups of Fe(II)- and 2-oxoglutarate-dependent oxygenases.[75] Co-crystal structure-guided modifications to CP2 were performed to generate a peptide that showed target engagement and stabilization in cells, at ~10-fold lower concentration than the parent peptide.[74] Interestingly, the cyclic peptide anchored into the active site pocket with Arg positioned where trimethyllysine in cognate histone substrate would bind. Methylation at this arginine converted this cyclic peptide CP2 into a substrate. This correlates with the arginine demethylation activity identified for this family of enzymes, including for KDM4A,[76] exemplifying how the cyclic peptides can also be used as tools to inform the biology of enzymes.

5.4.6 Strategies for Improving Cellular Permeability of "Probes": Bicyclic Cell-permeable Peptides

CPPs are often used to deliver otherwise impermeable peptides to intracellular targets. Trinh *et al.*[77] incorporated a CPP into a library of 5.7×10^6 bicyclic peptides on beads and used this library to identify a K-Ras inhibitor (Figure 5.6B). The peptides were comprised of one ring containing five random residues and a second ring made up of a CPP. The random sequence of amino acids was made up from a pool of 26 amino acids, namely 10 L-CAAs, six L-nCAAs and 10 D-nCAAs. L-Propargylglycine, an alkyne-containing amino acid, was included in the screen, and was later coupled to a known small-molecule Ras inhibitor using click chemistry. Peptides with single-digit μM K_D^{app} values were generated that had cellular uptake efficiencies of 25–56%; further optimization of the peptides resulted in a peptide with a K_D^{app} of 0.21 μM for K-Ras G12V, an IC_{50} of 3.4 μM for the Ras–Raf interaction and evidence of cellular activity. The results suggested that the CPP ring participates in binding, as the use of alternative CPPs reduced the binding affinity.[77]

5.4.7 OBOC-stapled Peptide

Although ribosomally synthesized peptide screening techniques can accommodate some nCAAs, chemical synthesis is still a convenient means to install a vast range of nCAAs. An OBOC approach was used by Rezaei Araghi *et al.*[78] to screen a library of stapled peptides against Mcl-1 that contained a mixture of CAAs and nCAAs. In this study, a truncated version of a known linear peptide inhibitor[79] identified using yeast display was used to develop a smaller stapled peptide with improved binding.

A library of 108 stapled peptide variants were prepared on resin and stapled, using Grubbs metathesis, and three amino acid positions were varied from the parent peptide in the library. Peptides were initially selected against biotinylated Mcl-1 in the presence of a 12.5-fold higher concentration of Bcl-x_L, a structurally homologous protein for which selectivity was desired against. The beads were washed and incubated with streptavidin-coated quantum dots; this allowed fluorescence-based sorting to isolate the brightest 5% of the beads and these went through to a second round of selection using a 50-fold higher concentration of Bcl-x_L. After further washing and incubation with streptavidin-coated quantum dots, the brightest 170 beads were isolated using fluorescence microscopy, cleaved from the resin and identified using matrix-assisted laser desorption/ionization mass spectrometry (MALDI-MS). From 108 initial sequences, six peptides were identified from the screen. The peptides identified had higher affinity for Mcl-1 than the parent and selectivity for Mcl-1 over similar proteins; they contained surprising substitutions. One of the peptides identified had Mcl-1 specific activity in cellular assays, with an EC_{50} of 27 nM.[78]

5.4.8 Peptides with Drug- or Natural Product-like Properties

In order to generate peptides with greater resistance to proteases, Baeriswyl and Heinis screened a peptide library that was subjected to proteases during the selection process.[80] They developed inhibitors of human plasma kallikrein, for which a biological and a bicyclic peptide inhibitor had been generated, but both were unstable to proteases. Peptides with improved stability to proteases and single-digit to sub-μM IC_{50} values were generated.[80]

Other methods have been used to bias peptide selections towards hits with properties more similar to those of drugs/natural products; Passioura *et al.* reprogrammed the genetic code used in RaPID screening so that it contained only hydrophobic amino acids, replacing seven of the most hydrophilic CAAs with 11 hydrophobic nCAAs.[81] Peptides were selected against the interleukin-6 receptor using both the reprogrammed library and a library comprised of the CAAs (without Met); hits from the radically reprogrammed library and CAA library were of comparably high affinity.[81]

Inspired by the few cases of cell-permeable and bioavailable peptides, Fouché *et al.*[21] set out to design a general cyclic decapeptide scaffold with desirable pharmacokinetic properties (Figure 5.6C). The resulting scaffold contained a β-hairpin structure, enabled by two opposite type II β-turns, which positioned backbone amides favourably for hydrogen bonding. Backbone NH groups not involved in intramolecular hydrogen bonding were methylated. Rigidification of the cyclic peptide and shielded hydrogen bond donors/acceptors improved passive cell permeability by reducing the energetic penalty for leaving the polar solvent around cells and entering the lipophilic cell membrane; stabilization of the β-turns also improved clearance times.[21]

The different strategies used in these studies provide clues and inspiration for designing cyclic peptides that will not only have desired affinities and functions, but can also be robustly used as chemical probes in cells and *in vivo*.

5.5 Conclusion

Cyclic peptides are an exciting class of compounds that are currently underexploited as chemical probes. The recent advances in cyclic peptide (bio)synthesis and screening technologies mean that *de novo* cyclic peptide ligands against a target of interest can now be efficiently generated. This is reflected in the many examples of *de novo* cyclic peptides for various targets reported during the last decade, including for challenging targets such as "difficult" PPIs. Although technical hurdles remain, most importantly reliable cell penetration, there is enormous potential for cyclic peptides to serve as chemical probes and tools for chemical biology. Although most cyclic peptides reported to date are functional inhibitors (*e.g.* catalysis, PPIs), they have also been successfully applied as investigative tools, as exemplified by cyclic peptides that can stabilize proteins (*e.g.* for crystallization),[74] induce dimerization[71] or act as new substrates,[74] informing their target function. Hence there is genuine excitement in this field – it is clear that we are just at the beginning of a new era of cyclic peptide discovery and we are likely to see more examples of cyclic peptides as chemical probes and their applications in target validation/therapeutics development in the future.

Abbreviations

CAA	Canonical amino acid
CPP	Cell-penetrating peptide
DEL	DNA-encoded library
HGF	Hepatocyte growth factor
HIF	Hypoxia-inducible factor
IVT	*In vitro* translation
KDM	Histone demethylase
nCAA	Non-canonical amino acid
NGS	Next-generation sequencing
OBOC	One bead, one compound
PfMATE	*Pyrococcus furiosus* multidrug and toxic compound extrusion
PPI	Protein–protein interaction
RaPID	Random non-standard Peptide Integrated Discovery
RTHS	Reverse two-hybrid system
SAR	Structure–activity relationship
SICLOPPS	Split-intein circular ligation of peptides and proteins
SPPS	Solid-phase peptide synthesis

Acknowledgements

We would like to thank Dr Albert Isidro Llobet for critical reading of and helpful suggestions on this chapter. L.S. is supported by the EPSRC Centre for Doctoral Training in Synthesis for Biology and Medicine (EP/L015838/1) and G.R. is supported by the Oxford–GSK–Crick Chemical Biology Centre for Doctoral Training Programme *via* the Systems Approaches to Biomedical Science (EP/R512333/1), EPSRC and GlaxoSmithKline. A.K. gratefully acknowledges the European Research Council Starting Grant (EPITOOLS-679479) for funding. We apologise for the incomplete citations and for research that we were not able to cite due to space constraints.

References

1. A. Henninot, J. C. Collins and J. M. Nuss, *J. Med. Chem.*, 2018, **61**, 1382–1414.
2. R. B. Merrifield, *J. Am. Chem. Soc.*, 1963, **85**, 2149–2154.
3. C. H. Arrowsmith, J. E. Audia, C. Austin, J. Baell, J. Bennett, J. Blagg, C. Bountra, P. E. Brennan, P. J. Brown, M. E. Bunnage, C. Buser-Doepner, R. M. Campbell, A. J. Carter, P. Cohen, R. A. Copeland, B. Cravatt, J. L. Dahlin, D. Dhanak, A. M. Edwards, S. V. Frye, N. Gray, C. E. Grimshaw, D. Hepworth, T. Howe, K. V. M. Huber, J. Jin, S. Knapp, J. D. Kotz, R. G. Kruger, D. Lowe, M. M. Mader, B. Marsden, A. Mueller-Fahrnow, S. Müller, R. C. O'Hagan, J. P. Overington, D. R. Owen, S. H. Rosenberg, B. Roth, R. Ross, M. Schapira, S. L. Schreiber, B. Shoichet, M. Sundström, G. Superti-Furga, J. Taunton, L. Toledo-Sherman, C. Walpole, M. A. Walters, T. M. Willson, P. Workman, R. N. Young and W. J. Zuercher, *Nat. Chem. Biol.*, 2015, **11**, 536–541.
4. M. Gao, K. Cheng and H. Yin, *Biopolymers*, 2015, **104**, 310–316.
5. S. Lee, J. Xie and X. Chen, *Biochemistry*, 2010, **49**, 1364–1376.
6. J. A. Wells and C. L. McClendon, *Nature*, 2007, **450**, 1001–1009.
7. T. A. F. Cardote and A. Ciulli, *ChemMedChem*, 2016, **11**, 787–794.
8. L. Nevola and E. Giralt, *Chem. Commun.*, 2015, **51**, 3302–3315.
9. T. E. McAllister, T.-L. Yeh, M. I. Abboud, I. K. H. Leung, E. S. Hookway, O. N. F. King, B. Bhushan, S. T. Williams, R. J. Hopkinson, M. Munzel, N. D. Loik, R. Chowdhury, U. Opperman, T. D. W. Claridge, Y. Goto, H. Suga, C. J. Schofield and A. Kawamura, *Chem. Sci.*, 2018, **9**, 4569–4578.
10. A. M. Ali, J. Atmaj, N. Van Oosterwijk, M. R. Groves and A. Dömling, *Comput. Struct. Biotechnol. J.*, 2019, **17**, 263–281.
11. R. Obexer, L. J. Walport and H. Suga, *Curr. Opin. Chem. Biol.*, 2017, **38**, 52–61.
12. J. Ottl, L. Leder, J. V. Schaefer and C. E. Dumelin, *Molecules*, 2019, **24**, 1629.
13. R. Behrendt, P. White and J. Offer, *J. Pept. Sci.*, 2016, **22**, 4–27.
14. H. Bruzzoni-Giovanelli, V. Alezra, N. Wolff, C. Z. Dong, P. Tuffery and A. Rebollo, *Drug Discovery Today*, 2018, **23**, 272–285.

15. M. Trellet, A. S. J. Melquiond and A. M. J. J. Bonvin, in *Computational Peptidology*, Humana Press, New York, 2015, pp. 221–239.
16. M. Ciemny, M. Kurcinski, K. Kamel, A. Kolinski, N. Alam, O. Schueler-Furman and S. Kmiecik, *Drug Discovery Today*, 2018, **23**, 1530–1537.
17. C. A. Lipinski, F. Lombardo, B. W. Dominy and P. J. Feeney, *Adv. Drug Delivery Rev.*, 2001, **46**, 3–26.
18. T. Dang and R. D. Süssmuth, *Acc. Chem. Res.*, 2017, **50**, 1566–1576.
19. D. Faulds, K. L. Goa and P. Benfield, *Drugs*, 1993, **45**, 953–1040.
20. C. A. Lipinski, *Adv. Drug Delivery Rev.*, 2016, **101**, 34–41.
21. M. Fouché, M. Schäfer, J. Berghausen, S. Desrayaud, M. Blatter, P. Piéchon, I. Dix, A. Martingarcia and H. J. Roth, *ChemMedChem*, 2016, **11**, 1048–1059.
22. A. Alex, D. S. Millan, M. Perez, F. Wakenhut and G. A. Whitlock, *MedChemComm*, 2011, **2**, 669–674.
23. P. M. Cromm, J. Spiegel and T. N. Grossmann, *ACS Chem. Biol.*, 2015, **10**, 1362–1375.
24. F. Bernal, A. F. Tyler, S. J. Korsmeyer, L. D. Walensky and G. L. Verdine, *J. Am. Chem. Soc.*, 2007, **129**, 2456–2457.
25. M. Pazgier, M. Liu, G. Zou, W. Yuan, C. Li, C. Li, J. Li, J. Monbo, D. Zella, S. G. Tarasov and W. Lu, *Proc. Natl. Acad. Sci. U. S. A.*, 2009, **106**, 4665–4670.
26. A. Galán, L. Comor, A. Horvatić, J. Kuleš, N. Guillemin, V. Mrljak and M. Bhide, *Mol. BioSyst.*, 2016, **12**, 2342–2358.
27. R. M. P. Siloto and R. J. Weselake, *Biocatal. Agric. Biotechnol.*, 2012, **1**, 181–189.
28. J. Shendure and H. Ji, *Nat. Biotechnol.*, 2008, **26**, 1135–1145.
29. P. Dickson and T. Kodadek, *Org. Biomol. Chem.*, 2019, **17**, 4676–4688.
30. G. P. Smith, *Science*, 1985, **228**, 1315–1317.
31. G. P. Smith and V. A. Petrenko, *Chem. Rev.*, 1997, **97**, 391–410.
32. I. Rentero Rebollo and C. Heinis, *Methods*, 2013, **60**, 46–54.
33. H. Qi, H. Lu, H. J. Qiu, V. Petrenko and A. Liu, *J. Mol. Biol.*, 2012, **417**, 129–143.
34. S. Bass, R. Greene and J. A. Wells, *Proteins: Struct., Funct., Genet.*, 1990, **8**, 309–314.
35. F. Breitling, S. Dübel, T. Seehaus, I. Klewinghaus and M. Little, *Gene*, 1991, **104**, 147–153.
36. S. J. McConnell, M. L. Kendall, T. M. Reilly and R. H. Hoess, *Gene*, 1994, **151**, 115–118.
37. C. Heinis, T. Rutherford, S. Freund and G. Winter, *Nat. Chem. Biol.*, 2009, **5**, 502–507.
38. X. S. Wang, P.-H. C. Chen, J. T. Hampton, J. M. Tharp, C. A. Reed, S. K. Das, D.-S. Wang, H. S. Hayatshahi, Y. Shen, J. Liu and W. R. Liu, *Angew. Chem., Int. Ed.*, 2019, **58**, 15904–15909.
39. R. W. Roberts and J. W. Szostak, *Proc. Natl. Acad. Sci. U. S. A.*, 1997, **94**, 12297–12302.
40. K. Josephson, A. Ricardo and J. W. Szostak, *Drug Discovery Today*, 2014, **4**, 388–399.

41. J. Ellman, D. Mendel, S. Anthony-Cahill, C. J. Noren and P. G. Schultz, *Methods Enzymol.*, 1991, **202**, 301–336.
42. C. Merryman and R. Green, *Chem. Biol.*, 2004, **11**, 575–582.
43. H. Saito, D. Kourouklis and H. Suga, *EMBO J.*, 2001, **20**, 1797–1806.
44. T. Passioura and H. Suga, *Chem. Commun.*, 2017, **53**, 1931–1940.
45. C. Heinis and G. Winter, *Curr. Opin. Chem. Biol.*, 2015, **26**, 89–98.
46. C. P. Scott, E. Abel-Santos, M. Wall, D. C. Wahnon and S. J. Benkovic, *Proc. Natl. Acad. Sci. U. S. A.*, 1999, **96**, 13638–13643.
47. T. M. Kinsella, C. T. Ohashi, A. G. Harder, G. C. Yam, W. Li, B. Peelle, E. S. Pali, M. K. Bennett, S. M. Molineaux, D. A. Anderson, E. S. Masuda and D. G. Payan, *J. Biol. Chem.*, 2002, **277**, 37512–37518.
48. J. A. Kritzer, S. Hamamichi, J. M. McCaffery, S. Santagata, T. A. Naumann, K. A. Caldwell, G. A. Caldwell and S. Lindquist, *Nat. Chem. Biol.*, 2009, **5**, 655–663.
49. A. Tavassoli, *Curr. Opin. Chem. Biol.*, 2017, **38**, 30–35.
50. T. S. Young, D. D. Young, I. Ahmad, J. M. Louis, S. J. Benkovic and P. G. Schultz, *Proc. Natl. Acad. Sci. U. S. A.*, 2011, **108**, 11052–11056.
51. J. R. Frost, N. T. Jacob, L. J. Papa, A. E. Owens and R. Fasan, *ACS Chem. Biol.*, 2015, **10**, 1805–1816.
52. A. Tavassoli and S. J. Benkovic, *Angew. Chem., Int. Ed.*, 2005, **44**, 2760–2763.
53. K. Lam, S. Salmon, E. Hersh, V. Hruby, W. Kazmierski and R. Knapp, *Nature*, 1991, **354**, 82–84.
54. R. A. Houghten, C. Pinilla, S. E. Blondelle, J. R. Appel, C. T. Dooley and J. H. Cuervo, *Nature*, 1991, **354**, 84–86.
55. R. Liu, X. Li, W. Xiao and K. S. Lam, *Adv. Drug Delivery Rev.*, 2017, **110–111**, 13–37.
56. T. Kodadek and K. Bachhawat-Sikder, *Mol. BioSyst.*, 2006, **2**, 25–35.
57. R. Liu and K. S. Lam, *Anal. Biochem.*, 2001, **295**, 9–16.
58. M. G. Paulick, K. M. Hart, K. M. Brinner, M. Tjandra, D. H. Charych and R. N. Zuckermann, *J. Comb. Chem.*, 2006, **8**, 417–426.
59. S. H. Joo, Q. Xiao, Y. Ling, B. Gopishetty and D. Pei, *J. Am. Chem. Soc.*, 2006, **128**, 13000–13009.
60. Z. Zhu, A. Shaginian, L. C. Grady, T. O'keeffe, X. E. Shi, C. P. Davie, G. L. Simpson, J. A. Messer, G. Evindar, R. N. Bream, P. P. Thansandote, N. R. Prentice, A. M. Mason and S. Pal, *ACS Chem. Biol.*, 2018, **13**, 53–59.
61. B. N. Tse, T. M. Snyder, Y. Shen and D. R. Liu, *J. Am. Chem. Soc.*, 2008, **130**, 15611–15626.
62. C. J. Brown, S. T. Quah, J. Jong, A. M. Goh, P. C. Chiam, K. H. Khoo, M. L. Choong, M. A. Lee, L. Yurlova, K. Zolghadr, T. L. Joseph, C. S. Verma and D. P. Lane, *ACS Chem. Biol.*, 2013, **8**, 506–512.
63. A. Zorzi, K. Deyle and C. Heinis, *Curr. Opin. Chem. Biol.*, 2017, **38**, 24–29.
64. Y. S. Chang, B. Graves, V. Guerlavais, C. Tovar, K. Packman, K. H. To, K. A. Olson, K. Kesavan, P. Gangurde, A. Mukherjee, T. Baker, K. Darlak, C. Elkin, Z. Filipovic, F. Z. Qureshi, H. Cai, P. Berry, E. Feyfant, X. E. Shi, J. Horstick, D. A. Annis, A. M. Manning, N. Fotouhi, H. Nash, L. T. Vassilev and T. K. Sawyer, *Proc. Natl. Acad. Sci. U. S. A.*, 2013, **110**, E3445–E3454.

65. Y. H. Lau, P. De Andrade, S. T. Quah, M. Rossmann, L. Laraia, N. Sköld, T. J. Sum, P. J. E. Rowling, T. L. Joseph, C. Verma, M. Hyvönen, L. S. Itzhaki, A. R. Venkitaraman, C. J. Brown, D. P. Lane and D. R. Spring, *Chem. Sci.*, 2014, **5**, 1804–1809.
66. A. L. Kung, S. Wang, J. M. Klco, W. G. Kaelin and D. M. Livingston, *Nat. Med.*, 2000, **6**, 1335–1340.
67. E. Miranda, I. K. Nordgren, A. L. Male, C. E. Lawrence, F. Hoakwie, F. Cuda, W. Court, K. R. Fox, P. A. Townsend, G. K. Packham, S. A. Eccles and A. Tavassoli, *J. Am. Chem. Soc.*, 2013, **135**, 10418–10425.
68. D. J. Asby, F. Cuda, F. Hoakwie, E. Miranda and A. Tavassoli, *Mol. BioSyst.*, 2014, **10**, 2505–2508.
69. I. N. Mistry and A. Tavassoli, *ACS Synth. Biol.*, 2017, **6**, 518–527.
70. W. Miao, K. Sakai, N. Ozawa, T. Nishiuchi, Y. Suzuki, K. Ito, T. Morioka, M. Umitsu, J. Takagi, H. Suga and K. Matsumoto, *Sci. Rep.*, 2018, **8**, 1–11.
71. K. Ito, K. Sakai, Y. Suzuki, N. Ozawa, T. Hatta, T. Natsume, K. Matsumoto and H. Suga, *Nat. Commun.*, 2015, **6**, 6373.
72. Y. Tanaka, C. J. Hipolito, A. D. Maturana, K. Ito, T. Kuroda, T. Higuchi, T. Katoh, H. E. Kato, M. Hattori, K. Kumazaki, T. Tsukazaki, R. Ishitani, H. Suga and O. Nureki, *Nature*, 2013, **496**, 247–251.
73. C. J. Hipolito, Y. Tanaka, T. Katoh, O. Nureki and H. Suga, *Molecules*, 2013, **18**, 10514–10530.
74. A. Kawamura, M. Münzel, T. Kojima, C. Yapp, B. Bhushan, Y. Goto, A. Tumber, T. Katoh, O. N. F. King, T. Passioura, L. J. Walport, S. B. Hatch, S. Madden, S. Müller, P. E. Brennan, R. Chowdhury, R. J. Hopkinson, H. Suga and C. J. Schofield, *Nat. Commun.*, 2017, **8**, 14773.
75. T. E. McAllister, K. S. England, R. J. Hopkinson, P. E. Brennan, A. Kawamura and C. J. Schofield, *J. Med. Chem.*, 2016, **59**, 1308–1329.
76. L. J. Walport, R. J. Hopkinson, R. Chowdhury, R. Schiller, W. Ge, A. Kawamura and C. J. Schofield, *Nat. Commun.*, 2016, **7**, 11974.
77. T. B. Trinh, P. Upadhyaya, Z. Qian and D. Pei, *ACS Comb. Sci.*, 2016, **18**, 75–85.
78. R. Rezaei Araghi, J. A. Ryan, A. Letai and A. E. Keating, *ACS Chem. Biol.*, 2016, **11**, 1238–1244.
79. G. W. Foight, J. A. Ryan, S. V. Gullá, A. Letai and A. E. Keating, *ACS Chem. Biol.*, 2014, **9**, 1962–1968.
80. V. Baeriswyl and C. Heinis, *Protein Eng. Des. Sel.*, 2013, **26**, 81–89.
81. T. Passioura, W. Liu, D. Dunkelmann, T. Higuchi and H. Suga, *J. Am. Chem. Soc.*, 2018, **140**, 11551–11555.
82. D. Hay, O. Fedorov, P. Filippakopoulos, S. Martin, M. Philpott, S. Picaud, D. S. Hewings, S. Uttakar, T. D. Heightman, S. J. Conway, S. Knapp and P. E. Brennan, *MedChemComm*, 2013, **4**, 140–144.
83. O. B. Cox, T. Krojer, P. Collins, O. Monteiro, R. Talon, A. Bradley, O. Fedorov, J. Amin, B. D. Marsden, J. Spencer, F. Von Delft and P. E. Brennan, *Chem. Sci.*, 2016, **7**, 2322–2330.

CHAPTER 6

Discovery and Use of Natural Products and Derivatives as Chemical Probes

MASAKI KITA*[a] AND HIDEO KIGOSHI*[b]

[a]Graduate School of Bioagricultural Sciences, Nagoya University, Furo-cho, Chikusa, Nagoya 464-8601, Japan; [b]Graduate School of Pure and Applied Sciences, University of Tsukuba, 1-1-1 Tennodai, Tsukuba 305-8571, Japan
*E-mail: mkita@agr.nagoya-u.ac.jp, kigoshi@chem.tsukuba.ac.jp

6.1 Introduction

A number of structurally and functionally diverse natural products have been discovered from both terrestrial and marine organisms. In the course of many years of biological evolution, natural products have specifically evolved to interact with biomacromolecules, especially proteins.[1] Owing to the highly specific interactions between ligands (small molecules) and target molecules, natural product scaffolds have contributed to the development of novel pharmacological tools and therapeutic agents. In fact, about 40% of therapeutic agents are derived from natural or biological sources.[2] "Chemical probes" are defined as selective small-molecule modulators of target biomolecule functions that answer mechanistic and phenotypic questions about their molecular targets in cell-based or animal studies. They are an essential component to be used to dissect complex biological processes, to regulate temporal and spatial control over biochemical pathways and to

Chemical Biology No. 16
The Discovery and Utility of Chemical Probes in Target Discovery
Edited by Paul Brennan and Saleta Vazquez Rodriguez

Published by the Royal Society of Chemistry, www.rsc.org

identify novel therapeutic targets.[3] Hence the utilization of natural product-based chemical probes is highly valuable and worth challenging.

Recently, protein–protein interactions (PPIs) regulated by small molecules have been of increasing interest in the fields of both basic research and drug discovery.[4] PPIs can be stabilized or disrupted to show potent biological and physiological effects. Some unique PPI modulators have been discovered from natural products. For example, the immunosuppressive macrolide FK506 (Prograf®) is a PPI inducer used in organ transplantation.[5] Whereas both FK506 and its structural congener rapamycin bind to the immunophilin FKBP12 in cells, their secondary targets are different [calcineurin or FKBP12–rapamycin complex-associated protein (FRAP, mTOR)], hence their effects on the cellular signalling pathways are also different. Since the protein complex interfaces are highly diverse, PPI inhibitors have higher specificity compared with inhibitors that bind to one target, such as enzyme active sites.[6] Therefore, from the point of view of the development of therapeutic drugs, the manipulation of PPIs has captured the attention of chemical biologists and medicinal scientists. The discovery and design of PPI modulators have been highlighted in several reviews.[7–10]

Various natural products that target cytoskeletons have been characterized. The representative cytoskeleton of eukaryotes is composed of three components: microfilaments (actin filaments), microtubules and intermediate filaments. These structures are constructed by the association of monomer proteins (actin, keratin, *etc.*). Regulation of PPIs on these cytoskeletal proteins permits the modulation of cytoskeletal dynamics. Microtubules are polymers composed of α- and β-tubulin heterodimers, which have many roles in cells, including signalling, migration and cell proliferation.[11] Several natural products that interfere with microtubule dynamics are used as antitumour agents.[12,13] For example, paclitaxel and epothilones enhance microtubule formation by stabilizing the interactions between the α/β-tubulin heterodimers. Vincristine and vinblastine (*Vinca* alkaloids) inhibit microtubule formation by suppressing the oligomerization of α/β-tubulin heterodimers. Eribulin mesylate (Halaven®) blocks the growth phase of microtubule dynamics by binding to the plus end of microtubules (see Figure 6.1).[14,15] Eribulin is a macrocyclic ketone derivative of halichondrin B, isolated from a marine sponge, *Halichondria okadai*. To date, eribulin is one of the most structurally complex, non-peptidic drugs. Dolastatin 10, a peptide of marine origin (the sea hare *Dolabella* sp.), has potent antineoplastic activity and interferes with tubulin assembly.[16] Brentuximab vedotin (Adcetris®), the mAb-targeted dolastatin 10 analogue, is utilized for Hodgkin's lymphoma and systemic anaplastic large-cell lymphoma.[17,18]

Among the natural products that interfere with cytoskeletal dynamics, a few are known to regulate PPIs. These unique interactions have been established by using activity-based chemical probes. In this chapter, we focus on natural product-derived chemical probes for target identification and mode of action studies, and also for binding position analysis using mass spectrometry.

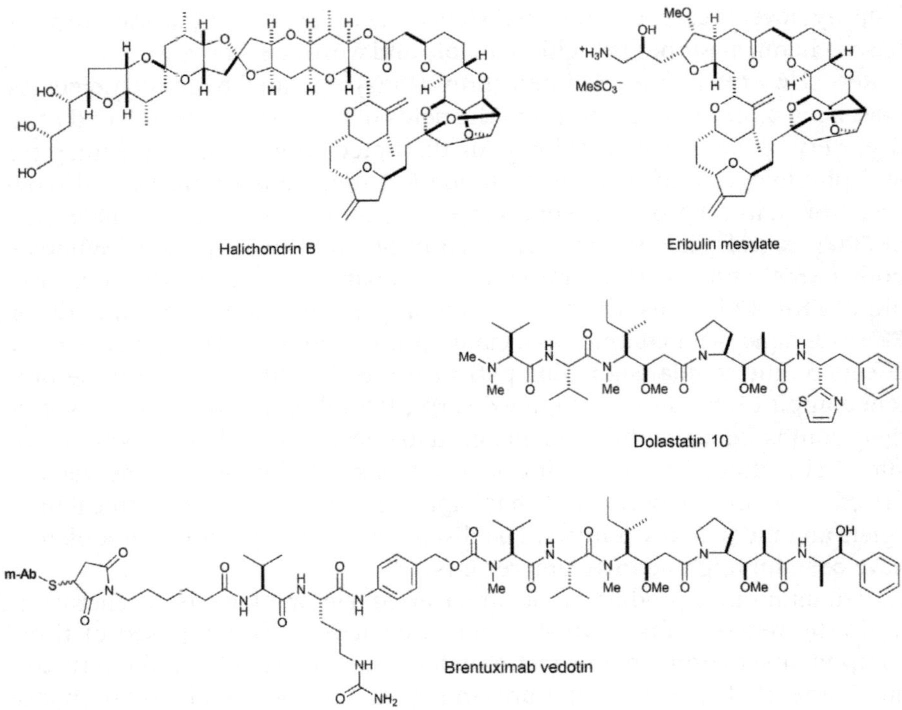

Figure 6.1 Structures of natural product-derived anti-tumour drugs and their original molecules.

6.2 Chemical Probes for Target Identification and Mode of Action Studies

Identification of the target molecules of bioactive small molecules (ligands) and precise analysis of their interactions are important in the fields of chemical biology and medicinal chemistry.[19,20] An effective approach to such binding-mode analysis is to utilize chemical probes, in which the bioactive ligands are derivatized with a tag that facilitates mechanistic analysis while the activity of the parent ligand is maintained.[21,22]

6.2.1 Actin and Actin-targeting Marine Natural Products

Actin is an abundant protein in eukaryotic cells and is important in regulating many cellular functions, such as cell division, muscle contraction and tumour migration.[23] The actin dynamics, including polymerization and depolymerization, are regulated by various actin-binding proteins, such as profilin and thymosin-β4. In addition to endogenous actin-binding proteins, various natural products have been shown to target actin (Figure 6.2) and some have potent cytotoxicity.[24–26] In the 1980s, latrunculins

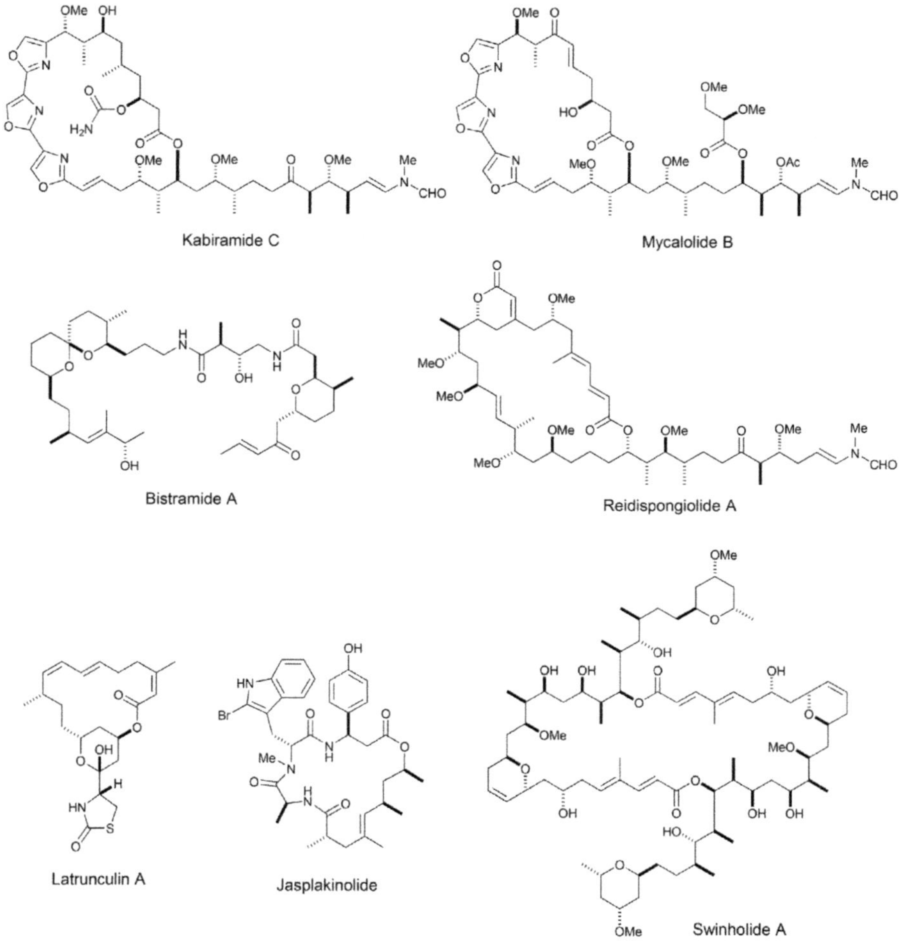

Figure 6.2 Structures of actin-targeting natural products of marine origin.

were characterized as the first actin-binding substances of marine origin.[27] To date, various actin-depolymerizing marine natural products have been reported. For example, ulapualides,[28] mycalolides,[29] kabiramides,[30] sphinxolides,[31] reidispongiolides[31] and bistramides[32,33] show potent cytotoxicity by inhibiting actin polymerization. Furthermore, several actin polymerization-stimulating or -blocking molecules induce apoptosis, including jasplakinolide,[34] latrunculin A[35] and mycalolide B.[36] They are useful tools for studying cellular actin dynamics and have potential as anti-tumour agents. A number of synthetic studies on these actin-targeting compounds have been reported, and also their use as effective chemical probes.

Although the producing organisms are diverse, several actin-binding macrolides have common structures, such as 24–26-membered macrocycles, aliphatic side chains and *N*-methylenamide moieties. They bind to actin to form a 1 : 1 complex and thus globular actins (G-actin) are sequestrated and filamentous actins (F-actin) are severed. In 2003, the crystal structure of the actin–kabiramide C complex was reported, in which kabiramide C bound to the hydrophobic cleft between subdomains (SDs) 1 and 3 of actin.[37] To date, 1 : 1 complexes of actin with several structurally related actin-binding macrolides, such as aplyronine A,[38] reidispongiolides A and C,[39] sphinxolide B,[39] swinholide A[40] and bistramide A,[41] have been reported and their interactions with actin molecule are very similar.[26] Hence it has been strongly suggested that their side-chain (tail) parts are important for binding strongly to actin.

In addition to monomeric lactones, swinholide A is a dimeric, 44-membered lactone isolated from the Japanese sponge *Theonella swinhoei*.[42] This molecule has two terminated side chains and forms a 2 : 1 actin–macrolide complex, where the side chains bind to the two different actin molecules.[40,43] As the congener of swinholide A, a 40-membered bislactone, bistheonellide A (or misakinolide A), was identified from the same sponge species.[44] Both bistheonellide A and swinholide A form a tertiary complex with two actin molecules and inhibit actin dynamics more potently than monomeric lactones.[45]

6.2.2 Aplyronine A, an Anti-tumour Actin-depolymerizing Macrolide

Over the past 40 years, a number of bioactive secondary metabolites, including anti-tumour compounds, have been discovered from the genera *Aplysia* and *Dolabella*.[46,47] Aplyronine A was isolated from the Japanese sea hare *A. kurodai* (see Figure 6.3 and Table 6.1).[48,49] Aplyronine A exists as an inseparable diastereomer mixture of two kinds of amino acid esters [*N*,*N*,*O*-trimethylserine (TMSer) and *N*,*N*-dimethylalanine (DMAla)]. Total synthesis of aplyronine A[50] and aplyronine B and C[51] established their structures and potent biological activity. Recently, second-generation total synthesis of aplyronine A using Ni/Cr-mediated cross-coupling was accomplished.[52] Also, Paterson and co-workers reported the total synthesis of aplyronine A, C and D.[53,54]

Aplyronine A–C show potent cytotoxicity against HeLa S3, a human cervical carcinoma cell line (IC_{50} = 0.48, 3.1 and 21 μg mL^{-1}, respectively).[55–57] Aplyronine A also shows anti-tumour activity *in vivo* against several tumours (such as P388 leukaemia and Lewis lung carcinoma). As with other related actin-binding macrolides, aplyronine A potently inhibits actin polymerization and forms a 1 : 1 complex with monomeric actin.[58] Furthermore, both the macrolide ring part and the side-chain part are essential for the potent activity of aplyronine A, whereas both the C1–C23 macrolactone and the C21–C34 side-chain parts alone showed little cytotoxicity.[59,60] Interaction between actin and the aplyronine A side-chain part was established by photoaffinity labelling experiments.[61]

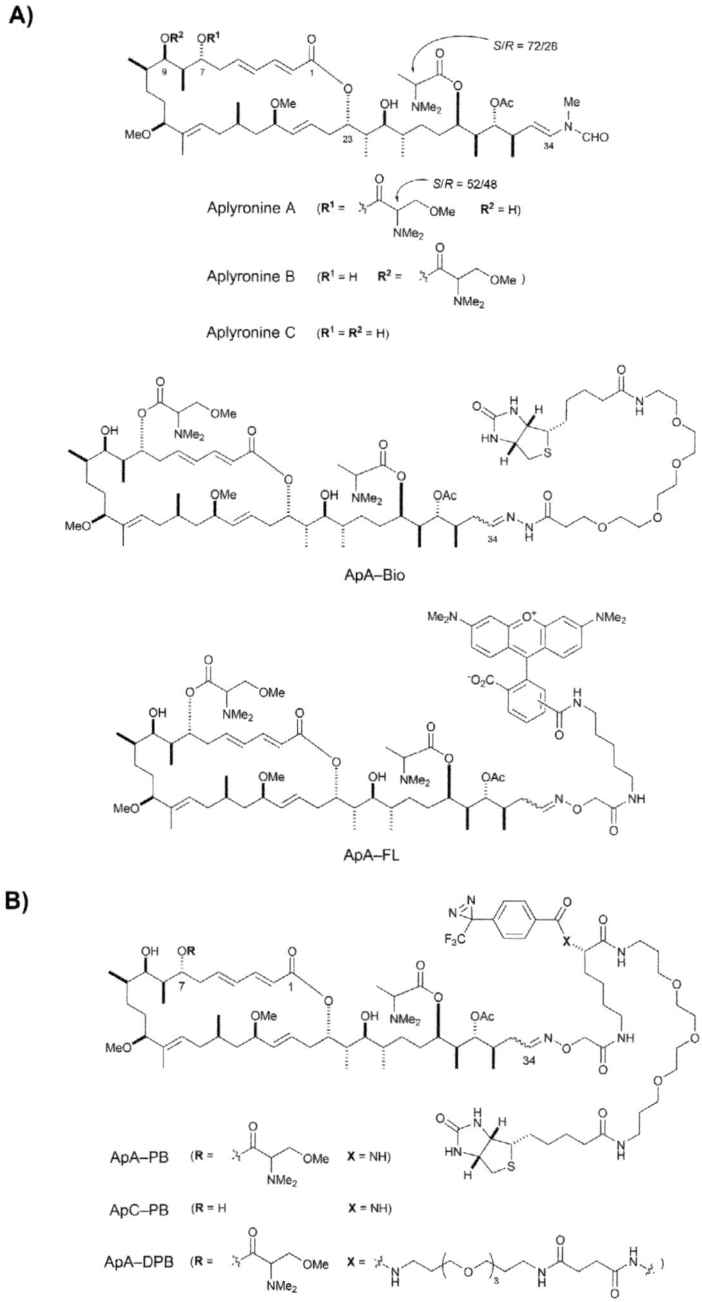

Figure 6.3 Structures of aplyronines and their derivatives for target ID studies.

Table 6.1 Biological activities of aplyronines and their derivatives. Reproduced from ref. 62, https://pubs.acs.org/doi/10.1021/acsomega.9b01099, with permission from American Chemical Society, Copyright 2019.

		Actin-depolymerizing activity[a]: EC_{50}/μM	
Compound	Cytotoxicity (HeLa S3): IC_{50}/nM	Ultracentrifugation method	Pyrenyl actin method
ApA	0.010 [0.45][b]	1.4	1.3 [1.6][d]
ApB	[2.9][b]	—[c]	—[c]
ApC	17 [22][b]	—[c]	1.4
ApA–Bio	0.096	—[c]	1.8
ApA–FL	0.37	<5	0.8
ApA–PB	1.2	—[c]	—[c]
ApC–PB	310	—[c]	—[c]
ApA–DPB	0.54	—[c]	—[c]
ApA–PF	13	—[c]	—[c]
ApA–PPF	28	—[c]	—[c]
ApA–PaP	0.66	<50	—[c]

[a]Values indicate the concentrations required to depolymerize F-actin (3 μM for monomer) to 50% of its control amplitude.
[b]The values calculated as free salts are given in square brackets.
[c]Not examined.
[d]The value using 3.7 μM for monomer actin is given in square brackets.

As with kabiramide C, aplyronine A binds to the hydrophobic cleft between SD 1 and 3 of actin by inserting its side chain (see Figure 6.4).[38] On the other hand, its C7 TMSer moiety has positive charge and protrudes towards the bulk solvent region. This TMSer ester is highly important for the cytotoxicity of aplyronine A, since aplyronine C, which does not have the C7 TMSer ester moiety, is much less cytotoxic than aplyronine A. Recent molecular modelling studies suggested that the macrolide part (C1–C23) of aplyronine A almost completely overlapped those in the actin–ApA complex and that the whole macrolactone moiety of aplyronine A contributed to fixing the C7 TMSer ester conformation.[62] The same calculation indicated that the C9 TMSer ester of aplyronine B similarly protruded from the surface of actin, but the orientation of the macrolactone part in aplyronine B was very different from that in the actin–ApA complex. Therefore, it was proposed that fixed conformations of the C7 TMSer ester in aplyronine A are important for binding to the secondary molecules to exhibit potent cytotoxicity.

Aplyronine A significantly reduces cellular viability and induces DNA fragmentation at 1 nM in several tumour cells.[36] In addition, aplyronine A activates caspase 3 in tumour cells at 1 nM, whereas aplyronine C does not, even at 100 nM.[63] This remarkable difference between aplyronine A and C in apoptosis-related activities was consistent with their cytotoxicity. Overall, it has been suggested that aplyronine A might first bind to actin to form a 1:1 actin–ApA complex, which then binds to secondary molecule(s) to show antitumour activity.

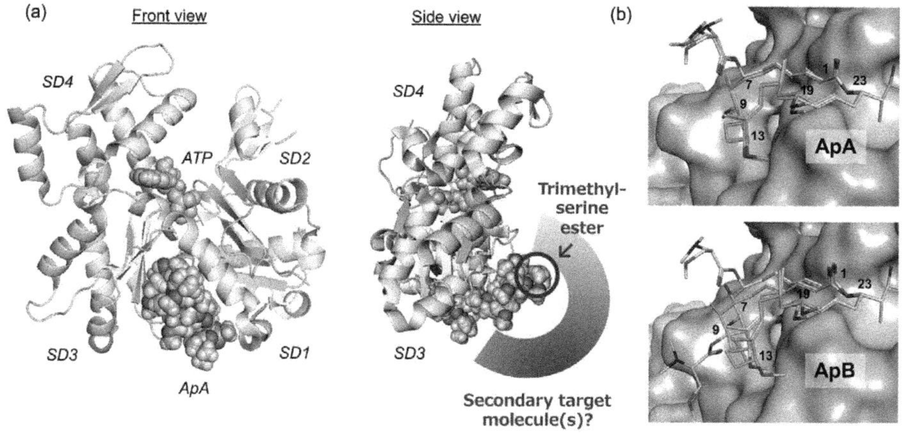

Figure 6.4 Structures of actin–aplyronine complexes. (a) X-ray structure (PDB code 1WUA). (b) Molecular modelling structures of ApA and ApB on actin, in which both actin and the C24–C34 side-chain parts of aplyronines were fixed. Note that the macrolide part (C1–C23) of ApA almost completely overlapped those in the actin–ApA complex (the ligand ApA is overlaid in both ApA and ApB models). Reproduced from ref. 62, https://pubs.acs.org/doi/10.1021/acsomega.9b01099, with permission from American Chemical Society, Copyright 2019.

6.2.3 Design and Synthesis of Aplyronine Chemical Probes

To identify the proposed secondary targets of aplyronine A, a number of chemical probes have been synthesized. According to the preparation of mycalolide and kabiramide biotin derivatives,[64] the *N*-methylenamide moiety of aplyronine A was hydrolysed under acidic conditions and condensed with hydrazide to afford an aplyronine A–biotin conjugate (ApA–Bio).[65] Similarly, condensation with an oxyamine yielded a tetramethylrhodamine (TAMRA) analogue of aplyronine A (ApA–FL).[63] ApA–Bio exhibited potent cytotoxicity, which was only *ca.* 10-fold less than that of natural aplyronine A against HeLa S3 cells (IC_{50} = 96 pM) (see Table 6.1). ApA–FL also exhibited potent cytotoxicity (IC_{50} = 370 pM). Both chemical probes showed potent actin-depolymerizing activity, comparable to that of aplyronine A. It was found that replacement of the C34 enamide moiety of aplyronines with hydrogen bond acceptors (*e.g.*, hydrazones and oximes) did not affect the potent activity of natural aplyronines.

When ApA–FL was incorporated into tumour cells, it was well retained and distributed throughout the cytoplasm. These results suggested that aplyronine A has potent cellular permeability and accumulation ability.[63] Since the concentration of intracellular actin is high and aplyronines have high affinity with actin, they might be trapped within cells as a 1 : 1 complex with actin. In addition, ApA–FL rapidly caused actin cytoskeleton disassembly in tumour cells.[65] These findings suggested that aplyronine A might interfere with cell-adherent systems or signalling pathways.

6.2.4 Target Identification of Aplyronine a Using Biotin Derivatives

Using ApA–Bio and NeutrAvidin agarose, interactions of aplyronine A with cellular proteins were investigated. Actin-related proteins 2 and 3 (Arp2 and Arp3) were affinity purified with a large quantity of actin from cell lysate.[65] Arp2 and Arp3 are actin-binding proteins that form an Arp2/3 complex and initiate the growth of additional actin filaments to stabilize a branched actin filament network.[66–68] The structures of G-actin, Arp2 and Arp3 are similar, especially around the hydrophobic clefts between SD 1 and 3. Therefore, it was proposed that aplyronine A might bind to Arp2 or Arp3 to inhibit the interactions between the Arp2/3 complex and F-actin. These properties could explain why aplyronine A exhibits potent disassembly of actin filaments.

Although Arp2 and Arp3 were identified as aplyronine A-binding proteins, their detailed molecular mechanism and binding modes were still unclear, since they have polymerization properties and are highly unstable as a complex *in vitro*. Hence an aplyronine A photoaffinity biotin (ApA–PB) and its aplyronine C derivative (ApC–PB) were synthesized (see Figure 6.3).[69] Although the amount of aplyronine C was very small in sea hare, this congener was successfully obtained by the hydrolysis of aplyronine A under mild basic conditions. ApA–PB showed potent cytotoxicity (IC_{50} = 1.2 nM) whereas ApC–PB was ~260-fold less cytotoxic than ApA–PB. Hence ApC–PB was used as a negative control in the target identification studies, like natural aplyronine C.

By carrying out photolabelling experiments using tumour cell lysate and affinity purification using ApA–PB and ApC–PB, Arp2 and Arp3 were obtained along with actin. However, neither photoaffinity probe formed covalent bonds with Arp2 or Arp3. It was suggested that Arp2 and Arp3 might indirectly bind to aplyronine A as adducts of the actin–ApA complex and thus they were not the target of aplyronine A to explain its potent anti-tumour activity.

Next, photolabelling experiments on tumour cells were carried out (see Figure 6.5). Notably, new target proteins were detected, following *in situ* photolabelling experiments with ApA–PB and affinity pull-down experiments.[70] Blotting analysis with streptavidin–HRP conjugate and peptide mass fingerprint (PMF) analysis established that ApA–PB covalently bound to β-tubulin. Meanwhile, ApC–PB photoreacted only with actin. Hence aplyronine A, but not aplyronine C, interacts with both actin and tubulin.

Since the distance between the aryltrifluoromethyldiazirine (photoreacting) group in ApA–PB and the side chain (C23–C34) of ApA is short, this moiety might also be close to the hydrophobic cleft of actin in the actin–probe complex. To increase the flexibility of the photoreacting group, an aplyronine A double-poly(ethylene glycol) (PEG)-linked photoaffinity biotin probe (ApA–DPB) was prepared. As expected, the amounts of both photolabelled tubulin and actin were increased by the treatment with ApA–DPB compared with ApA–PB.[70] Molecular modelling of actin–probe complexes suggested that a much greater area of the surface of actin was covered by the aryldiazirine in

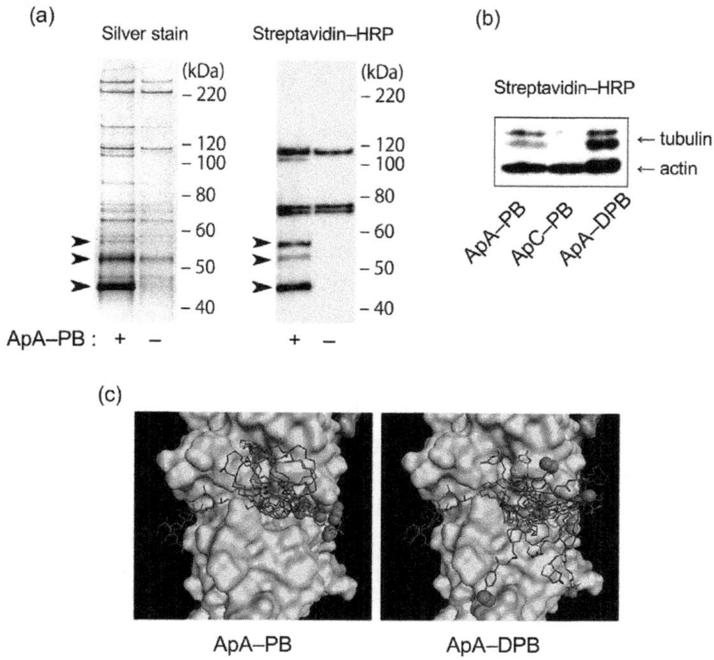

Figure 6.5 Actin–ApA complex binds to tubulin in cells and *in vitro*. (a) *In situ* photolabelling experiments with ApA–PB. Both actin (43–45 kDa) and tubulin (52–55 and 58 kDa) were photolabelled. (b) *In vitro* photolabelling experiments with rabbit actin and porcine tubulin with three kinds of photoaffinity biotin probes. (c) Molecular modelling of the complexes of actin–ApA photoaffinity biotin probes. Conformational analyses were performed by fixing both actin and the ApA moiety (C1–C32) in each probe. In each model, seven representative conformers of ApA probes are superimposed. The spheres represent the diazirine N atoms. Reproduced from ref. 70 with permission from American Chemical Society, Copyright 2013, and ref. 71 with permission from Elsevier, Copyright 2017.

ApA–DPB than in ApA–PB, which might explain why ApA–DPB bound more efficiently to tubulin.[71]

Whereas actin and aplyronine A alone did not affect tubulin dynamics, the 1 : 1 actin–ApA complex delayed the nucleation and growth phases and depolymerized microtubules.[70] It was also shown by gel-permeation HPLC analysis that the actin–ApA–tubulin heterodimer formed a 1 : 1:1 heterotrimeric complex (145 kDa). This synergistic binding between the actin–ApA complex and α/β-tubulin heterodimer was also established by surface plasmon resonance (SPR) analysis (K_D = 1.84 μM).[72] On the other hand, the actin–ApC complex did not bind to α/β-tubulin heterodimer or attenuate microtubule growth. Aplyronine A inhibited cell-cycle progression in the metaphase of tumour cells at 100 pM. By the same treatment, it caused irregular, multipolar spindle structures with unaligned chromosomes. In contrast, aplyronine C inhibited cell-cycle progression only at 100 nM, where the spindle formation was little affected. Therefore, the tubulin-binding properties of aplyronine A might clearly explain the differences in mitosis inhibition and cytotoxicity of aplyronine A and C.

6.2.5 Mode of Action of Aplyronine A as a PPI Inducer

We now consider the inhibitory effects on the spindle microtubule dynamics in cells caused by aplyronine A at extremely low concentration. The concentrations of intracellular actin and tubulin (*ca.* 100 and 20 μM, respectively) are high and the actin–ApA complex has high affinity with α/β-tubulin heterodimer. Thus, aplyronine A would form a 1 : 1 complex with actin in cytoplasm and further interact with liberated tubulin (see Figure 6.6).[9] It was shown that treatment with antimitotic agents at substoichiometric concentrations, such as vinblastine and colchicine, inhibits mitosis without depolymerization of spindle microtubules.[73] Similarly, paclitaxel also inhibits mitosis by stabilizing spindle microtubules but has little effect on the mass of polymerized microtubules.[74,75] Therefore, a few molecules of the actin–ApA–tubulin heterotrimeric complex could interact at the polymerization terminus of microtubules or be incorporated into the microtubule lattice to interfere with spindle microtubule dynamics.

A number of fundamental cellular processes, such as cell motility, cell division and neuronal pathfinding, are controlled by microtubule–actin interactions.[76] Since actin is a highly abundant protein in cells, actin-binding agents could interact with more multiple cellular targets than ever known. The findings on the unique PPI of aplyronine A established by chemical probes should provide important insights into the mechanisms of cytoskeleton-targeting natural products.

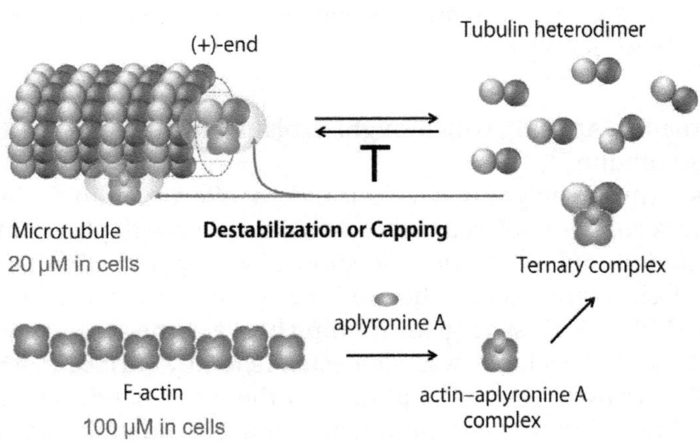

Figure 6.6 Proposed inhibition mechanism of microtubule assembly by an actin–aplyronine A complex. Reproduced from ref. 9 with permission from the Royal Society of Chemistry.

6.3 Chemical Probes for Binding Position Analysis

In the case of proteins, enzymatic digestion and peptide sequence analysis of labelled peptides enable us to decide the ligand-binding positions.[77] Chemical probe methods have the potential to complement other methods for analysing target molecule–ligand complex structures, such as NMR spectroscopy, X-ray crystallography and cryoelectron microscopy, especially when the interactions are weak or thermally unstable. Here we describe several attempts at target labelling of aplyronine A using photoaffinity and other reacting groups. Related mass-based binding position analysis for the avidin–biotin complex is also highlighted.

6.3.1 Development of Aplyronine Fluorescent Derivatives

While the formation of the actin–ApA–tubulin heterotrimeric complex has been established by gel-permeation and surface plasmon resonance (SPR) analyses, this complex was too unstable for X-ray crystallographic or NMR analyses to be conducted. To investigate the binding modes of the actin–ApA complex on tubulin, various aplyronine derivatives have been synthesized that possess detecting [biotin, tetramethylrhodamine (TAMRA) and amidopyrene (apy)] groups and reactive (photoaffinity and succinyl) groups. Among them, a TAMRA-conjugated ApA photoaffinity fluorescent probe (ApA–PF) and its PEG-linked derivative (ApA–PPF) showed moderate cytotoxicity (see Figure 6.7).[71] Both probes photoreacted with both actin and tubulin and this labelling was competitively inhibited by aplyronine A. However, the photolabelling yield of tubulin was too low (<1%) for their binding sites on tubulin to be determined by mass analysis. The reason for the low photolabelling yields was the addition of water molecules to the carbene species on the hydrophilic surface of actin, which was generated from an aryldiazirine moiety. Hence both the labelling yield and detectable sensitivity on mass analysis need to be improved to establish the binding position of ApA on the secondary target tubulin.

6.3.2 Development of Amidopyrene for Label-assisted Laser Desorption/Ionization Mass Spectrometry (LA-LDI-MS)

For PMF analysis on a minute scale, several kinds of mass analyses for biomacromolecules are useful, such as electrospray ionization (ESI)[78] and matrix-assisted laser desorption/ionization (MALDI).[79,80] These methods are highly sensitive and PMF can be used with proteins on the μg–ng scale. However, it is generally difficult to observe directly minute amounts of labelled peptides in the presence of excess reagents and other degraded products. LDI-MS without matrices was developed for the selective detection of labelled products. For example, molecular ions of aromatic amino acid-containing peptides[81] and polycyclic aromatic hydrocarbons (PAHs),[82–85] including anthracene and pyrene, were detected. In 2013, a

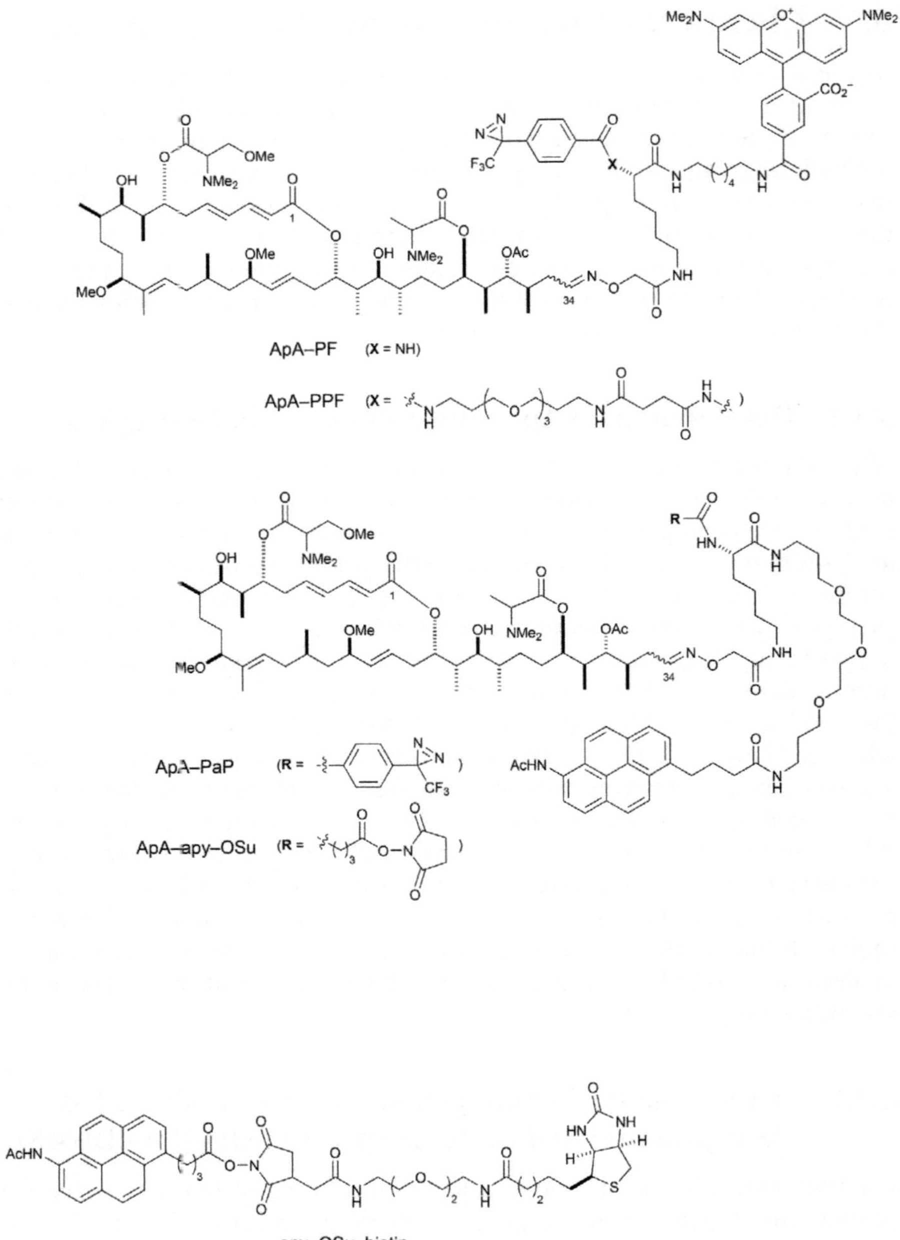

Figure 6.7 Structures of ApA fluorescent derivatives.

method for the selective detection of pyrene-conjugated compounds by LDI-MS in reaction mixtures was denoted label-assisted laser desorption/ionization mass spectrometry (LA-LDI-MS).[86] To date, several fluorescent tags have been applied as LA-LDI-MS tags, including PAHs and rhodamine fluorophores.[87]

Using LA-LDI-MS, commercially available 1-pyrenebutyric acid (**1**) was detected in amounts down to at least 100 pmol with an LDI-MS instrument equipped with a 355 nm Nd:YAG laser (see Figure 6.8). Since **1** has little absorption at 355 nm, nitrogen-substituted pyrenes were examined to improve the sensitivity on LDI-MS. Initially, the 6-aminopyrene methyl ester **2** with cationic properties was examined, but a homodimer (radical) cation was dominantly observed on LA-LDI-MS, which was oxidatively dimerized *in situ* by UV laser irradiation. Since acidic treatment or photoirradiation causes dimerization of aniline and related aromatic amines,[88] similar reactions might occur with **2** on LDI-MS. In contrast, in the case of the 6-amidopyrene methyl ester **3** with substantial absorption at 355 nm, a molecular ion peak $[M]^{+\bullet}$ and a fragment ion $[M-42]^{+\bullet}$ were solely observed using only a 10 fmol amount.[89] On electron ionization (EI)-MS, acetanilide fragments with the loss of ketene (CH_2=C=O) and thus aniline are generated.[90–92] Hence a fragment ion $[M-42]^{+\bullet}$ that corresponds to the molecular ion of **2** might be formed *in situ* from **3**.

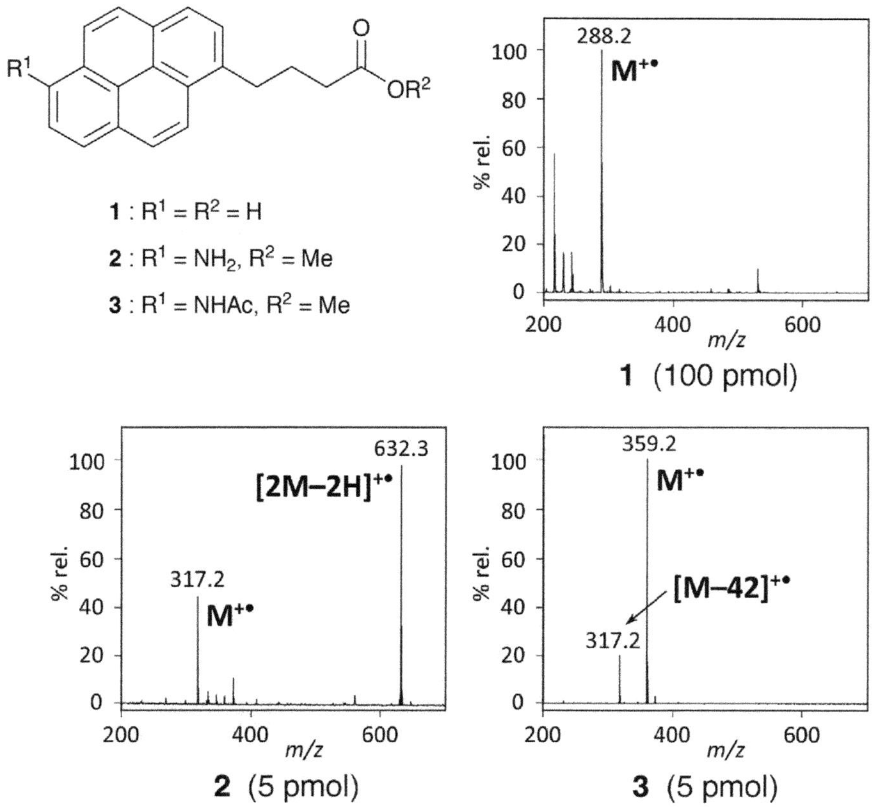

Figure 6.8 LA-LDI-MS of pyrenes. Adapted from ref. 89, https://doi.org/10.1038/srep17853, under the terms of a CC BY 4.0 license, https://creativecommons.org/licenses/by/4.0/.

6.3.3 Use of an Amidopyrene Tag for the Photoaffinity Probe of Aplyronine A

As mentioned, amidopyrene-conjugated compounds can be selectively detected by LA-LDI-MS without matrices. By using chemical probes that possess amidopyrene and diazirine groups, it was expected that pyrene-labelled peptide(s) could be solely detected by LDI-MS, after the photolabelling and enzymatic digestion of target proteins. If this approach is successful, labelled peptides would be observed without purification and the binding positions of ligands might be more easily determined. Hence a photoaffinity amidopyrene conjugate of aplyronine A (ApA–PaP) was prepared (see Figure 6.7). ApA–PaP showed strong cytotoxicity against HeLa S3 cells (IC_{50} = 0.66 nM). ApA–PaP caused abnormal multipolar spindles at 1 nM, as with the treatment with 0.1 nM ApA (see Figure 6.9). ApA–PaP also caused depolymerization of F-actin as with ApA. Thus, it was shown that modification with amidopyrene had little effect on the potent biological activity of aplyronine A.

To establish the ability of the photoaffinity group in the presence of amidopyrene, ApA–PaP was photoreacted with solvents. As a result, the MeOH and water adducts were obtained in 42% yields. In an LDI-MS analysis, oxime N–O bonds were selectively cleaved from both the ApA–PaP MeOH and water adducts, as confirmed by MS/MS analysis.[89] From the viewpoint of the selective detection of the conjugates with pyrenes, it is desirable to detect the fragment ions cleaved at the linker moiety.

Next, photolabelling experiments with ApA–PaP with actin were conducted, but the labelling yield was fairly low.[72] As with ApA–PF, the major side products might be hydrolysed products, where the carbene species reacted with a water molecule on the surface of actin.

6.3.4 Use of Amidopyrene for the Specific Detection of Labelled Peptides

To establish amidopyrene as a suitable detection group for chemical probes, it was conjugated with peptides and LDI-MS analysis of labelled peptides was examined. Actin was labelled with amidopyrene *N*-hydroxysuccinate

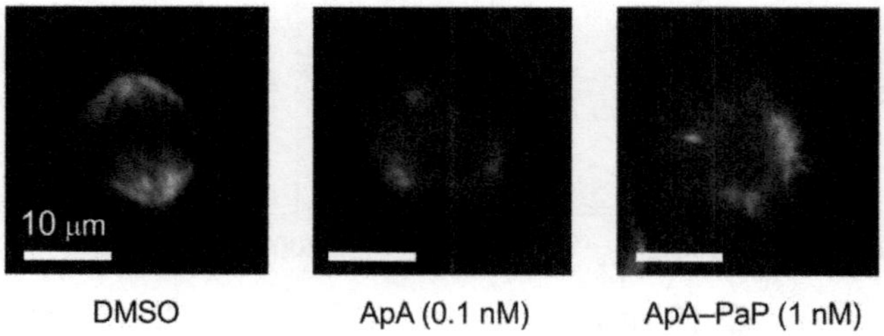

Figure 6.9 Confocal fluorescence images of HeLa S3 cells (metaphase) treated with ApA or ApA–PaP. Cells were stained with anti-α-tubulin and DAPI. Reproduced from ref. 89, https://doi.org/10.1038/srep17853, under the terms of a CC BY 4.0 license, https://creativecommons.org/licenses/by/4.0/.

(apy–OSu), which mainly reacts with the ε-amino group of the lysine (Lys) residue, which was digested with trypsin. In MALDI-MS, both apy-labelled and unlabelled actin peptides were detected with a similar intensity (see Figure 6.10). In contrast, eight apy-labelled peptides were predominantly detected by LA-LDI-MS.[93] With some apy-labelled peptides, 42 mu-reduced fragment ions were observed, which were assigned as aminopyrenes that had lost a ketene (CH_2=C=O) moiety, as mentioned earlier. Meanwhile, two non-labelled peptides were detected. They had several aromatic amino acids, such as phenylalanine, tyrosine and tryptophan, and might be excited by a UV_{355}

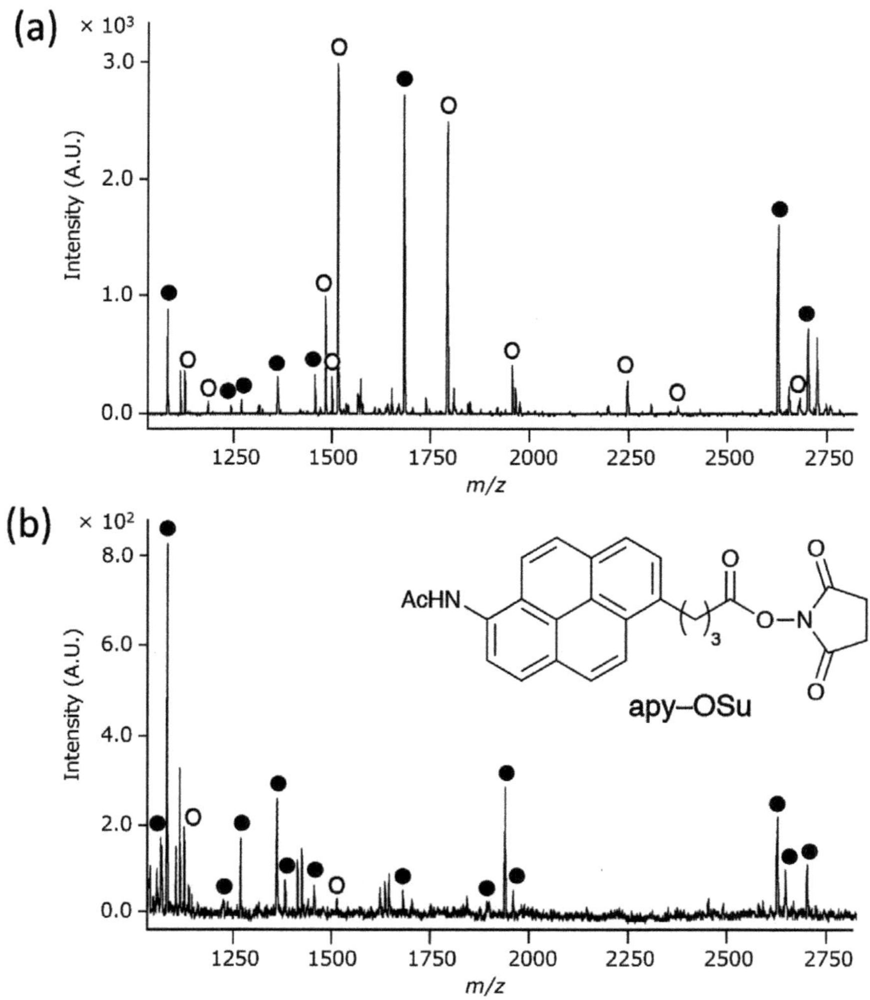

Figure 6.10 Mass spectra of the tryptic peptides of actin labelled with apy–OSu. (a) MALDI and (b) LA-LDI-MS. Solid and open circles indicate amidopyrene-labelled and unlabelled actin peptides, respectively. Reproduced from ref. 93 with permission from the Royal Society of Chemistry.

laser without matrices. Still, amidopyrene-labelled peptides were preferentially detected by LA-LDI-MS, which suggested that an apy-labelling approach is applicable for chemical probe methods to determine the binding sites of target proteins.

6.3.5 Binding Position Analysis of Actin with Aplyronine A Amidopyrene NHS Ester

Based on the findings with the ApA–PaP probe, an ApA–amidopyrene probe possessing an NHS ester (ApA–apy–OSu) was synthesized by condensation of a common amine intermediate of ApA–PaP with *N*,*N*′-disuccinimidyl glutarate (see Figure 6.7). This probe readily reacted with glycine under basic aqueous conditions, which established the high reactivity of ApA–apy–OSu with a primary amine.

As expected, actin was highly efficiently labelled with ApA–apy–OSu and this labelling was competitively inhibited by ApA, which suggested that this apy probe specifically bound to monomeric actin to form a 1 : 1 complex. MALDI-MS analysis of the digested actin peptides with trypsin and Glu-C showed that an apy-labelled peptide (m/z 2841.6, A108–R116 residues) was observed as the highest peak, in addition to several unlabelled peptides (see Figure 6.11). MALDI-MS/MS analysis established that the NHS ester in the probe dominantly reacted at K113. In the structure of the actin–ApA complex, A108–R116 are close to the binding site of the ApA side chain. Molecular modelling studies showed that the NHS ester of ApA–apy–OSu was closest to the K113 ε-amino group. These results suggested that the specific labelling of K113 by the ApA probe was reliable.

In the case of LA-LDI-MS, the same mass peak of apy-labelled peptide (m/z 2841.5) was observed as with MALDI-MS, but its intensity was low since the linker part was cleaved at the carbonyl α-position. In addition, several unlabelled peptides, especially those with aromatic amino acid residues, were detected. To detect apy-labelled peptides selectively, the detectable amount should be reduced below the picomole range to analyse various target protein–ligand interactions. Still, it is suggested that the LA-LDI-MS method using amidopyrene or other fluorescent tags could be improved as a new chemical probe method by further modifications of the detecting tags and the linker structures.

6.3.6 Ligand Dissociation-type Chemical Probes for Binding Mode Analysis

In the above studies on aplyronine A using chemical probes, the ligands, reacting groups and detecting groups were connected to an L-Lys linker, which was called a "ligand-bound type" probe. To improve the detection sensitivity of labelled products, "ligand-dissociation type" probes were designed, which have an *N*-hydroxysuccinimidyl (NHS) ester between the ligands and amidopyrene moieties (see Figure 6.12).[94] It was expected that if the nucleophilic

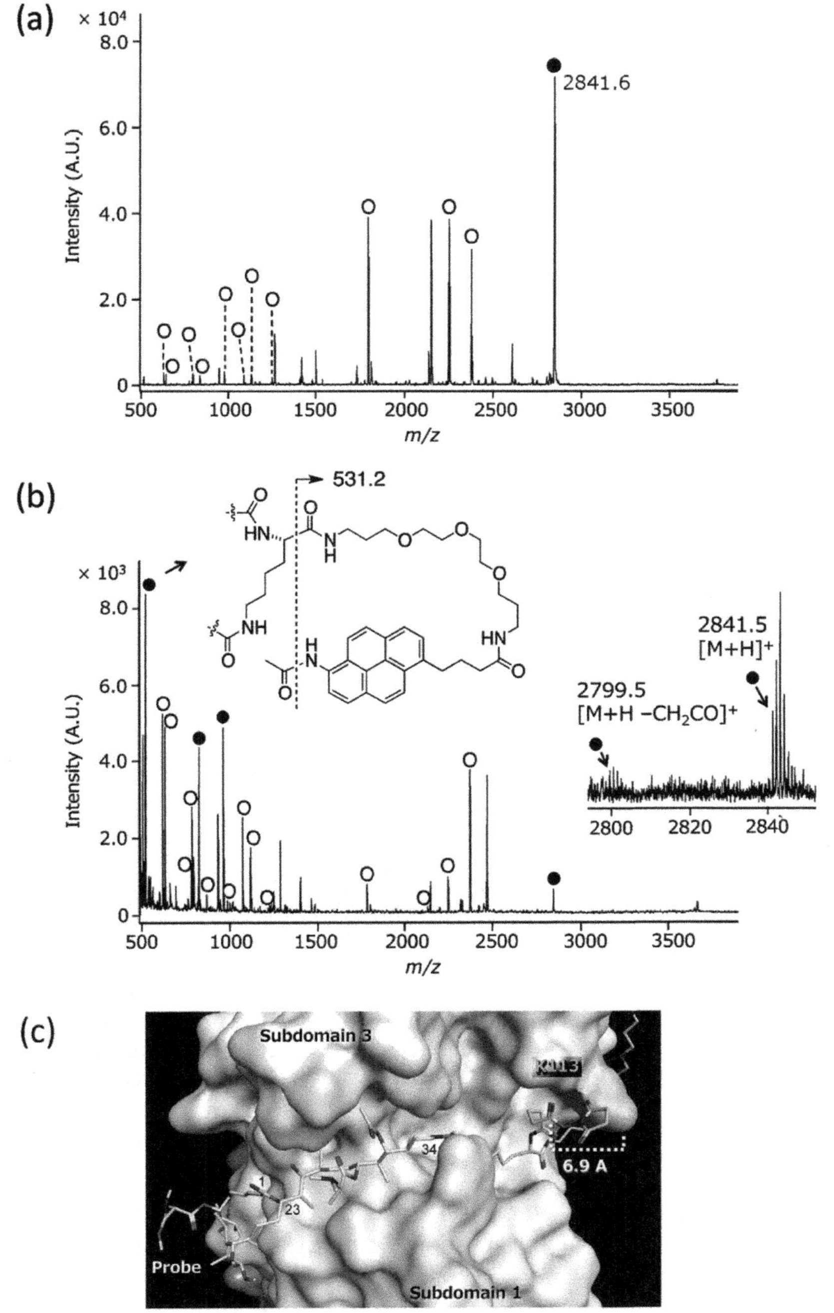

Figure 6.11 Mass spectra of the digested peptides of actin labelled with ApA–apy–OSu. (a) MALDI and (b) LA-LDI-MS. Solid and open circles indicate apy-labelled and unlabelled peptides, respectively. (c) Molecular modelling of the actin–ApA apy probe complex. A conformational analysis was performed by fixing actin and the ApA moiety (C1–C34). Reproduced from ref. 93 with permission from the Royal Society of Chemistry.

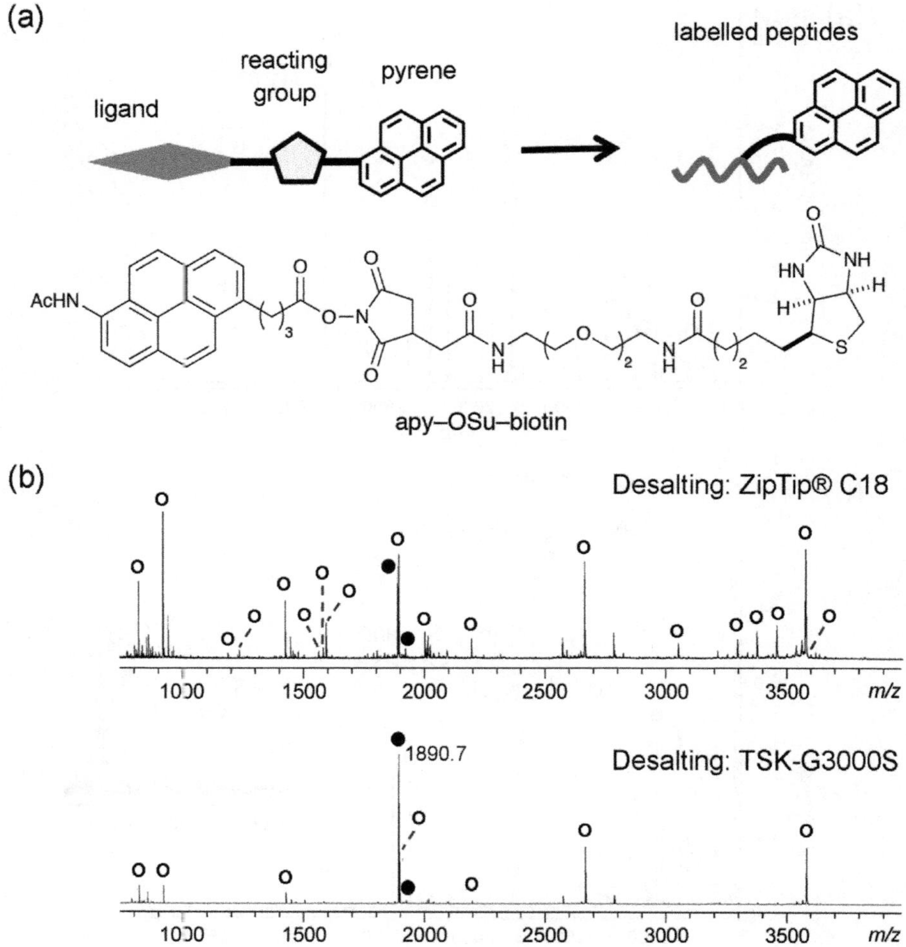

Figure 6.12 Ligand-dissociation type apy biotin probe. (a) Design concept. (b) MALDI-MS analysis of tryptic peptide fragments of avidin labelled with apy–OSu–biotin. Solid and open circles indicate apy-labelled and unlabelled peptides, respectively. Reproduced from ref. 94 with permission from the Royal Society of Chemistry.

attack of target molecules is specific, the covalent bond between the ligand and reacting group would be cleaved and only the apy moiety would remain on the nearest ligand-binding sites of the target proteins.

To confirm this proposal, an amidopyrene biotin derivative, apy–OSu–biotin, was designed as one of the ligand-dissociation type probes. In the avidin–biotin complex (PDB code 2avi),[95] K135 is closest to the ligand-binding site among the eight Lys residues. Furthermore, to interact more efficiently with the target molecules and to increase the flexibility of apy derivatives, the ligand biotin and the NHS ester moieties were connected *via* a PEG linker.

After labelling of avidin with apy–OSu–biotin, the labelled protein was denatured with 6 M guanidine·HCl, since avidin is very rigid and stable against enzymatic digestion. Subsequent reductive *S*-alkylation and trypsin digestion gave apy-labelled avidin peptides. To remove salts from tryptic peptides, a conventional reversed-phase ODS tip column (ZipTip® C_{18}) was initially used. On MALDI-MS, apy-labelled peptides were detected as expected, but their intensities were fairly low. Meanwhile, with a TSK-G3000S polystyrene gel batch column, apy-labelled peptides were more strongly absorbed than non-labelled peptides. By using stepwise aqueous MeOH elution, apy peptides were partially affinity purified and their relative mass intensities on MALDI-MS were improved.

As expected, the dominantly detected apy-labelled peptide (*m*/*z* 1890.7, S125–R138) had the K135 residue and its ε-amino group reacted with apy–OSu–biotin. Since an excess amount of apy–OSu unselectively reacted with five Lys residues in avidin, avidin labelling by the ligand-dissociation type biotin probe was highly site specific.

To explain the high labelling selectivity of the apy biotin probe at the K135 residue of avidin, docking simulation studies were performed. As expected, the biotin moiety in the apy–OSu–biotin highly overlapped that in the X-ray structure of the avidin–biotin complex in the most stable conformer (RMSD = 0.378 Å) (see Figure 6.13). In this complex, the NHS ester moiety was closest to the K135 ε-amino group and too far away from other Lys residues. These results supported the specific labelling of K135 by apy–OSu–biotin.

In general, proteins possess multiple hydrophobic ligand-binding sites, hence it is difficult to predict ligand-binding positions without laboratory-based experimental data. To predict ligand-binding sites precisely, covalent docking simulations were performed based on the results of LDI-MS

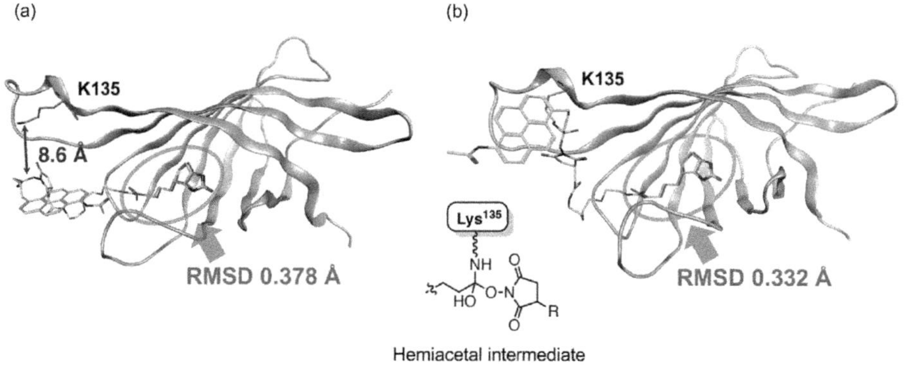

Figure 6.13 Molecular modelling of the complexes of avidin and apy–OSu–biotin. (a) Docking simulation (site: dummy atoms). (b) Covalent docking simulation of the hemiacetal intermediate by the reaction of NHS esters at K135 of avidin. Reproduced from ref. 94 with permission from the Royal Society of Chemistry.

analysis. For the ligand-dissociation type probes including apy–OSu–biotin, ligands are removed during the labelling reactions, so we postulated a hemiacetal intermediate, which is formed by the nucleophilic attack of the Lys ε-amino group of avidin on the NHS esters. As a result, the biotin moieties in the probe highly overlapped the original on the most stable model (RMSD = 0.332 Å). These results showed that LDI-MS analysis in combination with covalent docking simulations can promptly determine the ligand binding position on the target molecules.

6.4 Conclusion

We have focused on the target identification and binding position analysis of key natural products using chemical probes. To our knowledge, ApA is the first molecule that stabilizes PPI between actin and tubulin to inhibit microtubule dynamics. Furthermore, to analyse readily the binding mode of protein–ligand interactions, amidopyrene was prepared as a new LA-LDI-MS tag. In combination with LDI-MS, analyses of digested peptides and molecular modelling studies established the binding position of the ligands on the target proteins. This method may be applicable for analysing unknown various ligand-binding sites of biomacromolecules.

Since natural products have broad structural complexity and functional diversity, they can interact with multiple cellular targets in various ways. The use of PPI-modulating scaffolds should give perspectives for the design of simplified mimetics and functional molecules that tether different interactions. Further precise design, practical synthesis and utilization of chemical probes are important to gain better perspectives to develop new pharmacological tools and epoch-making therapeutic agents.

Abbreviations

Arp2/3	Actin-related protein 2/3
ApA	Aplyronine A
Apy	Amidopyrene
DTT	Dithiothreitol
EI	Electron ionization
ESI	Electrospray ionization
FKBP	FK506-binding protein
IC_{50}	Half-maximal inhibitory concentration
LA-LDI	Label-assisted laser desorption/ionization
Lys	Lysine
MALDI	Matrix-assisted laser desorption/ionization
MS	Mass spectrometry
NHS	*N*-hydroxysuccinyl
NMR	Nuclear magnetic resonance
ODS	Octadecylsilyl
PAH	Polycyclic aromatic hydrocarbon

PDB	Protein Data Bank
PEG	Poly(ethylene glycol)
PMF	Peptide-mass fingerprint
PPI	Protein–protein interaction
RMSD	Root mean square deviation
SD	Subdomain
SPR	Surface plasmon resonance
TAMRA	Tetramethylrhodamine
TMSer	*N*,*N*,*O*-trimethylserine
YAG	Yttrium aluminium garnet

Acknowledgements

We thank all of our collaborators for their work and valuable suggestions regarding chemical probe studies. This work was supported in part by JSPS grants (18H04613 and 19H02839 to M.K. and 18H04606 to H.K.) and the JSPS A3 Foresight Program. Support was also provided by PRESTO, JST (JPMJPR1535), NU–AIST Alliance Project, Yamada Science Foundation and the Naito Foundation.

References

1. J. Clardy and C. Walsh, *Nature*, 2004, **432**, 829.
2. D. J. Newman, G. M. Cragg and K. M. Snader, *J. Nat. Prod.*, 2003, **66**, 1022.
3. E. E. Carlson, *ACS Chem. Biol.*, 2010, **5**, 639.
4. S. Bovens and C. Ottmann, in *Chemical Biology: Learning through Case Studies*, ed. H. Waldmann and P. Jannig, Wiley-VCH Verlag GmbH, Weinheim, 2009, pp. 105–119.
5. D. M. Spencer, T. J. Wandless, S. L. Schreiber and G. R. Crabtree, *Science*, 1993, **262**, 1019.
6. J. A. Wells and C. L. McClendon, *Nature*, 2007, **450**, 1001.
7. L. G. Milroy, T. N. Grossmann, S. Hennig, L. Brunsveld and C. Ottmann, *Chem. Rev.*, 2014, **114**, 4695.
8. M. R. Arkin, Y. Tang and J. A. Wells, *Chem. Biol.*, 2014, **21**, 1102.
9. M. Kita and H. Kigoshi, *Nat. Prod. Rep.*, 2015, **32**, 534.
10. S. A. Andrei, E. Sijbesma, M. Hann, J. Davis, G. O'Mahony, M. W. D. Perry, A. Karawajczyk, J. Eickhoff, L. Brunsveld, R. G. Doveston, L.-G. Milroy and C. Ottmann, *Expert Opin. Drug Discovery*, 2017, **12**, 925.
11. M. A. Jordan and L. Wilson, *Nat. Rev. Cancer*, 2004, **4**, 253.
12. C. Dumontet and M. A. Jordan, *Nat. Rev. Drug Discovery*, 2010, **9**, 790.
13. D. G. Kingston, *J. Nat. Prod.*, 2009, **72**, 507.
14. M. J. Yu, W. Zheng and B. M. Seletsky, *Nat. Prod. Rep.*, 2013, **30**, 1158.
15. J. Cortes, J. O'Shaughnessy, D. Loesch, J. L. Blum, L. T. Vahdat, K. Petrakova, P. Chollet, A. Manikas, V. Diéras, T. Delozier, V. Vladimirov, F. Cardoso, H. Koh, P. Bougnoux, C. E. Dutcus, S. Seegobin, D. Mir, N. Meneses, J. Wanders and C. Twelves, *Lancet*, 2011, **377**, 914.

16. G. R. Pettit, in *International Oncology Updates: Marine Anticancer Compounds in the Era of Targeted Therapies*, ed. B. Chabner and H. Cortes-Funes, Permanyer, Barcelona, 2009, pp. 19–50.
17. J. A. Francisco, C. G. Cerveny, D. L. Meyer, B. J. Mixan, K. Klussman, D. F. Chace, S. X. Rejniak, K. A. Gordon, R. Deblanc, B. E. Toki, C. L. Law, S. O. Doronina, C. B. Siegall, P. D. Senter and A. F. Wahl, *Blood*, 2003, **102**, 1458.
18. P. D. Senter and E. L. Sievers, *Nat. Biotechnol.*, 2012, **30**, 631.
19. M. Schenone, V. Dančík, B. K. Wagner and P. A. Clemons, *Nat. Chem. Biol.*, 2013, **9**, 232.
20. U. Rix and G. Superti-Furga, *Nat. Chem. Biol.*, 2009, **5**, 616.
21. F. Kotzyba-Hibert, I. Kapfer and M. Geoldner, *Angew. Chem., Int. Ed. Engl.*, 1995, **34**, 1296.
22. T. Böttcher, M. Pitscheider and S. A. Sieber, *Angew. Chem., Int. Ed.*, 2010, **49**, 2680.
23. C. G. dos Remedios, D. Chhabra, M. Kekic, I. V. Dedova, M. Tsubakihara, D. A. Berry and N. J. Nosworthy, *Physiol. Rev.*, 2003, **83**, 433.
24. K. S. Yeung and I. Paterson, *Angew. Chem., Int. Ed.*, 2002, **41**, 4632.
25. G. Fenteany and S. Zhu, *Curr. Top. Med. Chem.*, 2003, **3**, 593.
26. J. S. Allingham, V. A. Klenchin and I. Rayment, *Cell. Mol. Life Sci.*, 2006, **63**, 2119.
27. I. Spector, N. R. Shocher, Y. Kashman and A. Groweiss, *Science*, 1983, **219**, 493.
28. E. Vincent, J. Saxton, C. Baker-Glenn, I. Moal, J. D. Hirst, G. Pattenden and P. E. Shaw, *Cell. Mol. Life Sci.*, 2007, **64**, 487.
29. S. Saito, H. Watabe, H. Ozaki, N. Fusetani and H. Karaki, *J. Biol. Chem.*, 1994, **269**, 29710.
30. J. Tanaka, Y. Yan, J. Choi, J. Bai, V. A. Klenchin, I. Rayment and G. Marriott, *Proc. Natl. Acad. Sci. U. S. A.*, 2003, **100**, 13851.
31. R. D. Perrins, G. Cecere, I. Paterson and G. Marriott, *Chem. Biol.*, 2008, **15**, 287.
32. A. V. Statsuk, R. Bai, J. L. Baryza, V. A. Verma, E. Hamel, P. A. Wender and S. A. Kozmin, *Nat. Chem. Biol.*, 2005, **1**, 383.
33. S. A. Rizvi, D. S. Courson, V. A. Keller, R. S. Rock and S. A. Kozmin, *Proc. Natl. Acad. Sci. U. S. A.*, 2008, **105**, 4088.
34. S. C. Posey and B. E. Bierer, *J. Biol. Chem.*, 1999, **274**, 4259.
35. H. Konishi, S. Kikuchi, T. Ochiai, H. Ikoma, T. Kubota, D. Ichiwaka, H. Fujiwara, K. Okamoto, C. Sakakura, T. Sonoyama, Y. Kokuba, J. Sasaki, T. Matsui and E. Otsuji, *Anticancer Res.*, 2009, **29**, 2091.
36. O. Ohno, M. Morita, T. Kitamura, T. Teruya, K. Yoneda, M. Kita, H. Kigoshi and K. Suenaga, *Bioorg. Med. Chem. Lett.*, 2013, **23**, 1467.
37. V. A. Klenchin, J. S. Allingham, R. King, J. Tanaka, G. Marriott and I. Rayment, *Nat. Struct. Biol.*, 2003, **10**, 1058.
38. K. Hirata, S. Muraoka, K. Suenaga, K. Kuroda, K. Kato, H. Tanaka, M. Yamamoto, M. Takata, K. Yamada and H. Kigoshi, *J. Mol. Biol.*, 2006, **356**, 945.

39. J. S. Allingham, A. Zampella, M. V. D. Auria and I. Rayment, *Proc. Natl. Acad. Sci. U. S. A.*, 2005, **102**, 14527.
40. V. A. Klenchin, R. King, J. Tanaka, G. Marriott and I. Rayment, *Chem. Biol.*, 2005, **12**, 287.
41. S. A. Rizvi, V. Tereshko, A. A. Kossiakoff and S. A. Kozmin, *J. Am. Chem. Soc.*, 2006, **128**, 3882.
42. M. Kobayashi, J. Tanaka, T. Katori and I. Kitagawa, *Chem. Pharm. Bull.*, 1990, **38**, 2960.
43. M. R. Bubb, I. Spector, A. D. Bershadsky and E. D. Korn, *J. Biol. Chem.*, 1995, **270**, 3463.
44. J. Tanaka, T. Higa, M. Kobayahshi and I. Kitagawa, *Chem. Pharm. Bull.*, 1990, **38**, 2967.
45. S. Saito, S. Watabe, H. Ozaki, M. Kobayashi, T. Suzuki, H. Kobayashi, N. Fusetani and H. Karaki, *J. Biochem.*, 1998, **123**, 571.
46. K. Yamada and H. Kigoshi, *Bull. Chem. Soc. Jpn.*, 1997, **70**, 1479.
47. M. Kita and H. Kigoshi, in *Handbook of Anticancer Drugs from Marine Origin*, ed. S.-K. Kim, Springer, 2015, pp. 701–739.
48. K. Yamada, M. Ojika, T. Ishigaki, Y. Yoshida, H. Ekimoto and M. Arakawa, *J. Am. Chem. Soc.*, 1993, **115**, 11020.
49. M. Ojika, H. Kigoshi, T. Ishigaki, I. Tsukada, T. Tsuboi, T. Ogawa and K. Yamada, *J. Am. Chem. Soc.*, 1994, **116**, 7441.
50. H. Kigoshi, M. Ojika, T. Ishigaki, K. Suenaga, T. Mutou, A. Sakakura, T. Ogawa and K. Yamada, *J. Am. Chem. Soc.*, 1994, **116**, 7443.
51. H. Kigoshi, K. Suenaga, T. Mutou, T. Ishigaki, T. Atsumi, H. Ishiwata, A. Sakakura, T. Ogawa, M. Ojika and K. Yamada, *J. Org. Chem.*, 1996, **61**, 5326.
52. I. Hayakawa, K. Saito, S. Matsumoto, S. Kobayashi, A. Taniguchi, K. Kobayashi, Y. Fujii, T. Kaneko and H. Kigoshi, *Org. Biomol. Chem.*, 2017, **15**, 124.
53. I. Paterson, S. J. Fink, L. Y. W. Lee, S. J. Atkinson and S. B. Blakey, *Org. Lett.*, 2013, **15**, 3118.
54. N. Anzǐcěk, S. Williams, M. P. Housden and I. Paterson, *Org. Biomol. Chem.*, 2018, **16**, 1343.
55. K. Yamada, M. Ojika, H. Kigoshi and K. Suenaga, in *Drugs from the Sea*, ed. N. Fusetani, Karger, Basel, 2000, pp. 59–73.
56. K. Yamada, M. Ojika, H. Kigoshi and K. Suenaga, *Proc. Jpn. Acad., Ser. B*, 2010, **86**, 176.
57. K. Yamada, M. Ojika, H. Kigoshi and K. Suenaga, *Nat. Prod. Rep.*, 2009, **26**, 27.
58. S. Saito, S. Watabe, H. Ozaki, H. Kigoshi, K. Yamada, N. Fusetani and H. Karaki, *J. Biochem.*, 1996, **120**, 552.
59. K. Suenaga, N. Kamei, Y. Okugawa, M. Takagi, A. Akao, H. Kigoshi and K. Yamada, *Bioorg. Med. Chem. Lett.*, 1997, **7**, 269.
60. H. Kigoshi, K. Suenaga, M. Takagi, A. Akao, K. Kanematsu, N. Kamei, Y. Okugawa and K. Yamada, *Tetrahedron*, 2002, **58**, 1075.
61. T. Kuroda, K. Suenaga, A. Sakakura, T. Handa, K. Okamoto and H. Kigoshi, *Bioconjugate Chem.*, 2006, **17**, 524.

62. K. Futaki, M. Takahashi, K. Tanabe, A. Fujieda, H. Kigoshi and M. Kita, *ACS Omega*, 2019, **4**, 8598.
63. M. Kita, K. Yoneda, Y. Hirayama, K. Yamagishi, Y. Saito, Y. Sugiyama, Y. Miwa, O. Ohno, M. Morita, K. Suenaga and H. Kigoshi, *ChemBioChem*, 2012, **13**, 1754.
64. S. Wada, S. Matsunaga, S. Saito, N. Fusetani and S. Watabe, *J. Biochem.*, 1998, **123**, 946.
65. M. Kita, Y. Hirayama, M. Sugiyama and H. Kigoshi, *Angew. Chem., Int. Ed.*, 2011, **50**, 9871.
66. R. D. Mullins, J. A. Heuser and T. D. Pollard, *Proc. Natl. Acad. Sci. U. S. A.*, 1998, **95**, 6181.
67. L. Blanchoin, K. J. Amann, H. N. Higgs, J.-B. Marchand, D. A. Kaiser and T. D. Pollard, *Nature*, 2000, **404**, 1007.
68. E. D. Goley and M. D. Welch, *Nat. Rev. Mol. Cell Biol.*, 2006, **7**, 713.
69. M. Kita, Y. Hirayama, K. Yamagishi, K. Yoneda, R. Fujisawa and H. Kigoshi, *J. Am. Chem. Soc.*, 2012, **134**, 20314.
70. M. Kita, Y. Hirayama, K. Yoneda, K. Yamagishi, T. Chinen, T. Usui, E. Sumiya, M. Uesugi and H. Kigoshi, *J. Am. Chem. Soc.*, 2013, **135**, 18089.
71. M. Kita, K. Yamagishi, K. Tsuchiya, Y. Seguchi, H. Nakane and H. Kigoshi, *Bioorg. Med. Chem.*, 2017, **25**, 6322.
72. Y. Hirayama, K. Yamagishi, T. Suzuki, H. Kawagishi, M. Kita and H. Kigoshi, *Bioorg. Med. Chem.*, 2016, **24**, 2809.
73. M. A. Jordan, D. Thrower and L. Wilson, *Cancer Res.*, 1991, **51**, 2212.
74. M. A. Jordan, R. J. Toso, D. Thrower and L. Wilson, *Proc. Natl. Acad. Sci. U. S. A.*, 1993, **90**, 9552.
75. M. A. Jordan, K. Wendell, S. Gardiner, W. B. Derry, H. Copp and L. Wilson, *Cancer Res.*, 1996, **56**, 816.
76. O. C. Rodriguez, A. W. Schaefer, C. A. Mandato, P. Forscher, W. M. Bement and C. M. Waterman–Storer, *Nat. Cell Biol.*, 2003, **5**, 599.
77. K. J. Pacholarz, R. A. Garlish, R. J. Taylor and P. E. Barran, *Chem. Soc. Rev.*, 2012, **41**, 4335.
78. J. B. Fenn, M. Mann, C. K. Meng, S. F. Wong and C. M. Whitehouse, *Science*, 1989, **246**, 64.
79. K. Tanaka, H. Waki, Y. Ido, S. Akita, Y. Yoshida and T. Yoshida, *Rapid Commun. Mass Spectrom.*, 1988, **2**, 151.
80. M. Karas, D. Bachman, U. Bahr and F. Hillenkamp, *Int. J. Mass Spectrom. Ion Processes*, 1987, **78**, 53.
81. M. Karas, D. Bachman and F. Hillenkamp, *Anal. Chem.*, 1985, **57**, 2935.
82. C. Robins and P. A. Limbach, *Rapid Commun. Mass Spectrom.*, 2003, **17**, 2839.
83. G. Petroselli, M. K. Mandal, L. C. Chen, G. T. Ruiz, E. Wolcan, K. Hiraoka, H. Nonami and R. Erra-Balsells, *J. Mass Spectrom.*, 2012, **47**, 313.
84. B. Martinez-Haya, A. R. Hortal, P. Hurtado, M. D. Lobato and J. M. Pedrosa, *J. Mass Spectrom.*, 2007, **42**, 701.
85. S. F. Macha, T. D. McCarley and P. A. Limbach, *Anal. Chim. Acta*, 1999, **397**, 235.

86. J. R. Cabrera-Pardo, D. I. Chai, S. Liu, M. Mrksich and S. A. Kozmin, *Nat. Chem.*, 2013, **5**, 423.
87. A. Mandal, M. Singha, P. S. Addy and A. Basak, *Mass Spectrom. Rev.*, 2019, **38**, 3and references therein.
88. W. Boszczyk and T. Latowski, *Z. Naturforsch., B*, 1989, **44**, 1589.
89. K. Yoneda, Y. Hu, M. Kita and H. Kigoshi, *Sci. Rep.*, 2015, **5**, 17853.
90. T. A. Lee, in *A Beginner's Guide to Mass Spectral Interpretation*, John Wiley & Sons, Chichester, England, 1998, pp. 106–116.
91. R. H. Shapiro and K. B. Tomer, *Org. Mass Spectrom.*, 1969, **2**, 579.
92. M. Ohashi, T. Kizaki, K. Tsujimoto, Y. Shida and Y. Yamada, *J. Mass Spectrom. Soc. Jpn.*, 1973, **21**, 85.
93. K. Yoneda, Y. Hu, R. Watanabe, M. Kita and H. Kigoshi, *Org. Biomol. Chem.*, 2016, **14**, 8564.
94. R. Watanabe, Y. Hu, K. Iio, K. Yoneda, A. Hattori, A. Arai, H. Kigoshi and M. Kita, *Org. Biomol. Chem.*, 2018, **16**, 7883.
95. O. Livnah, E. A. Bayer, M. Wilchek and J. L. Sussman, *Proc. Natl. Acad. Sci. U. S. A.*, 1993, **90**, 5076.

CHAPTER 7

Targeted Protein Degradation Chemical Probes

C. GALDEANO*[a] AND X. BARRIL[a,b]

[a]Department of Pharmacy, Pharmaceutical Technology and Physical Chemistry, Faculty of Pharmacy and Food Science, and Institute of Biomedicine (IBUB), University of Barcelona, 08028 Barcelona, Spain; [b]Catalan Institution for Research and Advanced Studies (ICREA), Passeig Lluis Companys 23, 08010 Barcelona, Spain
*E-mail: cgaldeano@ub.edu

7.1 Small Molecules that Hijack the Ubiquitin–Proteasome System

During normal cellular homeostasis, proteins are constantly synthesized and destroyed. The ubiquitin–proteasome system (UPS) pathway is ultimately responsible for the controlled degradation of 80% of the proteome *via* ubiquitination. Protein ubiquitination occurs through a highly organized and orchestrated cascade of enzymatic reactions, involving an E1 ubiquitin-activating enzyme, an E2 ubiquitin-conjugating enzyme and an E3 ubiquitin-ligase complex that covalently binds ubiquitin to the side chain of a lysine.[1] Repeated iterations of this ubiquitination process result in long chains of ubiquitin in a given target protein and its posterior degradation by the 26S proteasome. The therapeutic and biological potential of intervention in the UPS was demonstrated by proteasome inhibitors such as peptide boronic acids (bortezomib) and epoxyketones (carfilzomib), which have had success

Chemical Biology No. 16
The Discovery and Utility of Chemical Probes in Target Discovery
Edited by Paul Brennan and Saleta Vazquez Rodriguez

Published by the Royal Society of Chemistry, www.rsc.org

in the clinic for the treatment of multiple myeloma among other diseases, by deubiquitinating enzyme (DUB) inhibitors and by E1, E2 and E3 inhibitors.[2]

Instead of blocking the UPS components to achieve therapeutic benefit, in recent years novel approaches able to hijack the UPS specifically to induce proteasomal degradation of a desired protein have appeared.

7.1.1 Introducing PROTAC Molecules, Molecular Glues and Greasy Tags

Proteolysis-targeting chimera (PROTAC) molecules are heterobifunctional small molecules designed to induce intracellular protein degradation. The PROTAC approach works through a validated sub-stoichiometrically catalytic mechanism based on a dual interaction. One "warhead" of the PROTAC molecule binds to a protein of interest (POI), whereas the other warhead recruits a specific E3 ligase (see Figure 7.1), forming a ternary complex (E3 ligase–PROTAC POI) that allows ubiquitin transfer from the E2 or the E3 ligase (depending of the type of E3 ligase) to a lysine γ-amino group on the surface of the POI, which leads to degradation of the POI. As the role of the PROTAC molecule is simply to facilitate the interaction between the E3 ligase complex and the POI, the PROTAC molecule eventually dissociates and is free to mediate the ubiquitination and degradation of additional substrate molecules, *i.e.* the PROTAC molecules are acting as a true catalyst.

Distinct from PROTAC molecules, molecular glues are small molecules that insert into a naturally occurring protein–protein interaction (PPI), making interactions with both the substrate and the ligase with a unique molecular scaffold (see Figure 7.1). The protein degradation strategy based on molecular glues originated from the observation that immunomodulatory drugs (IMiDs, *e.g.* thalidomide, lenalidomide, pomalidomide) binding to the

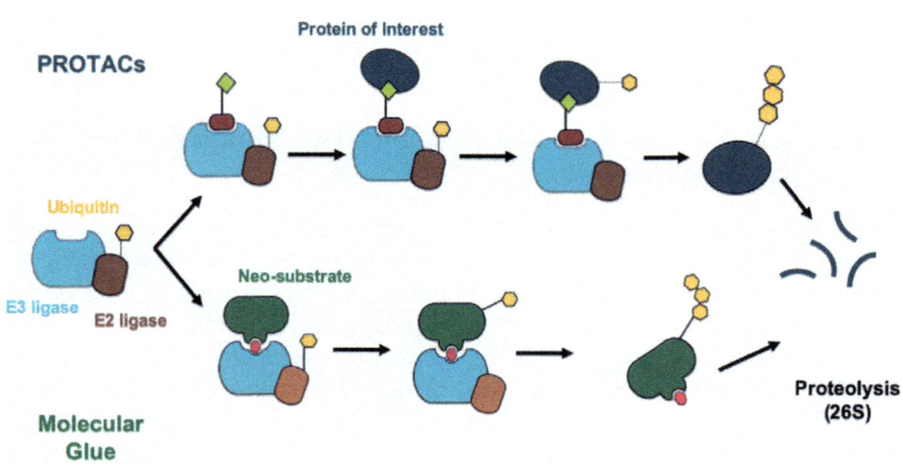

Figure 7.1 Mechanism of action of PROTAC molecules and molecular glues.

E3 ligase cereblon promoted the degradation of important non-native substrates, such as ikaros, aiolos, GSTP1 and CK1α, which structurally only share a common β-hairpin containing a glycine residue.[3] In addition to IMiDs, the use of PPI enhancer molecules to potentiate substrate–ligand interactions could also be useful in developing novel therapies and chemical probes targeting undruggable targets. Simonetta *et al.* reported the identification and rational design of molecular glues enhancing the interaction between the E3 ligase βTrCP and the transcription factor β-catenin.[4] These molecular glues potentiated the ubiquitination and posterior degradation of an engineered mutant β-catenin protein in cells, opening up new avenues in the development of specific mutation-dependent therapies.

Another approach to trigger selective protein degradation *via* the UPS is based on the observation that proteins showing large hydrophobic regions in their surfaces engage the cellular quality control machinery to induce its proteasomal degradation. Based on this observation, Craig Crews' laboratory at Yale University hypothesized that appending a hydrophobic moiety (adamantyl) in a covalent fashion to the surface of a protein would mimic the partially denatured state of a protein, triggering its corresponding degradation.[5] This approach has been employed to degrade a limited number of targets, including androgen receptor (AR)[6] and the pseudo-kinase Her3.[7]

From a chemical probe point of view, the "plug and play" modular approach of PROTAC molecules could present significant advantages compared with other targeted protein degradation approaches. In principle, the design of PROTAC molecules can be guided by the binary affinity of each protein and warhead. In contrast, a rational design for molecular glues remains elusive and challenging, since these molecules do not require any binary ligand affinity for the proteins outside of the E3 ligase–ligand–POI complex and have to promote *de novo* PPIs, restricting their utility for the development of chemical probes for part of the proteome. Finally, the mechanism of action of the hydrophobic tags remains unresolved, also hampering its application for the generation of chemical probes.

7.1.2 What Drives the Activity of PROTAC Molecules and Molecular Glues?

In contrast to traditional occupancy-driven pharmacology, where therapeutic efficacy and pharmacologically relevant inhibition are mostly achieved upon >90% target engagement,[8] the activity of PROTAC molecules does not depend exclusively on the affinity of the warhead ligands for their corresponding targets. The efficacy of PROTAC molecules, also molecular glues, depends strongly on their ability to form an E3 ligase–ligand–POI ternary complex (see Figure 7.2). Mathematically, the formation of this ternary complex depends on the concentration of the two protein counterparts, the affinity of the ligands towards these proteins and the cooperative effects.[9] Favourable and repulsive interactions between the E3 ligase and the POI affect the cooperativity and, in consequence, the formation of the ternary complex. Several studies, including a pioneering structural study at the laboratory of Alessio Ciulli (University of Dundee),[10] demonstrated that one of the major drivers of the potency of the PROTAC molecules is their ability to form optimal PPI between the E3 ligase and the POI. In line with these structural findings, seminal work at Crews' laboratory also demonstrated that what correlates best with the degradation potency is the ability to form the ternary complex and not the affinity of the binary complexes.[11]

Most PROTAC molecules suffer the pharmacological "hook effect". This effect occurs because once the ligand is in excess relative to its target protein partner, there is sufficient free ligand available to compete for the protein target separately, forming binary complexes (see Figure 7.2). The result is an autoinhibition activity at certain concentrations with a biphasic dose-dependent curve that can lead to misinterpretation of the biological phenotypes coming from the inhibition and the degradation of the POI. In addition, the hook effect makes dosing of PROTAC molecules in cells, animals and humans problematic. Strong cooperativity not only increases the PROTAC efficacy, but also eliminates the hook effect. Hence it is particularly desirable for *in vivo* dosing.[9]

Once the ternary complex is formed, ubiquitination of the POI must occur to initiate the actual degradation. Depending on the number of ubiquitin units, the type and length of the chains that they formed and by which one of

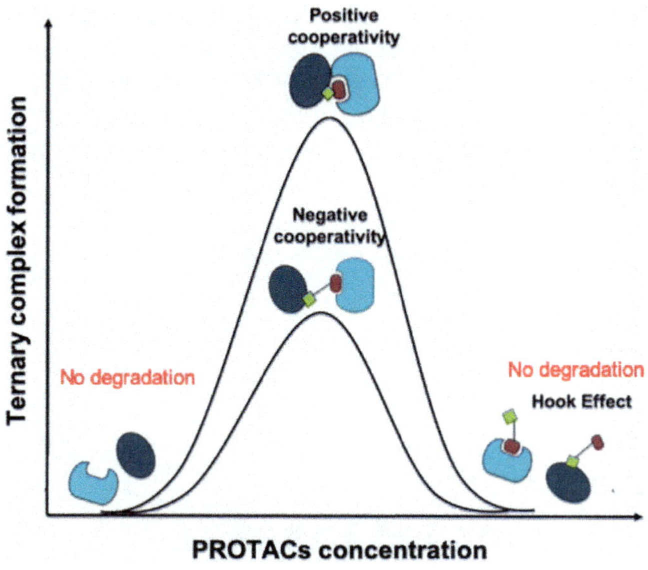

Figure 7.2 Bell-shaped dose–response and hook effect of PROTAC molecules.

the seven lysine residues the ubiquitins are linked (K6, K11, K27, K29, K33, K48, K63), the effect on the target degradation may differ greatly and can result in very different cellular responses.[12] In the case of PROTAC molecules, linear polyubiquitination (>4 units) in the K48 is preferred for 26S proteasome recognition and degradation. Finally, efficient degradation of a target protein requires the presence of a disordered region, *i.e.* a proteasome initiation region.[13]

7.1.3 Targeted Protein Degradation: From an Academic Exercise to Drug Candidates

The first proofs-of-concept of the PROTAC approach were published in 2001 and 2004 as a result of research collaboration between the groups of Craig Crews and Raymond Deshaies (Caltech).[14,15] These first-generation PROTAC molecules were mainly peptides either for the E3 ligase or for the POI warhead ligands, or both, and needed microinjection or polyarginine tags mimicking the HIV viral Tat protein for delivery inside cells; thereafter the PROTAC approach remained largely dormant for over a decade (see Figure 7.3), and only a handful of academic examples employing IAP and MDM2 E3 ligases were described in this period.[16,17] However, the development and discovery of specific drug-like ligands targeting E3 ligases impressively reactivated the PROTAC field (see Figure 7.3). On the one hand, in 2012 a collaboration between the laboratories of Craig Crews and Alessio Ciulli resulted in the publication of the first series of drug-like small molecules binding to the von Hippel–Lindau (VHL) E3 ligase. Subsequently, in 2014, Ciulli and co-workers described the first nanomolar binder of VHL (VH032), which retains sufficient potency and selectivity to be constructed as part of a PROTAC molecule. On the other hand, pushed by the finding of Hiroshi Handa's group (Tokyo Institute of Technology) that thalidomide binds to the E3 ligase cereblon,[18] in 2014 two independent teams were able

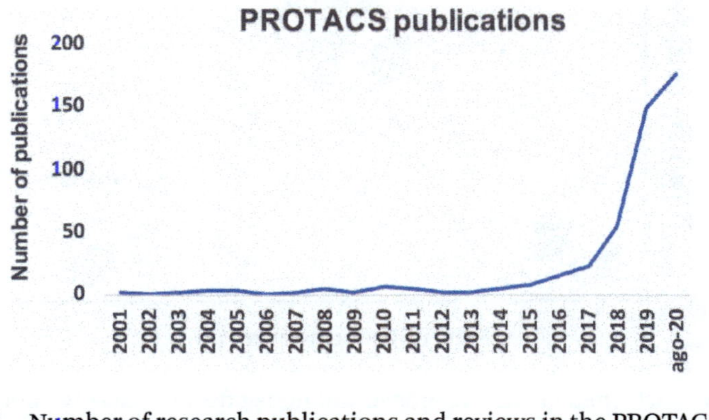

Figure 7.3 Number of research publications and reviews in the PROTAC field. Data extracted from PubMed.

to elucidate the X-ray structure of lenalidomide and thalidomide binding to cereblon, providing the structural basis for the design of PROTAC molecules and molecular glues.[19,20]

Employing VHL and cereblon as E3 ligases, in May and June 2015 three teams, led by Ciulli, Bradner (Dana-Farber Cancer Institute) and Crews, published several independent studies describing active PROTAC molecules with drug-like properties (see Figure 7.4).[21–25] These seminal studies unleashed the field in both academic and industrial settings. Finally, during 2019, two candidate PROTAC drugs entered clinical trials; ARV-110 and ARV-471, from Arvinas Inc., are orally available PROTAC molecules that target the androgen receptor (AR) and the estrogen receptor (ER) and are designed to treat metastatic castration-resistant prostate cancer and locally advanced or metastatic AR-positive breast cancer, respectively.

In parallel, in recent years the serendipitous discovery that several small molecules, including drugs in clinical trials (*e.g.* avadomide, indisulam and iberdomide), have a molecular glue mechanism opened up the horizon in the development of this class of molecules.[26]

7.1.4 Role of E3 Ligases in Small-molecule Degraders

E3 ubiquitin ligases confer substrate specificity for ubiquitination *via* PPI recognition. Recruitment by an E3 ligase is a mandatory step in PROTAC molecules and molecular glues. E3 ligases can be classified into three main types (RING, HECT and RBR), depending on the presence of characteristic domains and on the mechanism of ubiquitin transfer to the substrate protein.[27] RING E3s are the most abundant and include the cullin–RING ligases (CRLs), the most exploited subfamily for artificially redirecting protein degradation.

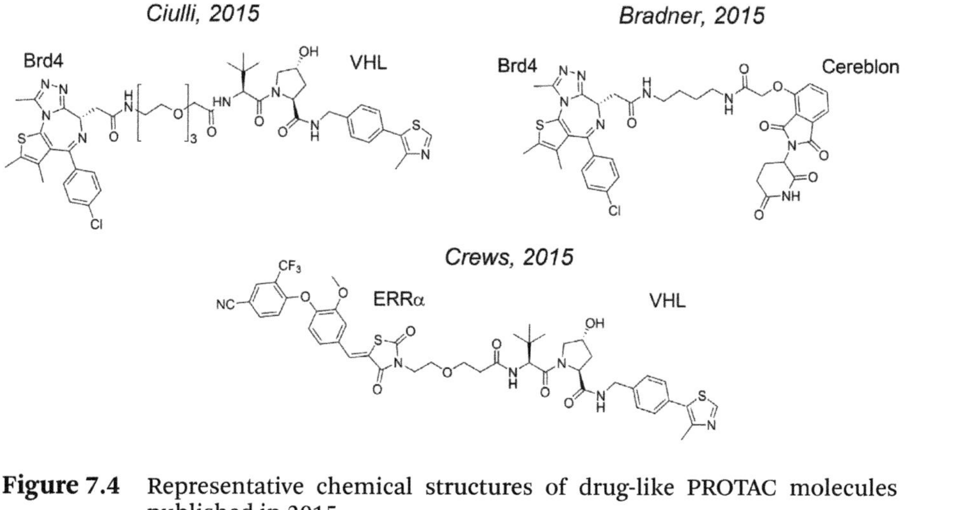

Figure 7.4 Representative chemical structures of drug-like PROTAC molecules published in 2015.

So far, only a handful of E3 ligases have been exploited for PROTAC molecules and molecular glues. The major hurdle for exploiting these E3 ligases is the lack of corresponding small-molecule ligands for them that could be incorporated into the PROTAC structure. Considering that E3 ligases represent a large gene family with more than 600 members, there is a clear mismatch between the number of members of the family and the number of E3 ligases exploited for targeted protein degradation purposes.

The two most exploited E3 ligases have been the CRLs VHL and cereblon. Although these E3 ligases have routinely demonstrated their potential, degrading dozens of different targets following a "plug and play" approach, several studies have raised concerns about the appearance of resistance.[28–30] Ottis *et al.*[28] and Zhang *et al.*[29] reported that cancer cells lose components of the UPS to develop resistance to PROTAC molecules. Specifically, VHL-recruiting PROTAC molecules showed resistance by loss of cullin 2 and cereblon-recruiting PROTAC molecules showed resistance by loss of the cereblon E3 ligase itself. The fact that cells generate resistance by mutating E3 ligases over other more general components of the ubiquitin proteasome complex e.g. cullin adaptors proteins, may relate to the redundancy or non-essentiality of these components for cell survival. In fact, exploiting E3 ligases that are essential to cells is a promising strategy to avoid this resistance mechanism.[31]

Other E3 ligases have been exploited with small molecules (IAP, MDM2, RNF114, RNF4, DCAF16 and AhR) or peptides (the very first βTRCP and Keap1) for the construction of PROTAC molecules. SNIPER (specific and non-genetic IAP-dependent protein eraser) molecules recruit the IAP ubiquitin ligase family and, remarkably, suffer from auto-ubiquitination and self-degradation. Taking into account that the level of IAP correlates with the degree of resistance to cancer therapy, this may be a desired effect.[32] However, this could also be a limitation to expanding the SNIPER approach to other targets and to developing chemical probes based on it. Nevertheless, in 2017 specific development and optimization of the IAP warhead resulted in SNIPER molecules that demonstrated proof-of-concept of degradation *in vivo* and circumvented the auto-ubiquitination effect.[33]

Like SNIPER molecules, MDM2-based PROTAC molecules could exert a synergistic and dual effect. The MDM2 E3 ligase is targeted by the nutlin family of small molecules. Because nutlin-based PROTAC molecules could upregulate the tumour suppressor p53, a very well-characterized MDM2 substrate, Crews' laboratory developed PROTAC molecules that are able to degrade BRD4 and at the same time upregulate p53, unlocking a new strategy in the targeted protein degradation field.[34]

By incorporating β-naphthoflavone as a warhead ligand, Ohoka *et al.* developed a novel class of PROTAC molecules that recruit the aryl hydrocarbon receptor (AhR) E3 ligase complex.[35] Although AhR is ubiquitously expressed, it has been shown that inflammatory stimuli and cytokines induce AhR expression during activation and differentiation of immune cells. AhR is also upregulated in several types of tumours, representing a source of disease-dependent selectivity for PROTAC molecules.[36]

Employing covalent ligands for the E3 ligase warhead, several PROTAC molecules have been developed. Daniel Nomura's group (UC Berkeley) was able to degrade BRD4 using covalent warhead ligands for RNF4[37] and RNF114.[38] In a similar fashion, Benjamin Cravatt and co-workers (Scripps Research Institute) were also able to degrade BRD4 by engaging DCAF16.[39]

Similarly to IMiDs binding to cereblon, indisulam (and similar molecules such as E7820) binds with weak affinity to DCAF15 (17 μM in the case of indisulam). However, this affinity is sufficient to target the degradation of RBM39, thanks to a large DCAF–RMB39 interface. Crucially, RMB39 does not bind to DCAF15 alone. From the X-ray structure of indisulan bound to DCAF15, steric clashes for DCAF15-recruiting PROTAC molecules were envisaged.[40–42]

The unique features of DCAF E3 ligases make the members of this family especially interesting for protein degradation purposes. DCAF E3 ligases show an exceptionally large ubiquitination zone and the cullin 4 ligase arm where DCAF ligases are mounted has unique flexibility compared with the other CRLs. However, there is limited understanding of the physiological function of DCAFs.[43,44]

7.1.4.1 Exploiting the Expression of E3 Ligases: Towards Specific Chemical Probes

E3 ligases have diverse temporal and spatial expression profiles. Importantly, the differences in tissue and cellular expression of a particular E3 ligase can be exploited to design unique and more selective chemical probes, since they will not be active in cells and tissues where the E3 ligase is not expressed. Based on proteomic data, Schapira *et al.* reviewed the expression of E3 ligases in 81 cell and tissue types.[31] Whereas some E3 ligases are widely expressed in tissues (*e.g.* MDM2 and cereblon) and can provide valid chemical probes in a variety of cellular contexts, others are specifically expressed in small amounts in tissues and provide the basis for tissue-specific chemical probes. From the 81 analyzed, 20 E3 ligases were identified to have a narrow window of expression in human tissues. For example, ASB9 is expressed only in pancreas and testicles, whereas FBXL16 is localized only in cerebral cortex and caudate.

PROTAC molecules also offer the possibility of developing organelle-specific chemical probes. As mentioned, following a chemical–proteomic approach, Benjamin Cravatt's laboratory was able to identify PROTAC molecules that bind covalently to the E3 ligase DCAF16 and, crucially, promote nuclear-restricted degradation of proteins.[39]

An illustrative example came from the laboratory of Guangrong Zheng (University of Florida). They developed cereblon-based PROTAC molecules with increased selectivity for BCL-X_L-dependent MOLT4 cells over human platelet cells. This selectivity originates from the low expression of cereblon in human platelets. In fact, the potential of BCL-X_L as an anticancer target has been hampered until now by the on-target platelet toxicity, because platelets depend on BCL-X_L to maintain their viability.[45]

7.1.5 Design Principles of PROTAC Molecules

The design of bifunctional molecules involves three main aspects: the substitution points on each individual ligand, the length and composition of the linker and the choice of E3 ligase. Regarding the first aspect, functionalization of ligands is a common strategy in chemical biology to incorporate fluorescent probes, reactive moieties, affinity tags or other groups that endow the ligand with additional capabilities without participating in the binding event. When chemical handles are not readily available, it will be necessary to identify a solvent-exposed part of the ligand and a growing vector where it is possible to introduce them without interfering with ligand binding.[46] This was the case for the development of BRD9-degrading PROTAC molecules.[47] The dimethylamine group of the inhibitor BI-7273 was identified as a solvent-exposed group not involved in interactions with the protein. Additionally, for synthetic reasons, the dimethylamine group of BI-7273 was replaced with a piperazine group. On the other hand, the chemistry for functionalizing many PROTAC warheads on the POI side has already been developed, *e.g.* the BRD4 inhibitor JQ1.[48] This is also true for the E3 ligase head ligands, as they were particularly developed with the PROTAC strategy in mind. In fact, some E3 ligase ligands can be substituted from more than one position, *e.g.* VHL ligands and intermediates for different chemistries are readily available.[47] This permits a "plug and play" approach, where pairs of ligands can be paired through common linkers using facile, reproducible and high-yield synthetic chemistry procedures. Although this approach can be successful, resulting in rapid elaboration of useful chemical probes, the development of optimal PROTAC molecules may require a more elaborate approach. Also, as the PROTAC toolbox expands, the number of options and combinations will drastically increase and a rational selection will become even more necessary.

Linker design is a PROTAC-specific problem that has just begun to be investigated. As mentioned, cooperativity is due to protein–protein interactions that occur in the presence of the PROTAC molecule. If the linker is too short, the proteins would clash, leading to strong *anti*-cooperativity ($\alpha < 1$). If the linker is sufficiently long and the proteins do not interact, there will be no cooperativity ($\alpha = 1$) (see Figure 7.5).[49] Between these two regions, it is possible to multiply the effectiveness of the PROTAC molecule through positive cooperativity ($\alpha > 1$). However, this will only occur under the double condition that (i) the E3 ligase and the POI exhibit chemical complementarity and (ii) the linker allows, or even promotes, optimal complementarity of the protein–protein dimer.

A curious example of the influence of the linker in the mechanism of action of PROTAC molecules was reported by Shaomeng Wang's group (University of Michigan). They showed that a simple structural modification in the linker of an MDM2 PROTAC molecules can convert a PROTAC molecule into a molecular glue, warning that the mechanism of action should be considered when developing IMiD-based degraders.[50]

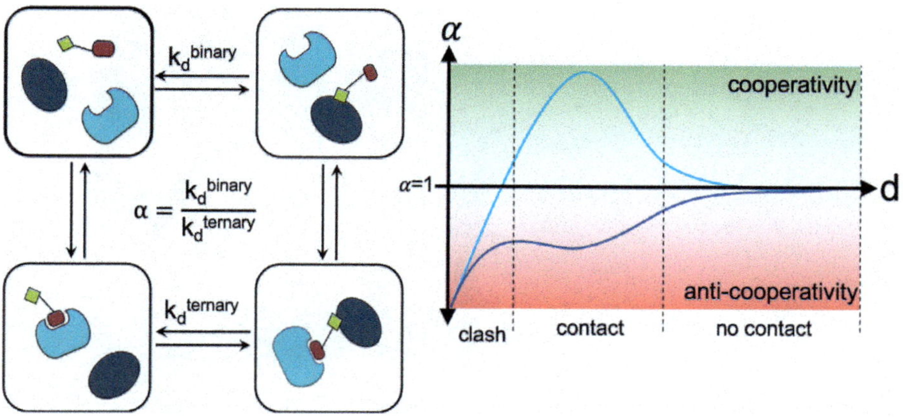

Figure 7.5 Influence of the linker in the cooperativity effect.

The 3D structures of PROTAC and glue molecules in a ternary complex reveal strong complementarity between the protein partners (see Figure 7.6). These favourable protein–protein intermolecular contacts contribute towards a cooperativity free energy term (ΔG_{coop}), which has an exponential relationship with α:

$$\alpha = \exp(-\Delta G_{coop}/RT)$$

Therefore, even small ΔG_{coop} values can boost the PROTAC effect by a large factor. For instance, two proteins with an intrinsic binding affinity of −4 kcal mol^{-1}, which would be experimentally undetectable ($K_d = 1$ mM), would translate into a 1000-fold α. The cooperativity values reported so far for PROTAC molecules are double digit at best, indicating that there is a large margin for improvement. In the case of monofunctional glue molecules, the cooperativity can be larger, which highlights both the challenge and the opportunity that this type of mechanism entails.[51]

7.1.5.1 *Optimization of PROTAC Molecules*

Although good pharmacokinetics are not essential for the development of chemical probes, it is obvious that the major part of PROTAC molecules does not fall into the "rule of five" proposed by Lipinski for oral drugs, molecular weights over 500 Da, relatively high lipophilicities (clog$P > 4$) and high polar surface areas (PSAs), representing a challenge for their testing (*i.e.* cellular permeation and solubility issues) and drug development [*i.e.* drug metabolism and pharmacokinetics (DMPK) optimization].[52] In contrast, molecular glues meet the historical guidelines regarding molecular weight and other properties, which represents a clear advantage to the progress of these

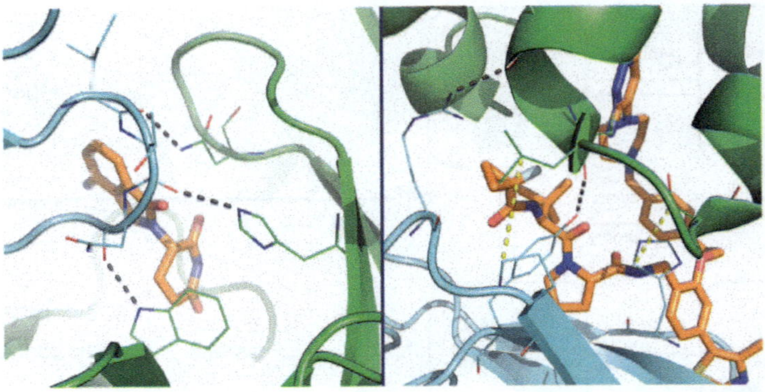

Figure 7.6 Two ligand-induced ternary complexes, showing protein–protein intermolecular contacts: hydrogen bonds in dark grey and hydrophobic contacts in yellow. Left: lenalidomide complex with cereblon (green) and CK1α (cyan) (PDB code 5FQD). Right: PROTAC complex with the bromodomain of SMARCA2 (green) and VHL (cyan) (PDB code 6HAX).

molecules into the clinic. However, challenging the dogma, in 2017 Arvinas Inc. reported the first orally bioavailable AR PROTAC degrader, effective at 1 mg kg^{-1} per day administered orally.

Several strategies to improve the properties of the PROTAC molecules and optimize them to design functional chemical probes have been developed. Researchers at Astex Pharmaceuticals described click-formed proteolysis-targeting chimeras (CLIPTACs).[53] CLIPTACs are PROTAC molecules formed in cells by a bio-orthogonal click reaction between a tetrazine-tagged thalidomide and a *trans*-cyclooctene-tagged ligand for the POI. The successful degradation of BRD4 and ERK1/2 validated the approach and CLIPTAC precursors showed a reduced molecular weight and PSA. Moreover, this approach can be used to screen POI ligands, linker lengths and E3 ligase warheads for PROTAC optimization.[54]

Based on the crystal structure of the ternary complex, Alessio Ciulli's laboratory developed a strategy to constrain the PROTAC molecule in its bioactive conformation *via* macrocyclization.[55] Despite its 12-fold weaker binary binding affinity at the BRD4 warhead, MacroPROTAC-1 has a very similar cellular activity to MZ1, the parent "free" BRD4 PROTAC degrader.

7.1.5.2 Computational Tools for the Design of PROTAC Molecules

Structure-based computational methods have a long and productive history in drug design. If the structure of the ternary complex is known (which is very rare nowadays), all pre-existing methods can be used for the rational design of optimized linkers. However, it should be noted that existing methods focus on protein–ligand dimeric interactions and will consider the protein–protein complex as a unified receptor. As such, these methods will fail to capture non-additive three-body effects, *i.e.* the reinforcement of the protein–protein complex caused by the presence of the ligand. To date, this effect has not been quantified in ternary complexes, but may not be negligible.

When the structure of the complex is not known, computational methods can be used to generate predictions of linker length and flexibility. Several schemes have been devised for this purpose. Fundamentally, they sample the configurational space available to the ternary complex, exploring the degrees of freedom of the linker, but maintaining the native binding mode of each head ligand. They may also incorporate a protein–protein docking engine to identify the best possible complementarity between proteins (within the constraints provided by PROTAC molecules). Although these methods use relatively crude approximations and explore the flexibility of the protein to only a very limited extent, the initial results are encouraging. However, in all cases this is a qualitative assessment at best.[56]

7.2 Small-molecule Degraders as Chemical Probes

Chemical probes represent one of the best ways to interrogate the function of a protein and to elucidate its importance as a drug target. The relevance of these small-molecule chemical probes to translate advances in genomics into better therapies has prompted several initiatives from the scientific community, *e.g.* Target 2035 and the Chemical Probes portal (https://www.chemicalprobes.org), to push their development.[57]

From a technical point of view, the discovery of chemical probes does not differ much from the discovery of initial hits in a drug discovery programmes.[58] However, their discovery must be extremely efficient, as the risks associated with the new mechanism of action rarely justify the long and costly hit identification campaigns that are typical in pharmaceutical companies.[59] However, as chemical probes explore new mechanisms of action, there is a lack of previous knowledge that makes discovery particularly challenging. At the same time, the chemical biology and drug discovery communities have come together to warn about the risks of low-quality tool compounds and have issued a set of strict guidelines that should be met by any chemical probe.[60] Several criteria have been proposed for chemical probes, with potency, selectivity, demonstrated target engagement and availability of a negative control being the most important. In this context, PROTAC molecules stand out from classical pharmacology with unique features for developing high-quality chemical probes.[61]

In this section, we discuss and exemplify some advantages that PROTAC molecules can offer in the generation of chemical probes (see Figure 7.7).

7.2.1 PROTAC Chemical Probes for Target Validation

The PROTAC approach emerges as a highly specific, robust and versatile approach to complement reverse genetic tools such as RNAi, CRISPR-CAS9 and other gene knockouts in the study of protein function and its target

	SiRNA/CRISPR	Chemical Probes Inhibitors	Targeted Protein Degradation
Removal of pathogenic proteins	✓	✗	✓
Catalytic mechanism of action	✓	✗	✓
Dosage Control	✗	✓	✓
Reversibility	✗	✓	✓
Intracellular targets	✓	✓	✓
Ability to develop high selectivity/potency	✓	Poor	✓
Ability to target proteins with scaffolding function	✓	✗	✓
Tissue penetration	Poor	✓	✓
Oral bioavailability	✗	✓	✓

Figure 7.7 Comparison of different approaches for target validation.

validation.[62] PROTAC molecules induce rapid, reversible, long-lasting, dose-dependent and profound protein degradation of a number of proteins, finally inducing a robust effect on downstream pathways. PROTAC molecules interfere at the post-translation level, although, being the least explored, this perhaps represents the cleanest, least invasive and most comprehensive approach to affect protein expression without any genetic manipulation of the organism. In fact, effective approaches for achieving protein depletion in large animals such as non-human primates are needed. Rao and co-workers reported that PROTAC technology depleted several target proteins with both oral and intraperitoneal administration in a range of animals, from mice to rhesus monkeys, in most tissues except the brain.[63]

Demonstrating intracellular protein target engagement is a mandatory condition for target validation and chemical probe generation. However, this is a very challenging step for non-catalytic proteins, *i.e.* transcription factor. Jones and co-workers used cereblon-based PROTAC molecules to demonstrate that a chemical probe that was discovered in a cell-based phenotypic screening binds the putative transcription factor pirin in living cells.[64]

When applying a chemical probe for target validation, it is strongly recommended to use a negative control.[65] A negative control is often an enantiomer or other close structural analogue of the original chemical probe, but inactive against the probe's primary target. VHL- and cereblon-based PROTAC molecules allow the efficient development of negative controls. In the case of VHL, this is by inversion of the stereocentre on the hydroxyproline moiety of the VHL. This change in the stereochemistry at the carbon bearing the hydroxyl group of the hydroxyproline ring is known to result in marked loss of binding.[66,67] In the case of cereblon, it is by methylation of the nitrogen on the glutarimide ring of pomalidomide and lenalidomide, causing steric hindrance in the binding pocket.[11,68]

Curiously, PROTAC molecules can validate as the target the E3 ligases that hijack, demonstrating the potential of PROTAC molecules as chemical probes also in the E3 ligase landscape. To address the question of whether VHL could degrade itself, Maniaci *et al.* developed a series of Homo-PROTACs, an approach to dimerize an E3 ligase, to trigger suicide-type chemical knockdown of VHL.[69] In the same fashion, Steinebach *et al.* described the chemical-induced degradation of cereblon by conjugation of two pomalidomide moieties, with no effect on other members of the same CRL ligase family.[70] Several groups have also investigated the hypothesis that the E3 ligases themselves could be hijacked against each other using heterodimerizing PROTAC molecules. Ciulli's group and Gütschow's laboratory (University of Bonn) also developed two independent series of VHL-cereblon PROTAC molecules. In both cases, preferential degradation of cereblon was observed.[71,72] Finally, by linking the MDM2 inhibitor MI-1061 to either a cereblon or VHL, Shaomeng Wang's group found that the corresponding PROTAC molecules were more efficacious than the parent MDM2 inhibitor to achieve tumour regression in a xenograft model.[73]

7.2.1.1 Genetic PROTAC Degradation for Target Validation

The dTAG system, developed by Bradner, Gray and colleagues at the Dana-Farber Cancer Institute, is an innovative approach that can be used for early target validation by combining the key advantages of genetic and pharmacological strategies.[74] The target protein is expressed as a chimera with mutant $FKBP12^{F36V}$, either by CRISPR/Cas9-mediated locus-specific knockin or lentiviral transgene expression. In parallel, a PROTAC molecule based on an E3 ligase warhead, *i.e.* cereblon linked to a highly selective $FKBP12^{F36V}$ already known ligand, was developed. The $FKBP12^{F36V}$ mutant presents a "hole" (F36V) to accept, with exceptional target specificity over the endogenous wild-type FKBP12, a described "bump" ligand. The resulting PROTAC molecule, also called dTAG, forms a ternary complex between the fusion protein and E3 ligase, causing polyubiquitination and degradation of the fused target protein (see Figure 7.8).

The dTAG PROTAC allows the validation of targets where there are no identified ligands available. The YEATS domain-containing protein ENL was previously identified in acute myeloid leukaemia (AML) as a key protein that supports pathogenesis *via* the maintenance of dysregulated gene expression; however, there were no known ligands available for ENL. To validate ENL as a target in AML, Erb *et al.* expressed ENL as an $FKBP12^{F36V}$ fusion protein in a human AML cell line and demonstrated selective degradation of ENL with nanomolar concentrations of dTAG-13.[75]

In a pioneering study using chemical genetic tools in the targeted protein degradation field, Crews and co-workers described HaloPROTAC molecules.[24] These molecules incorporate VHL ligands to degrade successfully HaloTag7 fusion proteins, so that the POI–HaloTag7 fusion recruited by VHL

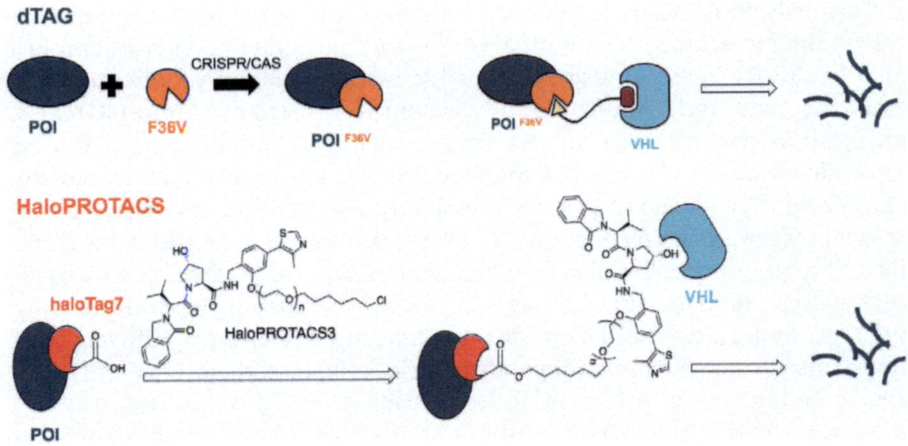

Figure 7.8 Graphical representation of dTAG and HaloPROTAC approaches.

is degraded *via* the UPS (see Figure 7.8). Similarly, HaloPROTAC molecules based on MDM2 were developed.[76] Proof-of-concept experiments demonstrated 90% green fluorescent protein (GFP)-HaloTag7 degradation in cells treated with nanomolar concentrations of the HaloPROTAC molecule. Crucially, plasmids containing HaloTag7 fused with 9000 human gene proteins are commercially available, allowing the degradation of numerous proteins in a single assay. However, it is important to note that the HaloTag7 system supplies overexpressed levels of target protein, which may disturb the normal protein turnover and degradation.

7.2.1.2 Expanding the PROTAC Approach for Target Validation

7.2.1.2.1 LYTACs A number of different targets, including enzymes, steroid receptors, transcription factors, BET proteins, several kinases and cytosolic parts of transmembrane receptors, have been degraded using PROTAC molecules. However, the "standard" PROTAC approach can only validate targets within the cells. In this context, two approaches were developed for the targeted degradation of extracellular secreted proteins and membrane proteins via the lysosomal pathway. On the one hand, lysosome-targeting chimeras (LYTACs) consist of an antibody fused to agonist glycopeptide ligands for the cation-independent mannose-6-phosphate receptor (CI-M6PR) (see Figure 7.9). CI-M6PR is a known lysosome-targeting receptor that directs proteins to the lysosome through an antibody specific for the protein. LYTAC molecules demonstrated efficient degradation of APOE4, epidermal growth factor receptor (EGFR), CD71 and PD-L1.[77]

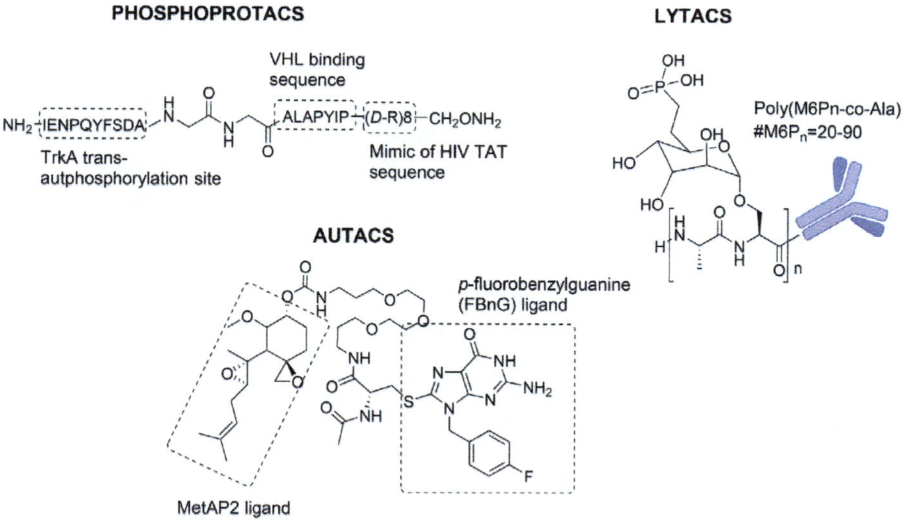

Figure 7.9 Representative chemical structures of a LYTAC, PhosphoPROTAC and AUTAC.

7.2.1.2.2 PhosphoPROTACs. Crews' group described an alternative approach to couple the conditional degradation of targeted proteins to the activation state of particular kinase-signalling pathways. Specifically, PhosphoPROTAC molecules were able to distinguish between different receptor tyrosine kinase (RTK) signalling pathways. Activation of the corresponding PhosphoPROTAC molecule and its ability to bind and degrade the POI occur only upon activation of the specific RTK with a suitable growth factor and posterior phosphorylation of the PhosphoPROTAC. On coupling a peptide sequence for VHL and incorporating different tyrosine phosphorylation sequences (TRKA or ERBB3), selective degradation was observed for the neurotrophic signalling effector fibroblast growth factor receptor substrate 2α or the survival-promoting phosphatidylinositol-3-kinase (see Figure 7.9).[78]

7.2.1.2.3 AUTACs. A wider range of substrates, including aggregated proteins and organelles, are degraded by autophagy. Autophagy-targeting chimeras (AUTACs) are bivalent molecules that downregulate specific cytosolic proteins by autophagy. On one side, AUTAC molecules contain an *S*-guanylation degradation moiety, whereas on the other side they contain a HaloTag-based or a small-molecule binder (*e.g.* MetAP2, FBKP12) warhead to provide target specificity, circumventing the lack of selectivity of the canonical autophagy (see Figure 7.8).[79] Actually, none of the available autophagy-inducing agents exhibit useful cargo selectivity. In a more specific approach, AUTAC molecules can also promote targeted mitochondrial turnover *via* mitophagy through binding to the mitochondria translocator protein (TSPO).[79]

7.2.2 Drugging the Undruggable with PROTAC Molecules

Only a small portion of the human proteome is pharmaceutically accessible based on traditional pharmacology approaches. The majority of proteins, for example transcription factors, scaffolding proteins and non-enzymatic proteins, are nowadays still considered undruggable. Only for 10% of the proteins it has it been possible to develop small molecules that are sufficiently bioactive,[80] for two major reasons: the lack of functional binding sites on these proteins[81] and the assumption of occupancy-driven pharmacology.[8]

PROTAC activity does not depend on the inhibition of an existing functional site in the target protein and, by definition, PROTAC molecules will elicit a response as long as a suitable head ligand binds to any site on the protein surface. PROTAC molecules only need to bring the target and the ubiquitin ligase together; it does not matter whether they bind to the active site or somewhere else on the target protein. This opens up the possibility of targeting proteins hitherto considered to be undruggable, especially transcription factors,[82] and proteins with druggable yet non-functional binding sites.

STAT3 has been considered a very attractive target in cancer and other inflammatory and autoimmune diseases. Previous efforts to target STAT3 have focused on disrupting upstream signalling molecules such as JAC and IL-6. Such upstream inhibitors have limited clinical activity as they are not able to block the STAT3 transcriptional activity completely. Bai *et al.* developed the first STAT3 PROTAC molecule, called SD-36.[83] SD-36 degrades >50% of STAT3 in leukaemia and lymphoma cell lines and this degradation is selective over other STAT proteins. Crucially, SD-36 also degrades STAT3 variants with somatic mutations that are present in haematological malignancies. Finally, *in vivo* treatment of leukaemia and lymphoma xenograft mice models with SD-36 resulted in complete and long-lasting tumour regression.

The lack of ordered structures of intrinsically disordered proteins (IDPs) is an inherent challenge for the generation of chemical probes based on standard structure-based drug discovery approaches. As a consequence, IDPs have been considered undruggable until now. In this context, the development of chemical probes based on PROTAC molecules can have a huge impact on targeting IDPs.

Tau protein, the major component of neurofibrillary tangles in Alzheimer's disease and other related tauopathies, is classified as an IDP. Several series of peptide-based PROTAC molecules containing tau recognition sequences and a VHL binding sequence[84] or a Keap1-dependent peptide[85] were developed. However, peptide-based PROTAC molecules may suffer from low brain permeability.

In July 2019, Arvinas Inc. presented the results of a preclinical study with tau-targeted PROTAC molecules. Following parenteral (peripheral) administration, this PROTAC degraded more than 95% of the pathological tau protein in the brain of a tauopathy mouse model. The data further indicated that the tau protein degradation is dose and concentration dependent, signifying that this PROTAC molecule effectively crosses the blood–brain barrier, reinforcing the idea that PROTAC molecules could degrade targets related to diseases of the central nervous system, including previously considered undruggable targets. In parallel, starting from a PET tracer that binds to the tau protein, Silva *et al.* designed a PROTAC molecule (QC-01-175) that engages both tau protein and cereblon.[86] QC-01-175 depleted tau in frontotemporal dementia patient-derived neuronal cell models.

7.2.3 Active Chemical Probes When Inhibition Fails: Exploiting Novel Biology

A PROTAC approach can repurpose abandoned small molecules for which effective phenotypic responses could not be observed by the inhibition of a particular target. This is especially interesting in multi-domain transcription regulators, where it is not always clear which domain it is most important to target, and frequently the more ligandable domain is not the one exerting the desired activity. In this respect, PROTAC molecules can recapture robustly the effect observed with genetic tools and provide relevant chemical probes. Indeed, several illustrative examples can be found in the literature.[87–89]

Small-molecule inhibitors targeting the SMARCA2 bromodomain are ineffective, as it is the ATPase domain that is functionally relevant. Selective suppression of SMARCA2 activity has been proposed as a therapeutic concept for SMARCA4-mutated cancers.[90] Using a bromodomain ligand that binds to SMARCA2 and SMARCA4, Ciulli's group developed PROTAC molecules that are able to degrade both subunits of the BAF complex.[87] Crucially, ACBI1, the best degrader of this series, induced anti-proliferative effects and cell death caused by SMARCA2 depletion in SMARCA4 mutant cancer cells and in AML cells dependent on SMARCA4 ATPase activity.

James Bradner's group were able to develop TRIM24 PROTAC molecules. Although potent and selective ligands of TRIM24 have been developed, they do not exert effective anti-proliferative responses. In contrast, the TRIM24 PROTAC (dTRIM24) represents a very useful tool to probe the function of TRIM24 by their depletion.[88]

Chemical inhibition of the PCAF/GCN5 bromodomains is insufficient to restore the diminished inflammatory response of PCAF-deficient immune cells. However, by generating the first PCAF/GCN5 PROTAC molecule, researchers at GlaxoSmithKline identified small molecules that are able to degrade PCAF/GCN5 and to potently modulate the expression of multiple inflammatory mediators in LPS-stimulated macrophages and dendritic cells.[89]

PROTAC molecules allow modulation of both enzymatic and scaffolding activity in kinase proteins, circumventing the problem of “kinome rewiring”,[91] whereby signalling leads to compensatory feedback activation of other kinases. This feature can also be exploited for other proteins with only scaffolding functions. Previous efforts to modulate FAK activity were limited to kinase inhibitors, with low success in clinical studies. Cromm *et al.* developed PROTAC molecules that are able to regulate both the kinase-dependent and the kinase-independent signalling of FAK, outperforming the original inhibitor defactinib.[92]

Many forms of cancer are driven by the hyperactivation of specific transmembrane RTKs due to mutations that confer growth factor-independent activation and resistance to kinase inhibitors. In contrast, VHL-based PROTAC molecules targeting RTKs were able to induce degradation of mutated RTK forms and target scaffolding functions of this protein, also showing that the PROTAC technology can avoid resistance mechanisms that are difficult or impossible to modulate using the inhibition approach.[93] This was also demonstrated by a PROTAC molecule able to degrade Bruton's tyrosine kinase (BTK) with the covalent kinase inhibitor ibrutinib as a warhead.[94,95] Ibrutinib was approved by the US Food and Drug Administration (FDA) in 2013 for the treatment of several types of non-Hodgkin's lymphoma. However, resistance to ibrutinib has been reported in various lymphomas due to a C481S BTK mutation, resulting in drastic reduction of its IC_{50} due to the lack of the covalent bond. Crucially, ibrutinib-based PROTAC molecules could degrade wild-type BTK and also ibrutinib-resistant BTK-C481S (total depletion at nanomolar concentrations).[94]

7.2.4 PROTAC Chemical Probes Allow Low-affinity Ligands

Occupancy-driven pharmacology is the default approach in drug discovery and chemical probe generation. It requires a strong binding affinity and high drug dosing, which results in lack of selectivity, potential off-target binding and exclusion of the part of the proteome that is considered undruggable, because potent molecules could not be developed. In fact, one of the main criteria for a good-quality chemical probe is that it should have an affinity of <100 nM, restricting the number of proteins for which chemical probes could be developed. The importance of the ternary complex for the PROTAC molecules' activity allows the use of molecules with affinities in the micromolar range.

Foratenib is a very promiscuous kinase inhibitor. Among all the kinases that it inhibits, foratenib shows a low binding affinity towards p38α. However, Smith *et al.* demonstrated that once built into a PROTAC molecule, foratenib can degrade, in an isoform-selective manner, p38α with nanomolar concentrations (DC_{50} = 210 nM, D_{max} = 91%).[96] Crucially, the authors demonstrated that the formation of the ternary complex mitigates the low binary affinity of foratenib. In addition, other examples can be found in the literature where PROTAC molecules enhance the activity in comparison with the binary activity,[97–99] demonstrating that high-activity PROTAC chemical probes (<100 nM) can be designed starting from low-affinity inhibitors (<1 μM).

PROTAC molecules have also demonstrated a catalytic mechanism of action and, unlike traditional inhibitors, their efficacy is not limited by equilibrium occupancy. A good example of this feature comes from a study by Crews' group where AR-targeting PROTAC molecules still block prostate tumour cell proliferation in the presence of competing androgen levels.[100]

7.2.5 Chemical Probes with Enhanced Selectivity

The selectivity criteria for chemical probes should be more stringent than those for drug candidates. In the case of PROTAC molecules, the need for an E3 ligase to degrade the target protein adds an additional layer of selectivity in comparison with traditional inhibitors. We discussed in Section 7.1.4.1 the potential of developing PROTAC-based tissue-specific molecules. However, the particular mechanism of action and the importance of the ternary complex formation can also provide enhanced selectivity compared with that provided by inhibition. Warheads recruiting different E3 ligases rule which protein can be degraded. An illustrative example of selectivity with this origin is provided by a seminal study by Crews' group. Whereas cereblon-based PROTAC molecules employing the kinase inhibitor bosutinib were able to degrade ABL/BCR-ABL, VHL-based PROTAC molecules with the same inhibitor were inactive towards the same protein. Additionally, although VHL-based PROTAC molecules employing dasatinib were selective degraders of ABL, no selectivity was observed between ABL and BCR-ABL with the cereblon-based PROTAC molecules using dasatinib (see Figure 7.10).[101]

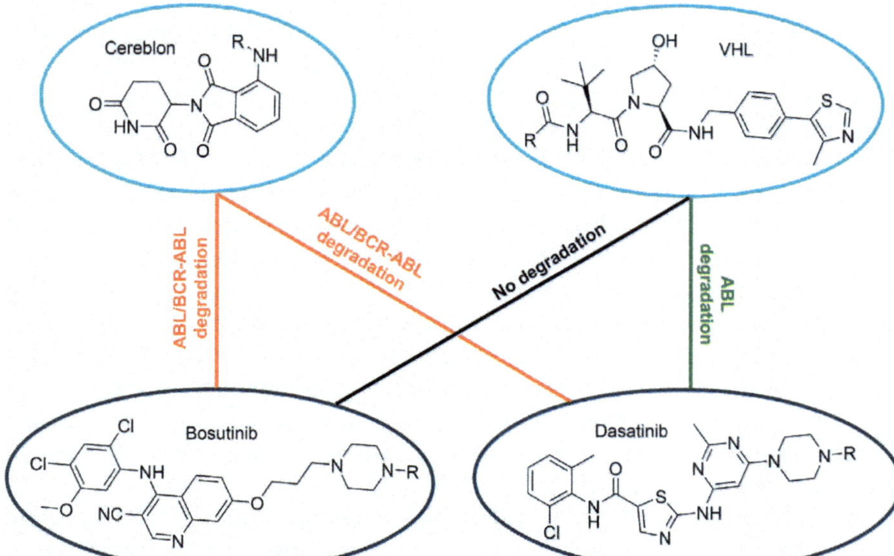

Figure 7.10 Scheme of the selectivity of ABL/BCR and VHL/cereblon PROTAC molecules.

Transforming a pan-selective or promiscuous chemical probe ligand into a more selective degrader probe allows new opportunities for improving target validation and could reduce off-target effects. Crews' group demonstrated that a promiscuous kinase inhibitor (foratinib) that binds >130 kinases, once built into a PROTAC system, could degrade <10 kinases.[11] Moreover, depending on which E3 ligase (VHL or cereblon) was employed, different kinases were degraded.

The development of a selective chemical probe for one of the bromodomain proteins inside the BET subfamily stimulated great interest in both academic and industrial settings. Ciulli's group reported that the PROTAC MZ1 induces reversible, long-lasting and selective removal of BRD4 over the highly homologous BRD2 and BRD3.[21] MZ1 is made by conjugating the pan-BET inhibitor (+)-JQ1 *via* a linker to a VHL ligand warhead. This selectivity window translated into a more BRD4-specific downstream transcriptional response in cancer cells treated with MZ1 compared with treatment with (+)-JQ1. Interestingly, initial similar studies with (+)-JQ1 built in a cereblon-based PROTAC molecule (dBET1)[23] or with ARV-825, also a cereblon PROTAC with the pan-BET inhibitor OTX015,[25] did not show noteworthy intra-BET selectivity. More recently, X-ray structures of BRD4 PROTAC molecules bound to VHL[48] and to cereblon (with a novel generation of dBET) were solved,[102] providing structural insights into how PROTAC molecules can induce target-specific interactions in ternary complex formation, governing the cooperativity between the counterparts. Moreover, these structures also showed that linker size and conformation and linking positions from both the E3 ligase warhead and target ligands are important factors in achieving potent target-specific degraders, due to the stabilization of different ternary complexes that results in distinct surface patches on the ligase and the target. Similarly, enhanced PROTAC selectivity in comparison with the binary ligands has been observed for other targets, including Mcl-1/Bcl-2,[103] HDACs,[104] TANK[99] and others.

Much interest has also been shown in identifying selective CDK6 over CDK4 inhibitors, a challenging endeavour as they share 94% amino acid sequence identity in their ATP-binding pockets. CDK4 and CDK6 are key regulators of the cell cycle and there are non-selective FDA-approved CDK4/6 inhibitors for treating patients with metastatic breast cancer. Using palbociclib, two independent teams were able to develop homologue-selective CDK6 and/or CDK4 degraders exploiting amino acid differences in the POI–E3 interface that induce differential ternary complex formation depending of the different linker lengths.[105,106] These selective PROTAC molecules represent perfect chemical probes for understanding, for example, the sensitivity of AML cell lines to selective CDK6 ablation.

In addition to selectivity for highly homologous proteins, PROTAC molecules can also achieve isoform selectivity, providing chemical probes with unique characteristics. Indeed, this was demonstrated by the already mentioned p38α PROTAC[96] and by SGK3-based PROTAC molecules.[107] SGK3–PROTAC1 did not degrade closely related SGK1 and SGK2 isoforms that are nevertheless engaged and inhibited by Sanofi 308-R, the original inhibitor.

7.2.5.1 Implications of Hijacking Natural Degradation

Although PROTAC molecules can enhance the homologue and isoform selectivity of traditional inhibitors, it is critical for the development of chemical probes based on PROTAC to monitor carefully off-target activities derived by the disruption of the natural degradation of the substrates or, in the case of molecular glues, by degradation of additional neo-substrates. Additionally, cereblon-binding molecules carry the safety risks of thalidomide, which caused an epidemic of severe birth defects in the 1950s. In studies aimed at understanding the molecular basis of thalidomide-induced teratogenicity, researchers at Celgene and the Dana-Farber Cancer Institute independently identified SALL4 as a thalidomide-dependent cereblon substrate whose degradation phenocopies genetic embryopathies caused by SALL4 mutation.[108,109] These findings were in line with the seminal discovery that drugs that bind to cereblon target for degradation two specific zinc finger (ZF) transcription factors, *i.e.* ikaros and aiolos. However, SALL4, ikaros and aiolos are not an exception. A research collaboration between the groups of Thomä (Friedrich Miescher Institute for Biomedical Research) and Ebert (Broad Institute of MIT and Harvard) showed that, in spite of the little sequence conservation (apart from residues that stabilize the ternary ZF fold), IMiDs cause cereblon-mediated degradation of a large number of ZF proteins. Additionally, thalidomide, altered with small chemical changes, allows the selective degradation of distinct ZF targets.[110]

The degradation of neo-substrates has attracted the interest of the targeted-protein degradation scientific community and it is becoming clear that cereblon-based PROTAC molecules should be carefully tracked and characterized since the activity of neo-substrates can be retained in PROTAC molecules, leading to undesired off-target degradation. Nowak *et al.* demonstrated that structural information can be very useful for fine-tuning the selectivity of neo-substrates. Changes in the linker composition of a BRD4 PROTAC degrader dictated the off-target degradation of the ZF ikaros.[102]

PROTAC molecules employ E3 ligase ligands that bind to the degrons of the natural substrates, with the possibility of disrupting their natural interaction. On the one hand, VHL is a tumour suppressor and acts as a regulator of the HIF hypoxia pathway by recognizing a post-translational hydroxylated form of HIF for degradation. Although it is difficult to determine the clinical consequences of disrupting this interaction with VHL-based PROTAC molecules, specifically, the levels of HIF proteins should be monitored in the development of these molecules. Frost *et al.* demonstrated that VH298, a very potent VHL ligand (K_d = 80 nM), stabilizes HIF-1α at micromolar concentrations, revealing an exploitable concentration window with PROTAC molecules acting at nanomolar concentrations.[67] On the other hand, as mentioned, whereas SNIPER molecules can trigger the degradation of the IAP E3 ligase, MDM2-based PROTAC could upregulate the tumour suppressor p53. Although IAP degradation and p53 upregulation can be desired effects in some therapeutic scenarios, they detract from their application for the development of chemical probes.

Protein downregulation after PROTAC treatment makes the proteomic approach very useful for identifying off-target proteins of the degraders. Quantitative tandem mass tag (TMT) labelling and posterior mass spectrometry have been used to assess changes in the proteome after PROTAC treatment.[11,87,111] Additionally, researchers at GSK established a powerful proteomic approach to discover the regulatory degradation mechanism of PROTAC molecules in biological systems.[112]

7.2.5.2 PhotoPROTAC, PHOTAC and Photocaged-PROTAC Chemical Probes

Being able to activate the degradation of a particular protein in a specific tissue and at a particular time has advantages in terms of selectivity and prevention of potential off-target effects. To do that, several approaches based on PROTAC molecules and photopharmacology have been developed, the two main approaches being photoswitches and photocaging. From a chemical probe point of view, photoswitches have some advantages over caging strategies, including their reversibility, dose control, the low intensities of light needed and the avoidance of dangerous by-products.

Two studies based on photoswitches have been published and more are envisaged. PhotoPROTAC contains an *o*-tetrafluoroazobenzene linker between two warheads. Whereas the azo-*cis*-isomer was inactive because the distance defined by the linker was very short, the longer azo-*trans*-isomer (3 Å longer) was able to degrade BRD4 in a persistent manner. Moreover, the highly biostable photoswitchable structural component led to reversible control over the topological distance between the two ligands, *i.e.* offering on/off protein degradation.[113] Similarly, PHOTAC contain a non-*ortho*-fluorinated linker in addition to cyclic azobenzene linkers.[114] Although this last strategy is also reversible, it necessitates continued light pulses to induce prolonged protein degradation. In contrast, PHOTAC molecules show colour dosing, *i.e.* the ability to control the concentration of the active species with the colour of the incident light. Finally, photo-removable blocking groups (caged) were added to a degrader of BRD4 PROTAC molecules. These photocaged PROTAC molecules were able to degrade BRD4 only after light irradiation.[115]

7.3 Final Remarks

Targeted-protein degradation has revolutionized the way in which we envisage therapies for targets that were previously considered undruggable. The field is growing exponentially and, quoting Alessio Ciulli, a key player in the PROTAC field: "the sky is the limit for PROTACs". However, these huge expectations come together with some challenges ahead, mainly on the design and development side: can a rational SAR be developed for PROTAC molecules and molecular glues? Can we go beyond serendipity in the generation of molecular glues? How can we predict ternary complex formation? Which parameters must be optimized in order to create orally bioavailable degraders? How can we include the importance of the ternary complex in PK/PD models? On what basis do we prioritize a specific E3 ligase for PROTAC construction? Concerning this last question, there is no doubt that in the near future much-needed E3 ligase small molecules will be identified, developed and exploited for protein degradation. In this context, allosteric ligands of E3 ligases would confer unique opportunities for chemical probes generation since they (i) do not compete with the natural substrates of E3 ligases and hence do not require a potent affinity, (ii) are less likely to interfere with the natural function of the E3 ligase and (iii) can be as selective as the ligands that bind to the substrate binding site. The use of allosteric ligands can increase the number of E3 ligases used for targeted-protein degradation purposes, pushing the development of tissue- and disease-specific PROTAC molecules.

The development of chemical probes based on targeted-protein degradation has an easier pathway than that for formal drugs, since DMPK concerns about PROTAC molecules do not apply for chemical probes. The substoichiometric mechanism of action and the relevance of the ternary complex in the PROTAC activity result in a unique opportunity to degrade proteins for which only a low-affinity ligand is available (see Figure 7.11). In this context, the use of E3 ligases to degrade their natural transcription factors, *e.g.* FBW7 to degrade c-MYC, in oncological scenarios when, for example, the degron site and/or the substrate is mutated, could take this property to the extreme, generating, at the same time, tumour-specific gain-of-function PROTAC molecules. Moreover, the potential to develop homologue- and isoform-selective and tissue- and organelle-specific chemical probes could not be matched with traditional inhibitor chemical probes (see Figure 7.11).

However, chemical probes based on targeted-protein degradation need to be characterized properly in all of the steps, *i.e.* target engagement and selectivity, ternary complex formation, ubiquitination and finally protein degradation. Several cell-based and biophysical approaches have been developed for these purposes. To conclude, we strongly believe that in the coming years a considerable number of chemical probes based on PROTAC will be developed for proteins previously considered undruggable, converting a large part of the undruggable proteome into a druggable one and delivering completely novel therapies to cure challenging diseases, such as cancer and Alzheimer's disease.

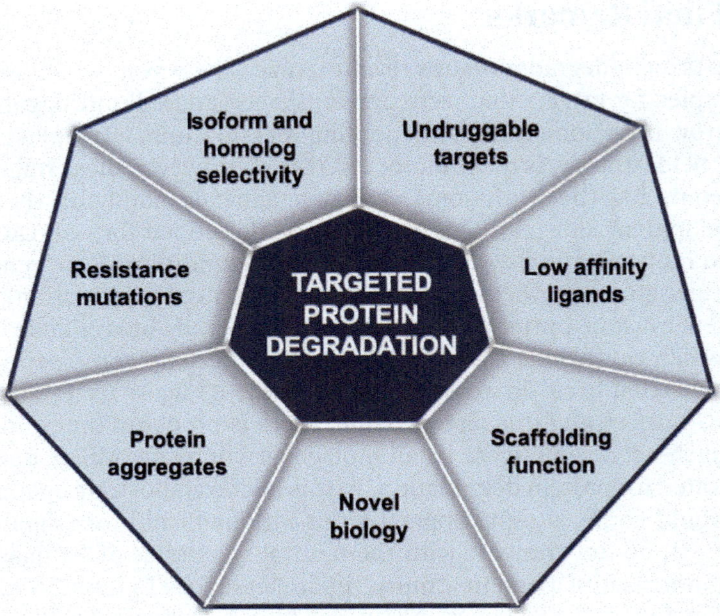

Figure 7.11 Representation of PROTACs' tenets.

Abbreviations

AML	Acute myeloid leukaemia
AhR	Aryl hydrocarbon receptor
AR	Androgen receptor
CI-M6PR	Cation-independent mannose-6-phosphate receptor
CRL	Cullin–RING ligase
DUB	Deubiquitinating enzyme
EGFR	Epidermal growth factor receptor
ER	Estrogen receptor
FDA	Food and Drug Administration
IAP	Inhibitor of apoptosis protein
IDP	Intrinsically disordered protein
IMiD	Immunomodulatory drug
POI	Protein of interest
PSA	Polar surface area
PPI	Protein–protein interaction
PROTAC	Proteolysis-targeting chimera
RTK	Receptor tyrosine kinase
SNIPER	Specific and non-genetic IAP-dependent protein eraser
UPS	Ubiquitin–proteasome system
VHL	Von Hippel–Lindau
ZF	Zinc finger

References

1. A. Hershko and A. Ciechanover, *Annu. Rev. Biochem.*, 1998, **67**, 425.
2. G. M. Burslem and C. M. Crews, *Chem. Rev.*, 2017, **117**, 11269.
3. J. Krönke, N. D. Udeshi, A. Narla, P. Grauman, S. N. Hurst, M. McConkey, T. Svinkina, D. Heckl, E. Comer, X. Li, C. Ciarlo, E. Hartman, N. Munshi, M. Schenone, S. L. Schreiber, S. A. Carr and B. L. Ebert, *Science*, 2014, **343**, 301.
4. K. R. Simonetta, J. Taygerly, K. Boyle, S. E. Basham, C. Padovani, Y. Lou, T. J. Cummins, S. L. Yung, S. K. Soly, F. Kayser, J. Kuriyan, M. Rape, M. Cardozo, M. A. Gallop, N. F. Bence, P. A. Barsanti and A. Saha, *Nat. Commun.*, 2019, **10**, 1402.
5. T. K. Neklesa, H. S. Tae, A. R. Schneekloth, M. J. Stulberg, T. W. Corson, T. B. Sundberg, K. Raina, S. A. Holley and C. M. Crews, *Nat. Chem. Biol.*, 2011, **7**, 538.
6. J. L. Gustafson, T. K. Neklesa, C. S. Cox, A. G. Roth, D. L. Buckley, H. S. Tae, T. B. Sundberg, D. B. Stagg, J. Hines, D. P. McDonnell, J. D. Norris and C. M. Crews, *Angew. Chem.*, 2015, **127**, 9795.
7. T. Xie, S. M. Lim, K. D. Westover, M. E. Dodge, D. Ercan, S. B. Ficarro, D. Udayakumar, D. Gurbani, H. S. Tae, S. M. Riddle, T. Sim, J. A. Marto, P. A. J. Anne, C. M. Crews and N. S. Gray, *Nat. Chem. Biol.*, 2014, **10**, 1006.
8. A. A. Adjei, *J. Clin. Oncol.*, 2006, **24**, 4054.
9. E. F. Douglass Jr, C. J. Miller, G. Sparer, H. Shapiro and D. A. Spiegel, *J. Am. Chem. Soc.*, 2013, **135**, 6092.
10. M. S. Gadd, A. Testa, X. Lucas, K.-H. Chan, W. Chen, D. J. Lamont, M. Zengerle and A. Ciulli, *Nat. Chem. Biol.*, 2017, **13**, 514.
11. D. P. Bondeson, B. E. Smith, G. M. Burslem, A. D. Buhimschi, J. Hines, S. Jaime-Figueroa, J. Wang, B. D. Hamman, A. Ishchenko and C. M. Crews, *Cell Chem. Biol.*, 2018, **25**, 78.
12. T. Kwon and A. Ciechanover, *Trends Biochem. Sci.*, 2017, **42**, 873.
13. K. Takahashi, A. Matouschek and T. Inobe, *ACS Chem. Biol.*, 2015, **10**, 2537.
14. K. M. Sakamoto, K. B. Kim, A. Kumagai, F. Mercurio, C. M. Crews and R. J. Deshaies, *Proc. Natl. Acad. Sci. U. S. A.*, 2001, **98**, 8554.
15. J. S. Schneekloth, F. N. Fonseca, M. Koldobskiy, A. Mandal, R. Deshaies, K. Sakamoto and C. M. Crews, *J. Am. Chem. Soc.*, 2004, **126**, 3748.
16. A. R. Schneekloth, M. Pucheault, H. S. Tae and C. M. Crews, *Bioorg. Med. Chem. Lett.*, 2008, **18**, 5904.
17. Y. Itoh, M. Ishikawa, M. Naito and Y. Hashimoto, *J. Am. Chem. Soc.*, 2010, **132**, 5820.
18. T. Ito, H. Ando, T. Suzuki, T. Ogura, K. Hotta, Y. Imamura and H. Handa, *Science*, 2010, **327**, 1345.
19. P. P. Chamberlain, A. Lopez-Girona, K. Miller, G. Carmel, B. Pagarigan, B. Chie-Leon, E. Rychak, L. G. Corral, Y. J. Ren, M. Wang, M. Riley, S. L. Delker, T. Ito, H. Ando, T. Mori, Y. Hirano, H. Handa, T. Hakoshima, T. O. Daniel and B. E. Cathers, *Nat. Struct. Mol. Biol.*, 2014, **21**, 803.

20. E. S. Fischer, K. Böhm, J. R. Lydeard, H. Yang, M. B. Stadler, S. Cavadini, J. Nagel, F. Serluca, V. Acker, G. M. Lingaraju, R. B. Tichkule, M. Schebesta, W. C. Forrester, M. Schirle, U. Hassiepen, J. Ottl, M. Hild, R. E. J. Beckwith, J. W. Harper, J. L. Jenkins and N. H. Thomä, *Nature*, 2014, **512**, 49.
21. M. Zengerle, K.-H. Chan and A. Ciulli, *ACS Chem. Biol.*, 2015, **10**, 1770.
22. D. P. Bondeson, A. Mares, I. E. D. Smith, E. Ko, S. Campos, A. H. Miah, K. E. Mulholland, N. Routly, D. L. Buckley, J. L. Gustafson, N. Zinn, P. Grandi, S. Shimamura, G. Bergamini, M. Faelth-Savitski, M. Bantscheff, C. Cox, D. A. Gordon, R. R. Willard, J. J. Flanagan, L. N. Casillas, B. J. Votta, W. den Besten, K. Famm, L. Kruidenier, P. S. Carter, J. D. Harling, I. Churcher and C. M. Crews, *Nat. Chem. Biol.*, 2015, **11**, 611.
23. G. E. Winter, D. L. Buckley, J. Paulk, J. M. Roberts, A. Souza, S. Dhe-Paganon and J. E. Bradner, *Science*, 2015, **348**, 1376.
24. D. L. Buckley, K. Raina, N. Darricarrere, J. Hines, J. L. Gustafson, I. E. Smith, A. H. Miah, J. D. Harling and C. M. Crews, *ACS Chem. Biol.*, 2015, **10**, 1831.
25. J. Lu, Y. Qian, M. Altieri, H. Dong, J. Wang, K. Raina, J. Hines, J. D. Winkler, A. P. Crew, K. Coleman and C. M. Crews, *Chem. Biol.*, 2015, **22**, 755.
26. P. P. Chamberlain and L. G. Hamann, *Nat. Chem. Biol.*, 2019, **15**, 937.
27. F. E. Morreale and H. Walden, *Cell*, 2016, **165**, 248.
28. P. Ottis, C. Palladino, P. Thienger, A. Britschgi, C. Heichinger, M. Berrera, A. Julien-Laferriere, F. Roudnicky, T. Kam-Thong, J. R. Bischoff, B. Martoglio and P. Pettazzoni, *ACS Chem. Biol.*, 2019, **14**, 2215.
29. L. Zhang, B. Riley-Gillis, P. Vijay and Y. Shen, *Mol. Cancer Ther.*, 2019, **18**, 1302.
30. R. Shirasaki, R. de Matos Simoes, S. Gandolfi, G. Matthews, D. Buckley, J. Raja, Q. Sievers, J. Bruggenthies, O. Dashevsky, H. Poarch, H. Tang, M. Bariteau, M. Sheffer, Y. Hu, S. Downey-Kopyscinski, P. Hengeveld, B. J. Glassner, E. Dhimolea, C. Ott, T. Zhang, N. Kwiatkowski, J. Laubach, R. Schlossman, A. Culhane, P. G. Richardson, R. Groen, E. Fischer, F. Vazquez, A. Tsherniak, W. Hahn, J. Levy, D. Auclair, J. Licht, J. Keats, L. Boise, B. Ebert, J. Bradner, N. Gray and C. Mitsiades, *Clin. Lymphoma, Myeloma Leuk.*, 2019, **19**, e134.
31. M. Schapira, M. F. Calabrese, A. N. Bullock and C. M. Crews, *Nat. Rev. Drug Discovery*, 2019, **18**, 949.
32. M. Naito, N. Ohoka and N. Shibata, *Drug Discovery Today Technol.*, 2019, **31**, 35.
33. N. Ohoka, K. Okuhira, M. Ito, K. Nagai, N. Shibata, T. Hattori, O. Ujikawa, K. Shimokawa, O. Sano, R. Koyama, H. Fujita, M. Teratani, H. Matsumoto, Y. Imaeda, H. Nara, N. Cho and M. Naito, *J. Biol. Chem.*, 2017, **292**, 4556.
34. J. Hines, S. Lartigue, H. Dong, Y. Qian and C. M. Crews, *Cancer Res.*, 2019, **79**, 251.
35. N. Ohoka, G. Tsuji, T. Shoda, T. Fujisato, M. Kurihara, Y. Demizu and M. Naito, *ACS Chem. Biol.*, 2019, **14**, 12, 2822.

36. N. T. Nguyen, H. Hanieh, T. Nakahama and T. Kishimoto, *Int. Immunol.*, 2013, **25**, 335.
37. C. C. Ward, J. I. Kleinman, S. M. Brittain, P. S. Lee, C. Y. S. Chung, K. Kim, Y. Petri, J. R. Thomas, J. A. Tallarico, J. M. McKenna, M. Schirle and D. K. Nomura, *ACS Chem. Biol.*, 2019, **14**, 2430.
38. J. N. Spradlin, X. Hu, C. C. Ward, S. M. Brittain, M. D. Jones, L. Ou, M. To, A. Proudfoot, E. Ornelas, M. Woldegiorgis, J. A. Olzmann, D. E. Bussiere, J. R. Thomas, J. A. Tallarico, J. M. McKenna, M. Schirle, T. J. Maimone and D. K. Nomura, *Nat. Chem. Biol.*, 2019, **15**, 747.
39. X. Zhang, V. M. Crowley, T. G. Wucherpfennig, M. M. Dix and B. F. Cravatt, *Nat. Chem. Biol.*, 2019, **15**, 737.
40. D. E. Bussiere, L. Xie, H. Srinivas, W. Shu, A. Burke, C. Be, J. Zhao, A. Godbole, D. King, R. G. Karki, V. Hornak, F. Xu, J. Cobb, N. Carte, A. O. Frank, A. Frommlet, P. Graff, M. Knapp, A. Fazal, B. Okram, S. Jiang, P.-Y. Michellys, R. Beckwith, H. Voshol, C. Wiesmann, J. Solomon and J. Paulk, *Nat. Chem. Biol.*, 2020, **16**, 15.
41. T. B. Faust, H. Yoon, R. X. A. P. Nowak, K. A. Donovan, Z. Li, Q. Cai, N. A. Eleuteri, T. Zhang, N. S. Gray and E. S. Fischer, *Nat. Chem. Biol.*, 2020, **17**, 7.
42. X. Du, O. Volkov, R. Czerwinski, H. Tan, C. Huerta, E. Morton, J. Rizzi, P. Wehn, R. Xu, D. Nijhawan and E. Wallace, *Structure*, 2019, **21**, 1625.
43. J. Cheng, J. Guo, B. J. North, K. Tao, P. Zhou and W. Wei, *Biochim. Biophys. Acta, Rev. Cancer*, 2019, **1871**, 138.
44. R. Chopra, A. Sadok and I. Collins, *Drug Discovery Today: Technol.*, 2019, **31**, 5.
45. X. Zhang, D. Thummuri, Y. He, X. Liu, P. Zhang, D. Zhou and G. Zheng, *Chem. Commun.*, 2020, **56**, 9866.
46. S. J. Hughes and A. Ciulli, *Essays Biochem.*, 2017, **61**, 505.
47. V. Zoppi, S. J. Hughes, C. Maniaci, A. Testa, T. Gmaschitz, C. Wieshofer, M. Koegl, K. M. Riching, D. L. Daniels, A. Spallarossa and A. Ciulli, *J. Med. Chem.*, 2018, **62**, 699.
48. M. G. J. Baud, E. Lin-Shiao, M. Zengerle, C. Tallant and A. Ciulli, *J. Med. Chem.*, 2016, **59**, 1492.
49. A. Zorba, C. Nguyen, Y. Xu, J. Starr, K. Borzilleri, J. Smith, H. Zhu, K. A. Farley, W. Ding, J. Schiemer, X. Feng, J. S. Chang, D. P. Uccello, J. A. Young, C. N. Garcia-Irrizary, L. Czabaniuk, B. Schuff, R. Oliver, J. Montgomery, M. M. Hayward, J. Coe, J. Chen, M. Niosi, S. Luthra, J. C. Shah, A. El-Kattan, X. Qiu, G. M. West, M. C. Noe, V. Shanmugasundaram, A. M. Gilbert, M. F. Brown and M. F. Calabrese, *Proc. Natl. Acad. Sci. U. S. A.*, 2018, **115**, 7285.
50. J. Yang, Y. Li, A. Aguilar, Z. Liu, C.-Y. Yang and S. Wang, *J. Med. Chem.*, 2019, **62**, 9471.
51. P. J. de Vink, S. A. Andrei, Y. Higuchi, C. Ottmann, L.-G. Milroy and L. Brunsveld, *Chem. Sci.*, 2019, **10**, 2869.
52. S. D. Edmondson, B. Yang and C. Fallan, *Bioorg. Med. Chem. Lett.*, 2019, **29**, 1555–1564.

53. H. Lebraud, D. J. Wright, C. N. Johnson and T. D. Heightman, *ACS Cent. Sci.*, 2016, **2**, 927–934.
54. H. Lebraud and T. D. Heightman, *Essays Biochem.*, 2017, **61**, 517–527.
55. A. Testa, S. J. Hughes, X. Lucas, J. E. Wright and A. Ciulli, *Angew. Chem., Int. Ed.*, 2020, **132**, 1744.
56. M. L. Drummond and C. I. Williams, *J. Chem. Inf. Model.*, 2019, **59**, 1634.
57. A. J. Carter, O. Kraemer, M. Zwick, A. Mueller-Fahrnow, C. H. Arrowsmith and A. M. Edwards, *Drug Discovery Today*, 2019, **24**, 2111.
58. P. Workman and I. Collins, *Chem. Biol.*, 2010, **17**, 561.
59. J. Blagg and P. Workman, *Cancer Cell*, 2017, **32**, 9.
60. A. Huston, C. H. Arrowsmith, S. Knapp and M. Schapira, *Nat. Chem. Biol.*, 2015, **11**, 542.
61. M. Moustakim, S. L. Felce, N. Zaarour, G. Farnie, F. E. McCann and P. E. Brennan, *Methods Enzymol.*, 2018, **610**, 27.
62. A. E. Campbell and D. Bennett, *Biochem. J.*, 2016, **473**, 2573.
63. X. Sun, J. Wang, X. Yao, W. Zheng, Y. Mao, T. Lan, L. Wang, Y. Sun, X. Zhang, Q. Zhao, J. Zhao, R.-P. Xiao, X. Zhang, G. Ji and Y. Rao, *Cell Discovery*, 2019, **5**, 10.
64. N. E. A. Chessum, S. Y. Sharp, J. J. Caldwell, A. E. Pasqua, B. Wilding, G. Colombano, I. Collins, B. Ozer, M. Richards, M. Rowlands, M. Stubbs, R. Burke, P. C. McAndrew, P. A. Clarke, P. Workman, M. D. Cheeseman and K. Jones, *J. Med. Chem.*, 2018, **61**, 918.
65. C. H. Arrowsmith, J. E. Audia, C. Austin, J. Baell, J. Bennett, J. Blagg, C. Bountra, P. E. Brennan, P. J. Brown, M. E. Bunnage, C. Buser-Doepner, R. M. Campbell, A. J. Carter, P. Cohen, R. A. Copeland, B. Cravatt, J. L. Dahlin, D. Dhanak, A. M. Edwards, M. Frederiksen, S. V. Frye, N. Gray, C. E. Grimshaw, D. Hepworth, T. Howe, K. V. M. Huber, J. Jin, S. Knapp, J. D. Kotz, R. G. Kruger, D. Lowe, M. M. Mader, B. Marsden, A. Mueller-Fahrnow, S. Müller, R. C. O'Hagan, J. P. Overington, D. R. Owen, S. H. Rosenberg, R. Ross, B. Roth, M. Schapira, S. L. Schreiber, B. Shoichet, M. Sundström, G. Superti-Furga, J. Taunton, L. Toledo-Sherman, C. Walpole, M. A. Walters, T. M. Willson, P. Workman, R. N. Young and W. J. Zuercher, *Nat. Chem. Biol.*, 2015, **11**, 536.
66. C. Loenarz, J. Mecinović, R. Chowdhury, L. A. McNeill, E. Flashman and C. J. Schofield, *Angew. Chem., Int. Ed.*, 2009, **48**, 1784.
67. J. Frost, C. Galdeano, P. Soares, M. S. Gadd, K. M. Grzes, L. Ellis, O. Epemolu, S. Shimamura, M. Bantscheff, P. Grandi, K. D. Read, D. A. Cantrell, S. Rocha and A. Ciulli, *Nat. Commun.*, 2016, **7**, 13312.
68. A. A. Akuffo, A. Y. Alontaga, R. Metcalf, M. S. Beatty, A. Becker, J. M. McDaniel, R. S. Hesterberg, W. E. Goodheart, S. Gunawan, M. Ayaz, Y. Yang, M. R. Karim, M. E. Orobello, K. Daniel, W. Guida, J. A. Yoder, A. M. Rajadhyaksha, E. Schonbrunn, H. R. Lawrence, N. J. Lawrence and P. K. Epling-Burnette, *J. Biol. Chem.*, 2018, **293**, 6187.
69. C. Maniaci, S. J. Hughes, A. Testa, W. Chen, D. J. Lamont, S. Rocha, D. R. Alessi, R. Romeo and A. Ciulli, *Nat. Commun.*, 2017, **8**, 830.

70. C. Steinbach, S. Lindner, N. D. Udeshi, D. C. Mani, H. Kehm, S. Köpff, S. A. Carr, M. Gütschow and J. Krönke, *ACS Chem. Biol.*, 2018, **13**, 2771.
71. M. Girardini, C. Maniaci, S. J. Hughes, A. Testa and A. Ciulli, *Bioorg. Med. Chem.*, 2019, **27**, 2466.
72. C. Steinebach, H. Kehm, S. Lindner, L. P. Vu, S. Köpff, Á. López Mármol, C. Weiler, K. G. Wagner, M. Reichenzeller, J. Krönke and M. Gütschow, *Chem. Commun.*, 2019, **55**, 1821.
73. Y. Li, J. Yang, A. Aguilar, D. McEachern, S. Przybranowski, L. Liu, C.-Y. Yang, M. Wang, X. Han and S. Wang, *J. Med. Chem.*, 2018, **62**, 448.
74. B. Nabet, J. M. Roberts, D. L. Buckley, J. Paulk, S. Dastjerdi, A. Yang, A. L. Leggett, M. A. Erb, M. A. Lawlor, A. Souza, T. G. Scott, S. Vittori, J. A. Perry, J. Qi, G. E. Winter, K.-K. Wong, N. S. Gray and J. E. Bradner, *Nat. Chem. Biol.*, 2018, **14**, 431.
75. M. A. Erb, T. G. Scott, B. E. Li, H. Xie, J. Paulk, H.-S. Seo, A. Souza, J. M. Roberts, S. Dastjerdi, D. L. Buckley, N. E. Sanjana, O. Shalem, B. Nabet, R. Zeid, N. K. Offei-Addo, S. Dhe-Paganon, F. Zhang, S. H. Orkin, G. E. Winter and J. E. Bradner, *Nature*, 2017, **543**, 270.
76. S. Tomoshige, M. Naito, Y. Hashimoto and M. Ishikawa, *Org. Biomol. Chem.*, 2015, **13**, 9746.
77. S. Banik, K. Pedram, S. Wisnocsky, N. Riley and C. Bertozzi, *Nature*, 2020, **584**, 291.
78. J. Hines, J. D. Gough, T. W. Corson and C. M. Crews, *Proc. Natl. Acad. Sci. U. S. A.*, 2013, **110**, 8942.
79. D. Takahashi, J. Moriyama, T. Nakamura, E. Miki, E. Takahashi, A. Sato, T. Akaike, K. Itto-Nakama and H. Arimoto, *Mol. Cell*, 2019, **76**, 1.
80. T. I. Oprea, C. G. Bologa, S. Brunak, A. Campbell, G. N. Gan, A. Gaulton, S. M. Gomez, R. Guha, A. Hersey, J. Holmes, A. Jadhav, L. J. Jensen, G. L. Johnson, A. Karlson, A. R. Leach, A. Ma'ayan, A. Malovannaya, S. Mani, S. L. Mathias, M. T. McManus, T. F. Meehan, C. von Mering, D. Muthas, D.-T. Nguyen, J. P. Overington, G. Papadatos, J. Qin, C. Reich, B. L. Roth, S. C. Schürer, A. Simeonov, L. A. Sklar, N. Southall, S. Tomita, I. Tudose, O. Ursu, D. Vidovic, A. Waller, D. Westergaard, J. J. Yang and G. Zahoránszky-Köhalmi, *Nat. Rev. Drug Discovery*, 2018, **17**, 317.
81. A. K. Arakaki, W. Tian and J. Skolnick, *BMC Genomics*, 2006, **7**, 315.
82. J. H. Bushweller, *Nat. Rev. Cancer*, 2019, **19**, 611.
83. L. Bai, H. Zhou, R. Xu, Y. Zhao, K. Chinnaswamy, D. McEachern, J. Chen, C.-Y. Yang, Z. Liu, M. Wang, L. Liu, H. Jiang, B. Wen, P. Kumar, J. L. Meagher, D. Sun, J. A. Stuckey and S. Wang, *Cancer Cell*, 2019, **36**, 498.
84. T.-T. Chu, N. Gao, Q.-Q. Li, P.-G. Chen, X.-F. Yang, Y.-X. Chen, Y.-F. Zhao and Y.-M. Li, *Cell Chem. Biol.*, 2016, **23**, 453.
85. M. Lu, T. Liu, Q. Jiao, J. Ji, M. Tao, Y. Liu, Q. You and Z. Jiang, *Eur. J. Med. Chem.*, 2018, **146**, 251.

86. M. C. Silva, F. M. Ferguson, Q. Cai, K. A. Donovan, G. Nandi, D. Patnaik, T. Zhang, H.-T. Huang, D. E. Lucente, B. C. Dickerson, T. J. Mitchison, E. S. Fischer, N. S. Gray and S. J. Haggarty, *eLife*, 2019, **8**, e45457.
87. W. Farnaby, M. Koegl, M. J. Roy, C. Whitworth, E. Diers, N. Trainor, D. Zollman, S. Steurer, J. Karolyi-Oezguer, C. Riedmueller, T. Gmaschitz, J. Wachter, C. Dank, M. Galant, B. Sharps, K. Rumpel, E. Traxler, T. Gerstberger, R. Schnitzer, O. Petermann, P. Greb, H. Weinstabl, G. Bader, A. Zoephel, A. Weiss-Puxbaum, K. Ehrenhöfer-Wölfer, S. Wöhrle, G. Boehmelt, J. Rinnenthal, H. Arnhof, N. Wiechens, M.-Y. Wu, T. Owen-Hughes, P. Ettmayer, M. Pearson, D. B. McConnell and A. Ciulli, *Nat. Chem. Biol.*, 2019, **15**, 672.
88. L. N. Gechijian, D. L. Buckley, M. A. Lawlor, J. M. Reyes, J. Paulk, C. J. Ott, G. E. Winter, M. A. Erb, T. G. Scott, M. Xu, H.-S. Seo, S. Dhe-Paganon, N. P. Kwiatkowski, J. A. Perry, J. Qi, N. S. Gray and J. E. Bradner, *Nat. Chem. Biol.*, 2018, **14**, 405.
89. Z. I. Bassi, M. C. Fillmore, A. H. Miah, T. D. Chapman, C. Maller, E. J. Roberts, L. C. Davis, D. E. Lewis, N. W. Galwey, K. E. Waddington, V. Parravicini, A. L. Macmillan-Jones, C. Gongora, P. G. Humphreys, I. Churcher, R. K. Prinjha and D. F. Tough, *ACS Chem. Biol.*, 2018, **13**, 2862.
90. B. G. Wilson, K. C. Helming, X. Wang, Y. Kim, F. Vazquez, Z. Jagani, W. C. Hahn and C. W. M. Roberts, *Mol. Cell. Biol.*, 2014, **34**, 1136.
91. H. J. Donnella, J. T. Webber, R. S. Levin, R. Camarda, O. Momcilovic, N. Bayani, K. N. Shah, J. E. Korkola, K. M. Shokat, A. Goga, J. D. Gordan and S. Bandyopadhyay, *Nat. Chem. Biol.*, 2018, **14**, 768.
92. P. M. Cromm, K. T. G. Samarasinghe, J. Hines and C. M. Crews, *J. Am. Chem. Soc.*, 2018, **140**, 17019.
93. G. M. Burslem, B. E. Smith, A. C. Lai, S. Jaime-Figueroa, D. C. McQuaid, D. P. Bondeson, M. Toure, H. Dong, Y. Qian, J. Wang, A. P. Crew, J. Hines and C. M. Crews, *Cell Chem. Biol.*, 2018, **25**, 67.
94. Y. Sun, X. Zhao, N. Ding, H. Gao, Y. Wu, Y. Yang, M. Zhao, J. Hwang, Y. Song, W. Liu and Y. Rao, *Cell Res.*, 2018, **28**, 779.
95. A. D. Buhimschi, H. A. Armstrong, M. Toure, S. Jaime-Figueroa, T. L. Chen, A. M. Lehman, J. A. Woyach, A. J. Johnson, J. C. Byrd and C. M. Crews, *Biochemistry*, 2018, **57**, 3564.
96. B. E. Smith, S. L. Wang, S. Jaime-Figueroa, A. Harbin, J. Wang, B. D. Hamman and C. M. Crews, *Nat. Commun.*, 2019, **10**, 131.
97. G. M. Burslem, J. Song, X. Chen, J. Hines and C. M. Crews, *J. Am. Chem. Soc.*, 2018, **140**, 16428.
98. C. Qin, Y. Hu, B. Zhou, E. Fernandez-Salas, C.-Y. Yang, L. Liu, D. McEachern, S. Przybranowski, M. Wang, J. Stuckey, J. Meagher, L. Bai, Z. Chen, M. Lin, J. Yang, D. N. Ziazadeh, F. Xu, J. Hu, W. Xiang, L. Huang, S. Li, B. Wen, D. Sun and S. Wang, *J. Med. Chem.*, 2018, **61**, 6685.
99. A. P. Crew, K. Raina, H. Dong, Y. Qian, J. Wang, D. Vigil, Y. V. Serebrenik, B. D. Hamman, A. Morgan, C. Ferraro, K. Siu, T. K. Neklesa, J. D. Winkler, K. G. Coleman and C. M. Crews, *J. Med. Chem.*, 2017, **61**, 583.

100. J. Salami, S. Alabi, R. R. Willard, N. J. Vitale, J. Wang, H. Dong, M. Jin, D. P. McDonnell, A. P. Crew, T. K. Neklesa and C. M. Crews, *Commun. Biol.*, 2018, **1**, 100.
101. A. C. Lai, M. Toure, D. Hellerschmied, J. Salami, S. Jaime-Figueroa, E. Ko, J. Hines and C. M. Crews, *Angew. Chem.*, 2015, **128**, 818–821.
102. R. P. Nowak, S. L. DeAngelo, D. Buckley, Z. He, K. A. Donovan, J. An, N. Safaee, M. P. Jedrychowski, C. M. Ponthier, M. Ishoey, T. Zhang, J. D. Mancias, N. S. Gray, J. E. Bradner and E. S. Fischer, *Nat. Chem. Biol.*, 2018, **14**, 706.
103. Z. Wang, N. He, Z. Guo, C. Niu, T. Song, Y. Guo, K. Cao, A. Wang, J. Zhu, X. Zhang and Z. Zhang, *J. Med. Chem.*, 2019, **62**, 8152.
104. K. Yang, Y. Song, H. Xie, H. Wu, Y.-T. Wu, E. D. Leisten and W. Tang, *Bioorg. Med. Chem. Lett.*, 2018, **28**, 2493.
105. S. Rana, M. Bendjennat, S. Kour, H. M. King, S. Kizhake, M. Zahid and A. Natarajan, *Bioorg. Med. Chem. Lett.*, 2019, **29**, 1375.
106. M. Brand, B. Jiang, S. Bauer, K. A. Donovan, Y. Liang, E. S. Wang, R. P. Nowak, J. C. Yuan, T. Zhang, N. Kwiatkowski, A. C. Müller, E. S. Fischer, N. S. Gray and G. E. Winter, *Cell Chem. Biol.*, 2019, **26**, 300.
107. H. Tovell, A. Testa, H. Zhou, N. Shpiro, C. Crafter, A. Ciulli and D. R. Alessi, *ACS Chem. Biol.*, 2019, **14**, 2024.
108. M. E. Matyskiela, S. Couto, X. Zheng, G. Lu, J. Hui, K. Stamp, C. Drew, Y. Ren, M. Wang, A. Carpenter, C.-W. Lee, T. Clayton, W. Fang, C.-C. Lu, M. Riley, P. Abdubek, K. Blease, J. Hartke, G. Kumar, R. Vessey, M. Rolfe, L. G. Hamann and P. P. Chamberlain, *Nat. Chem. Biol.*, 2018, **14**, 981.
109. K. A. Donovan, J. An, R. P. Nowak, J. C. Yuan, E. C. Fink, B. C. Berry, B. L. Ebert and E. S. Fischer, *eLife*, 2018, **7**, e38430.
110. Q. L. Sievers, G. Petzold, R. D. Bunker, A. Renneville, M. Słabicki, B. J. Liddicoat, W. Abdulrahman, T. Mikkelsen, B. L. Ebert and N. H. Thomä, *Science*, 2018, **362**, 558.
111. H.-T. Huang, D. Dobrovolsky, J. Paulk, G. Yang, E. L. Weisberg, Z. M. Doctor, D. L. Buckley, J.-H. Cho, E. Ko, J. Jang, K. Shi, H. G. Choi, J. D. Griffin, Y. Li, S. P. Treon, E. S. Fischer, J. E. Bradner, L. Tan and N. S. Gray, *Cell Chem. Biol.*, 2018, **25**, 88.
112. M. M. Savitski, N. Zinn, M. Faelth-Savitski, D. Poeckel, S. Gade, I. Becher, M. Muelbaier, A. J. Wagner, K. Strohmer, T. Werner, S. Melchert, M. Petretich, A. Rutkowska, J. Vappiani, H. Franken, M. Steidel, G. M. Sweetman, O. Gilan, E. Y. N. Lam, M. A. Dawson, R. K. Prinjha, P. Grandi, G. Bergamini and M. Bantscheff, *Cell*, 2018, **173**, 260.
113. P. Pfaff, K. T. G. Samarasinghe, C. M. Crews and E. M. Carreira, *ACS Cent. Sci.*, 2019, **5**, 1682.
114. M. Reynder, B. Matsuura, M. Bérouti, D. Simoneschi, A. Marzio, M. Pagano and D. Trauner, *Sci. Adv.*, 2020, **6**, eaay5064.
115. G. Xue, K. Wang, D. Zhou, H. Zhong and Z. Pan, *J. Am. Chem. Soc.*, 2019, **141**, 18370.

CHAPTER 8

Chemical Probes for Kinases

XIAOYUN LU[a], JEFF B. SMAILL[b] AND KE DING*[a]

[a]International Cooperative Laboratory of Traditional Chinese Medicine Modernization and Innovative Drug Discovery of the Chinese Ministry of Education (MCE), School of Pharmacy, Jinan University, 601 Huangpu Avenue West, Guangzhou 510632, China; [b]MWC–JNU International Joint Laboratory, Maurice Wilkins Centre for Molecular Biodiscovery, University of Auckland, Private Bag 92019, Auckland 1142, New Zealand
*E-mail: dingke@jnu.edu.cn

8.1 Introduction

Protein kinases, referred to here simply as kinases, are a family of enzymes that catalyse the transfer of phosphate groups from ATP to proteins, thus causing protein phosphorylation and in turn regulating survival, growth, differentiation, migration and cell death.[1] To date, approximately 518 kinases encoding about 2.5% of the human genome have been discovered.[2] Based on the nature of the residue that is phosphorylated, kinases can be broadly classified into two classes: protein serine/threonine kinases (PS/TKs) and protein tyrosine kinases (PTKs). PTKs can be further classified into two groups: receptor tyrosine kinases and non-receptor tyrosine kinases lacking extracellular binding domains. Because overexpression, dysregulation or mutations of kinases are directly associated with many diseases, such as oncogenesis, inflammatory-related diseases, immune diseases and infectious diseases, kinases have been widely investigated.[3,4] Hence drug-like chemical probes (defined as selective inhibitors with certain potency, selectivity and

Chemical Biology No. 16
The Discovery and Utility of Chemical Probes in Target Discovery
Edited by Paul Brennan and Saleta Vazquez Rodriguez

Published by the Royal Society of Chemistry, www.rsc.org

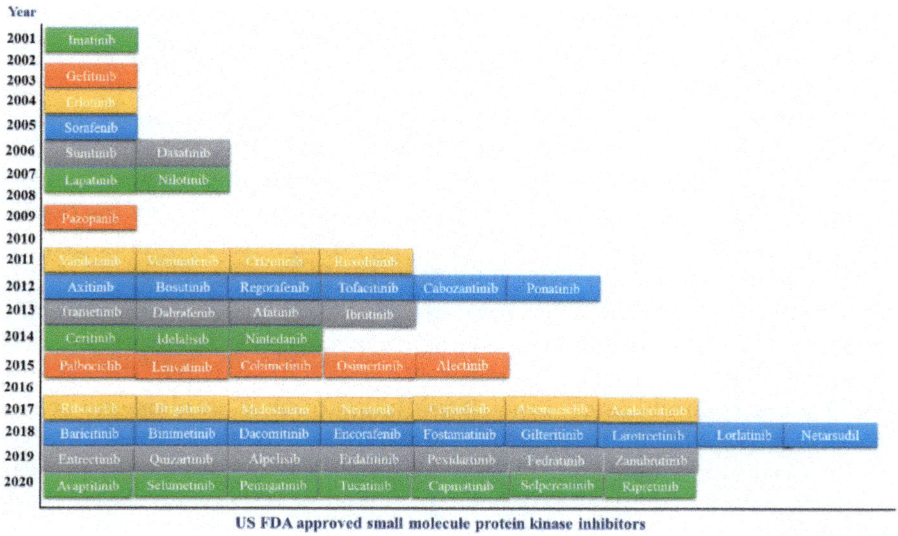

Figure 8.1 Small-molecule kinase inhibitors approved by the FDA from 2001 to May 2020.

cell activity criteria) and antibodies have attracted great attention in recent decades from researchers in the pharmaceutical industries and academic institutes. Since imatinib, the first kinase chemical probe, was approved in 2001, a further 61 small-molecule kinase inhibitors (SMKIs) have been approved by the US Food and Drug Administration (FDA) as of May 2020 (Figure 8.1).[5,6] These inhibitors have shown significant clinical benefits against human malignancies and other diseases. In this chapter, we briefly review the recent advances and challenges of some kinase inhibitors.

8.2 Binding Modes of Chemical Probes

Kinases have a highly conserved three-dimensional structure, particularly in the catalytically active ATP-binding domain. This structure includes a C-terminal-lobe region composed of α-helices, an N-terminal-lobe region composed of β-sheets and a hinge region that is an ATP-binding domain formed between the two lobes.[7] It contains a flexible activation loop starting with a conserved amino acid sequence, Asp–Phe–Gly (DFG), which can control access to the ATP-binding site. When the DFG side chains point into the ATP-binding site and control the Mg^{2+}/ATP, this is termed a DFG-in conformation. When the side chains face outwards from the ATP-binding pocket, this corresponds to a DFG-out conformation.[8] In addition, a regulatory αC-helix in the C-terminus can adopt active (αC_{in}) or inactive (αC_{out}) orientations. An important amino acid that controls access to the hydrophobic pocket in the ATP site is called the gatekeeper residue, which can directly prevent or weaken the interaction with inhibitors within mutation.

The chemical probes of kinases are categorized by binding modes and are mainly classified as reversible or irreversible inhibitors (Table 8.1). Reversible kinase inhibitors are further classified into five main types (types I-IV and I½) based on the DFG motif and position of the binding site in the protein (Figure 8.2).[3,9] Type I inhibitors are ATP-competitive inhibitors that bind to active kinases with the DFG-in conformation (Figure 8.2A). There are also some chemical probes that bind to inactive kinases (αC_{out} orientation) with a DFG-in conformation. These are called type I½ inhibitors (Figure 8.2B).[10,11] Type II inhibitors are also ATP competitive but bind to the kinase in a DFG-out conformation (Figure 8.2C). Allosteric inhibitors are inhibitors that bind outside the ATP-binding site and do not interact with the hinge region of the kinase. Based on the distance from the ATP pocket, these can be divided into type III and IV inhibitors.[12] Type III inhibitors bind to an allosteric pocket adjacent to the ATP-binding pocket (Figure 8.2D). Type IV inhibitors bind to an allosteric site remote from the ATP binding pocket (Figure 8.2E). None of the SMKIs approved by the FDA belongs to type IV, but several are currently in clinical trials.[13]

Table 8.1 Binding characteristic of kinase chemical probes.

SMKI	Type	Binding mode
Reversible	I	Bind to the active kinases with a DFG-in conformation
	I½	Bind to the inactive kinases in an αC_{out} orientation with a DFG-in conformation
	II	Bind to the inactive kinase with a DFG-out conformation
	III	Bind to an allosteric pocket adjacent to the ATP-binding pocket
	IV	Bind to an allosteric site remote from the ATP-binding pocket
Irreversible	V	Covalently bonded with a cysteine residue proximal to the ATP-binding site

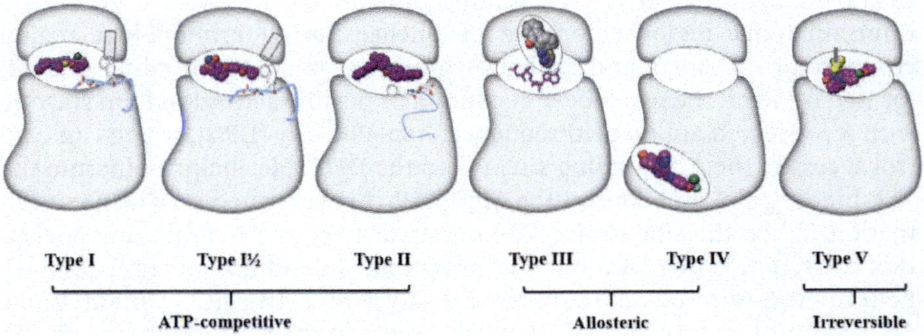

Figure 8.2 The binding modes of kinase chemical probes.

Most of the irreversible kinase inhibitors form a covalent bond with a cysteine residue proximal to the ATP-binding site and these are categorized as type V inhibitors. A few kinase chemical binding agents have been observed to bind to distinct regions of the kinase domain[14] or the non-kinase domain to regulate kinase activity.[15] Like allosteric chemical probes, these probes have higher selectivity and the ability to overcome resistance to ATP-binding probes. In general, the majority of the approved SMKIs that bind to the ATP-binding site are known as type I or II; a few of them, such as cabozantinib and cobimetinib, bind to the allosteric sites and are categorized as type III allosteric inhibitors (Table 8.2).

8.3 Chemical Probes for Representative Kinases

A single kinase is often associated with different types of chemical inhibitors. An important reason for this is that mutation of a target has become a significant driver for continuing development of new chemical probes to overcome clinical inactivity. In contrast, one inhibitor, a so-called multi-target kinase inhibitor, may also target multiple kinases. As of May 2020, the FDA-approved SMKIs have been classified based on types of kinase targets and binding modes (Table 8.2). It has been shown that the most common drug kinase targets include PTKs (Bcr-Abl, EGFR, VEGFR and ALK) and PS/TKs (B-Raf, MEK1/2, CDK4 and -6 and PI3K). This chapter provides an insight into the development of relatively selective chemical probes for the representative PTKs (Bcr-Abl and EGFR) and PS/TKs (B-Raf and MEK1/2) and highlights examples of successful applications overcoming the resistance induced by mutation of the gatekeeper.[16,17] We also explore the opportunities and challenges of this area.

8.3.1 Bcr-Abl

Bcr-Abl, a non-receptor tyrosine kinase family, has been validated as a therapeutic target for the treatment of Philadelphia chromosome-positive (Ph^+) chronic myelogenous leukaemia (CML) or acute lymphoblastic leukaemia (ALL).[18,19] Currently, five small-molecule Bcr-Abl inhibitors, imatinib (**1**), nilotinib (**2**) (Figure 8.3A), dasatinib (**3**), bosutinib (**4**) (Figure 8.4A) and ponatinib (**5**) (Figure 8.6A), have been approved by the FDA. A novel allosteric type IV inhibitor, ABL001, with improved kinase selectivity and enhanced capability to combat the gatekeeper Thr315 mutation resistance is also under clinical investigation.[13]

Imatinib, the first FDA-approved Bcr-Abl inhibitor, is a type II chemical probe optimized with the phenylaminopyrimidine (PAP) scaffold using a rational drug design strategy.[20] Imatinib shows high potency against Bcr-Abl (IC_{50} = 38 nM) and against K652 cells (IC_{50} = 280 nM). Owing to its high target selectivity, the therapeutic efficiency of imatinib is considerable and side effects are reduced, thus beginning a new era of targeted cancer therapy.[21] The X-ray

Table 8.2 FDA-approved SMKIs based on types of kinase targets and binding modes.

Kinase group	Kinase name	Drug	Binding mode[a]	Targeted disease
PTKs	Bcr-Abl	Imatinib	II	Philadelphia chromosome-positive CML or ALL
		Dasatinib	I	CML
		Bosutinib	II	CML
		Nilotinib	II	Philadelphia chromosome-positive CML
		Ponatinib	II	Philadelphia chromosome-positive CML or ALL
	EGFR	Gefitinib	I	NSCLC
		Erlotinib	I or I½	NSCLC, pancreatic cancer
		Lapatinib	I½	HER2-positive breast cancer
		Afatinib	V	NSCLC
		Osimertinib	V	NSCLC
		Dacomitinib	V	NSCLC
	ALK	Crizotinib	I½	ALK or ROS1-positive NSCLC
		Ceritinib	I½	ALK-positive NSCLC resistant to crizotinib
		Alectinib	I½	ALK-positive NSCLC
		Brigatinib	I	ALK-positive NSCLC
		Lorlatinib	I	ALK-positive NSCLC
	VEGFR	Pazopanib	I	Renal cell carcinoma, soft tissue sarcomas
		Sunitinib	II	Gastrointestinal stromal tumours, renal cell carcinoma
		Sorafenib	II	Hepatocellular carcinoma, renal cell carcinoma, thyroid cancer
		Vandetanib	I	Medullary thyroid cancer
		Axitinib	II	Advanced renal cell carcinoma
		Regorafenib	ND	Colorectal cancer, hepatocellular carcinoma
		Lenvatinib	I	Thyroid cancer
	Flt3	Midostaurin	I	Acute myelogenous leukaemia, mastocytosis, mast cell leukaemia
		Gilteritinib	ND	Acute myelogenous leukaemia
		Quizartinib	II	Childhood acute myeloid leukaemia, myeloid malignancies

	BTK	Ibrutinib	V	Chronic lymphocytic leukaemia, mantle cell lymphoma
		Acalabrutinib	V	Mantle cell lymphoma
	TRK	Larotrectinib	ND	Solid tumours with NTRK fusion proteins
		Entrectinib	I	Solid tumours with NTRK fusion proteins
	ErbB2	Neratinib	I	HER2-positive breast cancer
	FGFR	Nintedanib	I	Idiopathic pulmonary fibrosis
		Erdafitinib	I	Metastatic bladder cancer with FGFR mutations
	Flt3, CSF1R and KIT	Pexidartinib	II	Advanced solid tumours
	JAK	Ruxolitinib	ND	Myelofibrosis, polycythaemia vera
	JAK1 and -2	Baricitinib	ND	Rheumatoid arthritis
	JAK3	Tofacitinib	I	Rheumatoid arthritis
	RET and others	Cabozantinib	ND	Advanced medullary thyroid cancer
	Syk	Fostamatinib	I	Chronic immune thrombocytopenia
PS/TKs	B-Raf	Vemurafenib	I½	$BRAF^{V600E}$ mutant melanoma
		Dabrafenib	I½	$BRAF^{V600E}$ mutant melanoma
		Encorafenib	I½	$BRAF^{V600E/K}$ melanoma/NSCLC
	MEK1/2	Trametinib	III	Melanoma/NSCLC
		Cobimetinib	III	BRAF mutation-positive melanoma in combination with vemurafenib
		Binimetinib	III	Melanoma
	CDK4 and -6	Palbociclib	I½	Estrogen receptor- and HER2-positive breast cancer
		Abemaciclib	I½	Breast cancer
		Ribociclib	I½	Combination therapy for breast cancer
	PI3K	Alpelisib	ND	Advanced or metastatic breast cancer
		Copanlisib	I	Relapsed follicular lymphoma
		Idelalisib	I	Relapsed or refractory CLL and B cell and small lymphocytic lymphoma
	Rho	Netarsudil	ND	Glaucoma

[a]ND: no reported X-ray crystal complex structure.

crystal structure of imatinib with Bcr-Abl (Figure 8.3B) shows that it binds to the DFG-out conformation of the protein and the pyridine of imatinib forms a hydrogen bond with Met318 of the hinge region. The pyridine ring and the pyrimidine ring engage in π–π stacking with Phe317 and Phe382, respectively. The amine group forms a hydrogen bond with the gatekeeper Thr315 and the amide forms hydrogen bonds with Glu286 and Asp381. The piperazinyl group of the inhibitor is located in the hydrophobic cavity on the surface of the protein and forms hydrogen bonds with Ile360. Unfortunately, the success of imatinib was overshadowed by the rapid emergence of drug resistance.[22–24] The most frequently acquired mutations are G250E, Y253H, E255K(V), T315I, F317L and M351T, which led to the development and approval of the second- and third-generation Bcr-Abl inhibitors.

Nilotinib (**2**), designed by replacing the benzyl-4-methylpiperazinyl side chain in imatinib with 4-methyl-1-[(3-trifluoromethyl)phenyl]-1*H*-imidazole, is also a type II Bcr-Abl inhibitor but is more potent than imatinib with IC_{50} values of 12 and 25 nM against K562 and Ba/F3 cells harbouring Bcr-Abl WT, respectively.[25,26] As a second-generation inhibitor, nilotinib is active against a number of imatinib-resistant Bcr-Abl mutants, with the exception of the gatekeeper T315I mutant. The X-ray co-crystal structure of nilotinib and Abl provides a structural explanation for the differential activity of nilotinib and imatinib against imatinib-resistant Bcr-Abl. Nilotinib binds to the inactive conformation of Abl and maintains a network of protein contacts similar to those observed with imatinib except for helix E hook.[25] The (3-trifluoromethyl) phenyl group of nilotinib binds deeply into the DFG-out hydrophobic pocket of

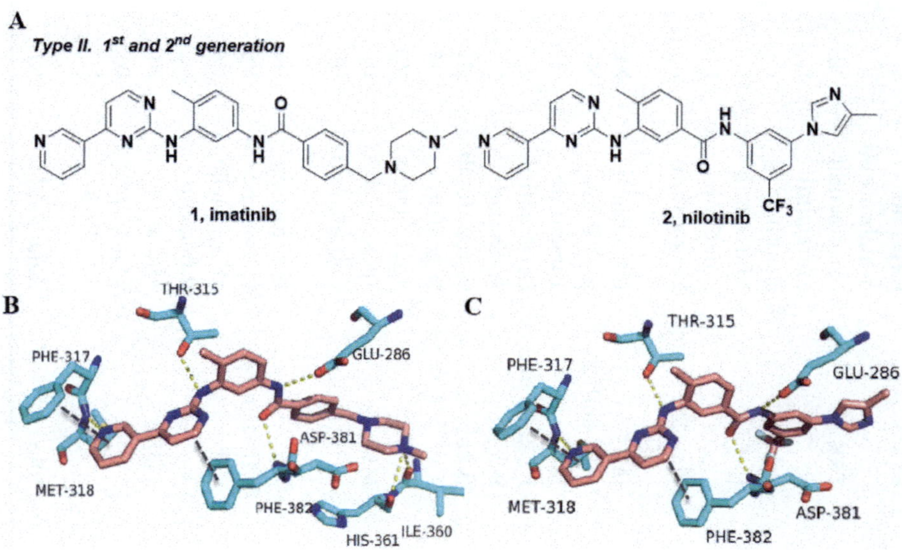

Figure 8.3 (A) Type II Bcr-Abl inhibitors imatinib and nilotinib. The type II binding modes of Abl and imatinib (B) (PDB code 2HYY) and nilotinib (C) (PDB code 3CS9). The key residues are represented as cyan sticks, hydrogen bonds are shown as yellow dashes and π–π stacking interactions are shown as black dashes.

the protein (Figure 8.3C). Nilotinib significantly prolongs survival in imatinib-resistant CML mouse models[26] and was approved by the FDA to treat certain types of CML in adults and children aged 1 year and older or CML that could not be treated successfully with imatinib.

Dasatinib (**3**), as a type I Bcr-Abl inhibitor, is about 325- and 16-fold more potent than imatinib and nilotinib, respectively, against Bcr-Abl WT, with IC_{50} = 0.8 nM.[27] However, like nilotinib, dasatinib is active against a number of imatinib-resistant Bcr-Abl mutants, with the exception of the gatekeeper T315I mutant. An X-ray crystal structure (Figure 8.4B) shows that dasatinib binds to the DFG-in conformation and its aminothiazole ring forms two hydrogen bonds with Met318 of the hinge region.[28] The 2-chloro-6-methylphenyl ring of dasatinib is oriented towards the hydrophobic pocket formed by Ile313, Met290 and Glu286 and the piperazine group is directed towards the surface-exposed portion of the hinge region. In addition, the amide nitrogen of dasatinib also forms a hydrogen bond with the gatekeeper Thr315, which explains why dasatinib is ineffective against the T315I mutant. Dasatinib potently inhibits the proliferation of Bcr-Abl-positive bone marrow progenitor cells from patients with imatinib-sensitive or imatinib-resistant CML and also effectively prolongs the survival of mice harbouring Bcr-abl WT.[29] Dasatinib was approved in 2006 for use in CML patients.

Bosutinib is a type II ATP-competitive Bcr-Abl inhibitor with an additional inhibitory effect on Src family kinases including Src, Lyn, Hck and Yes.[30]

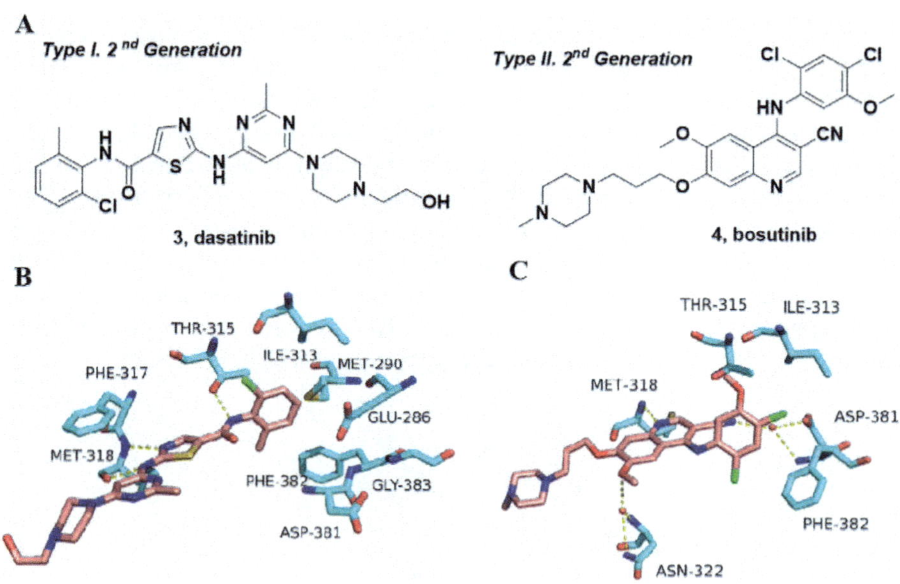

Figure 8.4 (A) Second-generation Bcr-Abl inhibitors dasatinib and bosutinib. The binding modes of Abl and dasatinib (B) (PDB code 2GQG) and bosutinib (C) (PDB code 3UE4). The key residues are represented as cyan sticks and hydrogen bonds are shown as yellow dashes.

In vitro and *in vivo* studies showed that bosutinib is active against multiple imatinib-resistant Bcr-Abl mutants except those with T315I gatekeeper mutation and has demonstrated efficacy against imatinib-resistant CML. The binding mode with Abl is very similar to that observed with Src, where the nitrogen of the quinoline forms a key hydrogen bond with Met318.[31] The 2,4-dichloro-5-methoxyaniline group is oriented at ~65° to the plane of the quinoline core and, similarly to the 2-chloro-6-methylphenyl ring of dasatinib, binds to a hydrophobic pocket. The flexible *N*-propyl-*N*-methylpiperazine group is directed towards the surface-exposed portion of the hinge region. The methoxy oxygen participates in a hydrogen-bonding interaction with Asn322, mediated by a water molecule (Figure 8.4C). Whereas imatinib, nilotinib and dasatinib all form hydrogen bonds with Thr315, the nitrile group of bosutinib makes van der Waals contacts with Thr315 and forms a hydrogen bond with Asp381, mediated by a water molecule. Bosutinib, like nilotinib and dasatinib, is still inactive against the common T315I mutant.

Although the second-generation Bcr-Abl inhibitors are active against many imatinib-resistant mutants of Bcr-Abl, they are unable to inhibit the most notable Bcr-Abl T315I gatekeeper mutation.[32] This T315I mutation alters the 3D structure of the ATP pocket of Bcr-Abl to eliminate a fundamental hydrogen bond, causing a steric clash in the case of imatinib (Figure 8.5).[23,33] To overcome the T315I mutation, considerable efforts have been made to develop third-generation Bcr-Abl inhibitors.[16,34] Ponatinib (**5**, Figure 8.6A), developed from an Abl/Src inhibitor, AP23464, by introducing rigid, rod-shaped alkyne groups to reduce the steric clash with Ile315,[35,36] was approved in December 2012 for the treatment of imatinib-resistant CML. Ponatinib exhibits low nanomolar activity in a biochemical and cellular proliferation assay against wild-type (WT) Bcr-Abl and Bcr-Abl T315I. The co-crystal structure of ponatinib with Bcr-Abl T315I demonstrates that ponatinib binds to Bcr-Abl in a type II binding mode similar to that adopted by imatinib (Figure 8.6B). Specifically, the imidazo[1,2-*b*]pyridazine core occupies the adenine pocket and the amide forms two hydrogen bonds with Glu286 and Asp381. The trifluoromethylphenyl group is bound deep in the DFG-out hydrophobic pocket. Compared with imatinib, a potential contributing factor to overcoming the T315I mutation was that the alkyne makes favourable van der Waals interactions with the gatekeeper Ile315 without any steric clash. To date, only ponatinib has been approved for the treatment of CML carrying the Bcr-Abl T315I mutation.

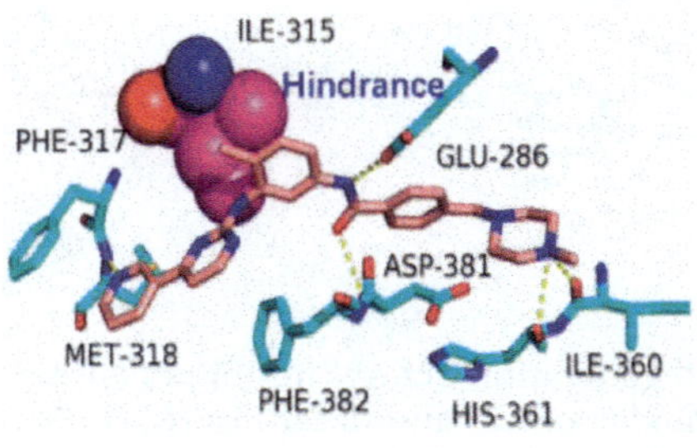

Figure 8.5 The steric clash between imatinib and Abl caused by the gatekeeper T315I mutation.

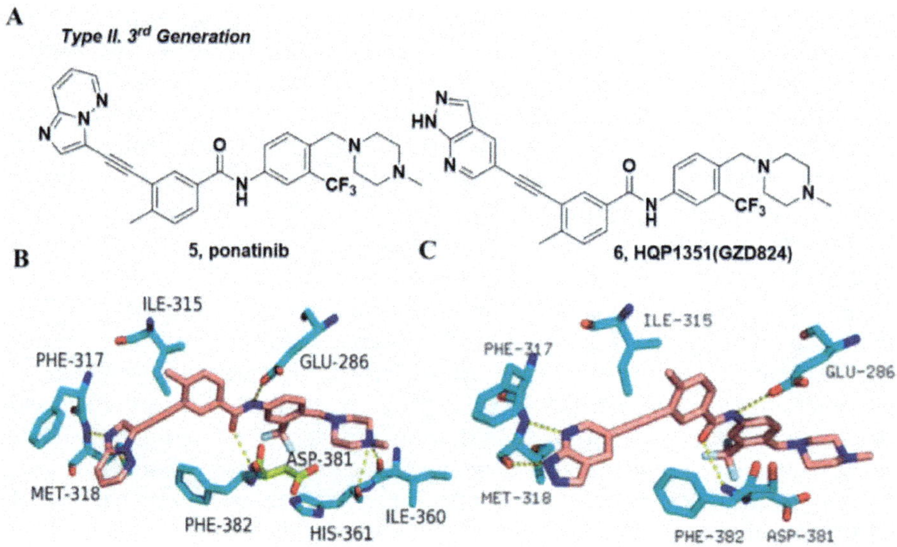

Figure 8.6 Third-generation Bcr-Abl inhibitors ponatinib and GZD824 (A). The type II binding modes of Abl and ponatinib (B) (PDB code 3IK3) and GZD824 (C) (docking model). The key residues are represented as cyan sticks and hydrogen bonds are shown as yellow dashes.

Our group reported a series of 1*H*-pyrazolo[3,4-*b*]pyridine-based compounds as third-generation type II inhibitors.[37] The representative compound GZD824 exhibits a broad-spectrum reaction with Bcr-Abl and also with mutants, including T315I. Similarly to ponatinib, GZD824 exhibits low nanomolar activities in biochemical and cellular proliferation assays. It displays ideal pharmacokinetic (PK) properties and significantly induces tumour regression in mouse xenograft tumour models driven by Bcr-Abl WT, Bcr-Abl T315I and other mutants. GZD824 maintains a network of protein contacts similar to that of imatinib (Figure 8.6C) and is currently in phase II clinical trials in China.

As mentioned, another method to override the Bcr-Abl T315I gatekeeper mutation is by targeting the allosteric site of kinase. The first allosteric Bcr-Abl inhibitor, the 4,6-disubstituted pyrimidine-based GNF-2 (7, Figure 8.7A), binds to the myristate-binding pocket of Bcr-Abl and was discovered in a fluorescence-based proliferation assay.[38] GNF-2 specifically inhibits the proliferation of the Bcr cells expressing Abl with an IC_{50} of 138 nM and displays no toxicity against non-transformed cells. Further genetic experiments and NMR and X-ray crystallographic studies demonstrated that GNF-2 binding to the myristate pocket does indeed interfere with the peptide substrate binding to Bcr-Abl.[39,40] X-ray crystallographic studies revealed that GNF-2 combined with imatinib shows a synergistic effect in inhibiting the kinase activity of Abl carrying the SH3 and SH2 domains (Figure 8.7B). As a type IV inhibitor, GNF-2 binds to the allosteric myristate site with its CF_3 group buried deeply in the hydrophobic pocket formed by Leu448, Ala452 and Leu360, while the amide group is directed towards to the surface of the protein. The nitrogen atom of the pyrimidine forms a hydrogen bond with Tyr454 mediated by a water molecule. In addition, the amine group forms a hydrogen-bond network with Ala452 and Glu481, also mediated by a water molecule. These unique binding interactions determine the remarkable selectivity of GNF-2 towards Bcr-Abl. Further lead optimization studies resulted in a GNF-2 analogue, GNF-5 (**8**, Figure 8.7A), which exhibits activity similar to that of GNF-2 and also more favourable PK properties. However, in cellular assays, both GNF-2 and GNF-5 failed to show significant potency against the BaF3 cells harbouring the Bcr-Abl T315I mutant, with GI_{50} values >10 μM. A combination of GNF-5 with nilotinib led to additive inhibitory activity in biochemical and cellular assays against Bcr-Abl T315I and displayed ideal *in vivo* efficacy in a murine bone-marrow transplantation study. These results suggested that the allosteric myristate-binding site inhibitors combined with ATP-competitive inhibitors can effectively overcome the Bcr-Abl T315I mutant resistance to either agent when used alone.

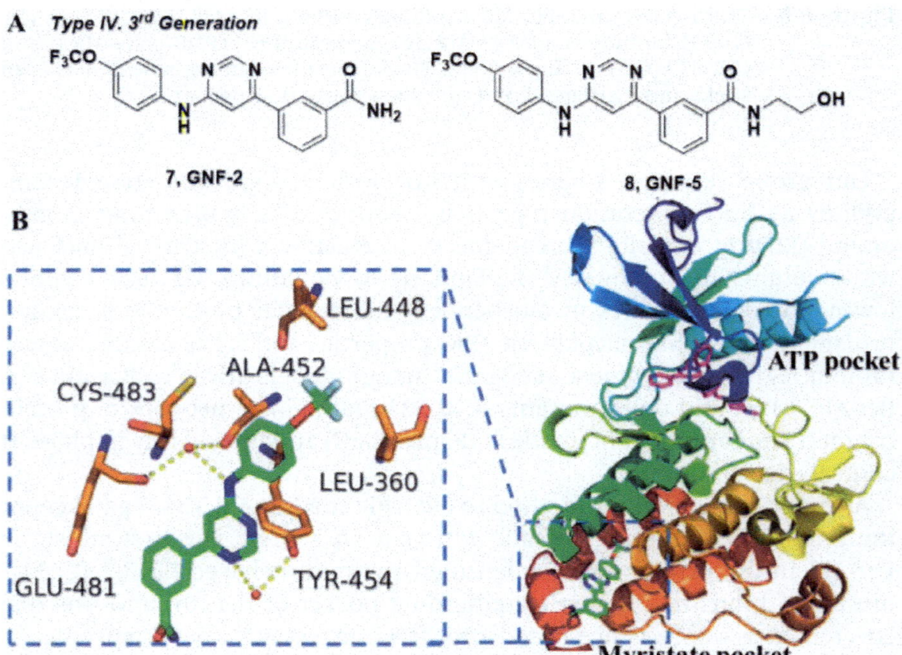

Figure 8.7 Type IV Bcr-Abl inhibitors GNF-2 and GNF-5 (A) and the X-ray crystal structure of GNF-2 with Abl kinase domain (B) (PDB code 3K5V). Details of the binding interactions between GNF-2 and the myristate-binding pocket are shown in the inset on the left. The key residues are represented as sticks and labelled and hydrogen bonds are shown as yellow dashes.

Further optimization aimed at improving the potency, selectivity and *in vivo* PK character led to the allosteric clinical candidate ABL001 (**9**, Figure 8.8A) for the treatment of CML and Ph$^+$-ALL.[13,41] Like GNF-2, ABL001 mimics the substrate myristate and potently binds to the myristoyl pocket of Bcr-Abl with a K_d value of ~0.5–0.8 nM and more than 100-fold potency over that of GNF-2. More importantly, ABL001 is active in the low nanomolar concentration range against all catalytic ATP-site mutations, including the gatekeeper T315I mutation. Studies of a combination of ABL001 with nilotinib indicated that the mixture maintains activity against SMKI-resistant ATP-binding site mutations and the components have no shared resistance for each other. Hence SMKIs have been suggested to be effective against ABL001-resistant mutations.[42] The findings suggest that combinations of ABL001 with ATP-competitive SMKIs suppress the emergence of resistance caused by the Bcr-Abl-driven mutation. ABL001 (asciminib) was the first allosteric type IV kinase inhibitor to be entered into clinical trials. X-ray crystallographic studies showed that ABL001 binds to the myristate pocket of Bcr-Abl with contacts similar to those enjoyed by GNF-2 (Figure 8.8B).

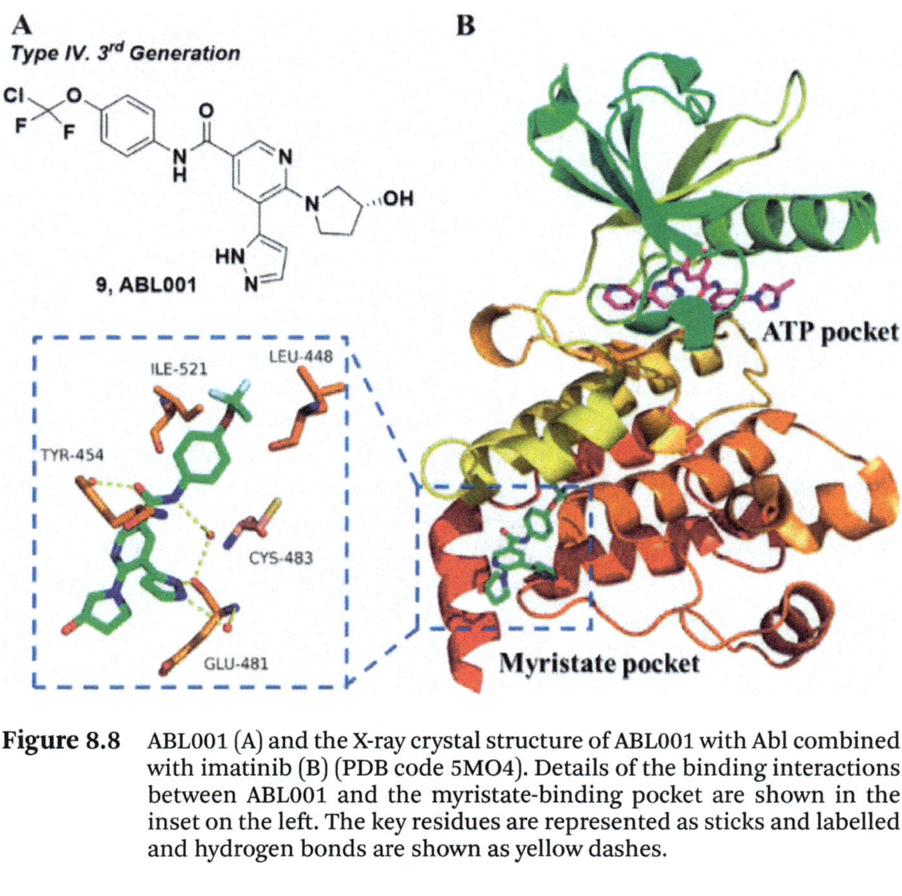

Figure 8.8 ABL001 (A) and the X-ray crystal structure of ABL001 with Abl combined with imatinib (B) (PDB code 5MO4). Details of the binding interactions between ABL001 and the myristate-binding pocket are shown in the inset on the left. The key residues are represented as sticks and labelled and hydrogen bonds are shown as yellow dashes.

8.3.2 EGFR

The epidermal growth factor (EGFR, ErbB1, HER1) receptor tyrosine kinase is one of the most validated molecular targets in non-small cell lung cancer (NSCLC) drug discovery.[43,44] Currently, six small-molecule EGFR inhibitors, gefitinib (**10**), erlotinib (**11**), lapatinib (**12**), afatinib (**13**), dacomitinib (**14**) and osimertinib (**15**), have received FDA approval and have achieved significant clinical benefits in NSCLC and breast cancer patients. Another covalent third-generation EGFR L858R/T790M inhibitor, olmutinib (**16**), was launched in Korea in 2016 for the treatment of patients with locally advanced or metastatic EGFR T790M-positive NSCLC. However, a tertiary Cys797 to Ser797 (C797S) point mutation disturbs covalent bond formation with the covalent third-generation inhibitors and creates the need to develop fourth-generation EGFR inhibitors.

The first-generation inhibitors gefitinib, erlotinib and lapatinib (Figure 8.9A), classified as anilinoquinazoline-based derivatives, are more active for EGFR (L858R or delE746-A750) than EGFR WT.[43,45] Gefitinib was the first-approved EGFR inhibitor and exhibits good selectivity against EGFR with mutations (IC_{50} = 33 nM) compared with other protein kinases. The X-ray crystal structure of gefitinib with EGFR demonstrates that it binds to an active conformation of EGFR kinase domain in the type I binding mode (Figure 8.9B).[46] The quinazoline N1 of gefitinib forms a direct hydrogen bond with the hinge residue Met793 and the N3 of the quinazoline forms a hydrogen-bond network with Thr790 and Thr854, mediated by a water molecule. Similarly to the 2,4-dichloro-5-methoxyaniline group of bosutinib with Abl, the 3-chloro-4-fluorophenyl group is oriented at ~30° to the plane of the quinoline core and binds in a hydrophobic pocket formed by Leu788 and Met766. The flexible propoxymorpholine group is oriented towards the surface-exposed portion of the hinge region. However, it was found that acquired drug resistance to gefitinib generally occurs after 9–14 months of clinical therapy.

Erlotinib was approved by the FDA in 2004 for the treatment of NSCLC. The X-ray crystal structure of erlotinib and EGFR demonstrates that erlotinib can bind to both the active conformation (type I – DFG_{in}–αC_{in}) and the inactive conformation (type I½ – DFG_{in}–αC_{out}) of the EGFR kinase domain (Figure 8.9C and D).[47,48] In both structures, erlotinib displays a binding conformation and contacts with EGFR similar to those observed with gefitinib.

Lapatinib is a dual inhibitor of EGFR and ErbB-2 and has been approved for the treatment of ErbB-2/HER2-postitive breast cancer.[49] It is the only approved EGFR inhibitor that was not approved for the treatment of NSCLC. Lapatinib has a unique mechanism of action with a very slow off-rate from EGFR and ErbB-2 compared with gefitinib and erlotinib.[50] The X-ray crystal structure of lapatinib with EGFR demonstrates that lapatinib binds to an inactive conformation (type I½ – DFG_{in}–αC_{out}) of EGFR (Figure 8.9E). N1 of the quinazoline in lapatinib forms a key hydrogen bond with Met793 and N3 participates in a water-mediated hydrogen-bond network with Thr854. However, unlike gefitinib and erlotinib, the chlorophenyl group forms a hydrophobic contact with gatekeeper Thr790 and the ethoxyfluorophenyl group

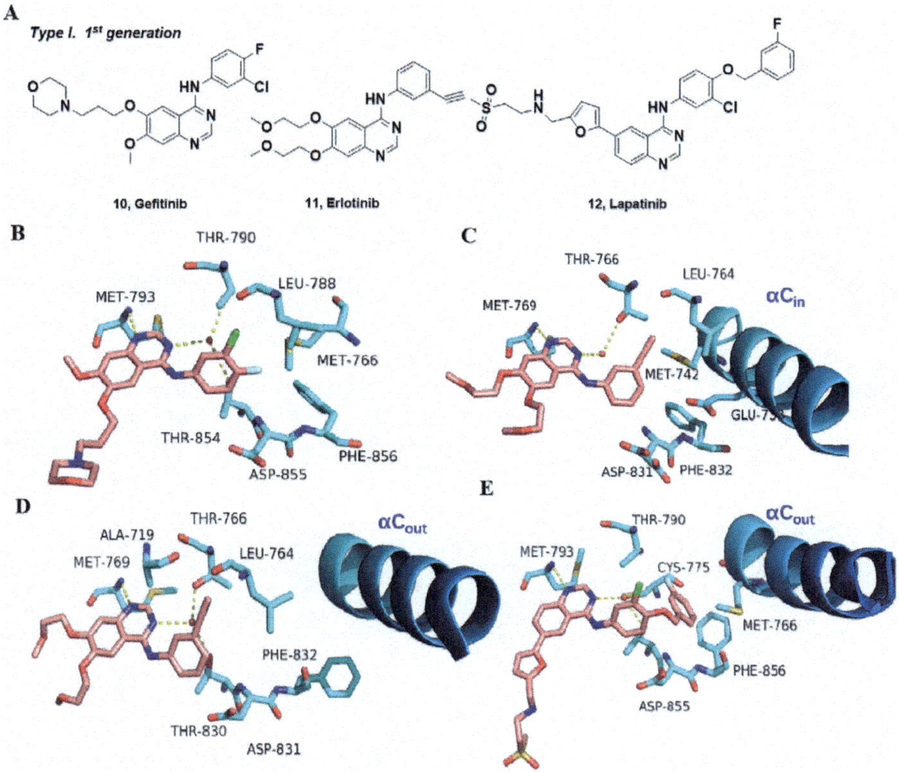

Figure 8.9 Type I EGFR inhibitors gefitinib, erlotinib and lapatinib (A) and X-ray crystal structures of gefitinib (B) (PDB code 2ITY), erlotinib (C) (PDB code 1M17) and (D) (PDB code 4HJO) and lapatinib (E) (PDB code 1XKK) with EGFR. The key residues are represented as cyan sticks and hydrogen bonds are shown as yellow dashes.

binds to a rear hydrophobic pocket formed by Cys775, Met766 and Phe856 with the αC-helix in an out conformation. The flexible methyl sulfone linker is oriented towards the surface-exposed portion of the hinge region.

The gatekeeper T790M mutation occurs in 55–70% of resistant NSCLC patients during the use of the first-generation inhibitors of EGFR (gefitinib, erlotinib and lapatinib) in the clinic. The T790M mutation increases the affinity of ATP for EGFR, therefore increasing competition between ATP and first-generation reversible inhibitors, restoring the enzymatic activity at physiologically achievable EGFR–SMKI concentrations (Figure 8.10A).[51] Thus, second-generation EGFR inhibitors with irreversible Cys797-targeting moieties to compete with the high ATP-binding ability have been developed to overcome EGFR T790M mutation-mediated resistance. These include afatinib (**13**) and dacomitinib (**14**) (Figure 8.11A), both of which have been approved. Afatinib and dacomitinib, type V covalent inhibitors, have a binding conformation similar to that of gefitinib, which has a covalent bond between the acrylamide and Cys797 (Figure 8.11B). Like lapatinib, dacomitinib binds to the inactive conformation of EGFR with an αC helix in the out

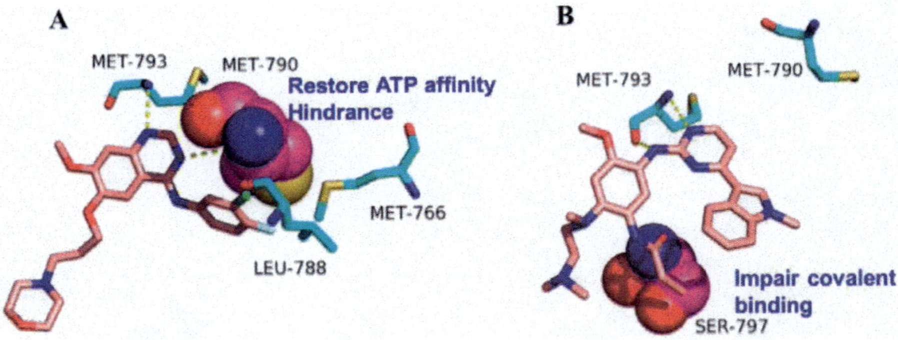

Figure 8.10 (A) The steric clash between gefitinib and EGFR caused by the gatekeeper T790M mutation, along with ATP affinity restored. (B) The disturbance of the covalent bonding between osimertinib and EGFR caused by the C797S mutation.

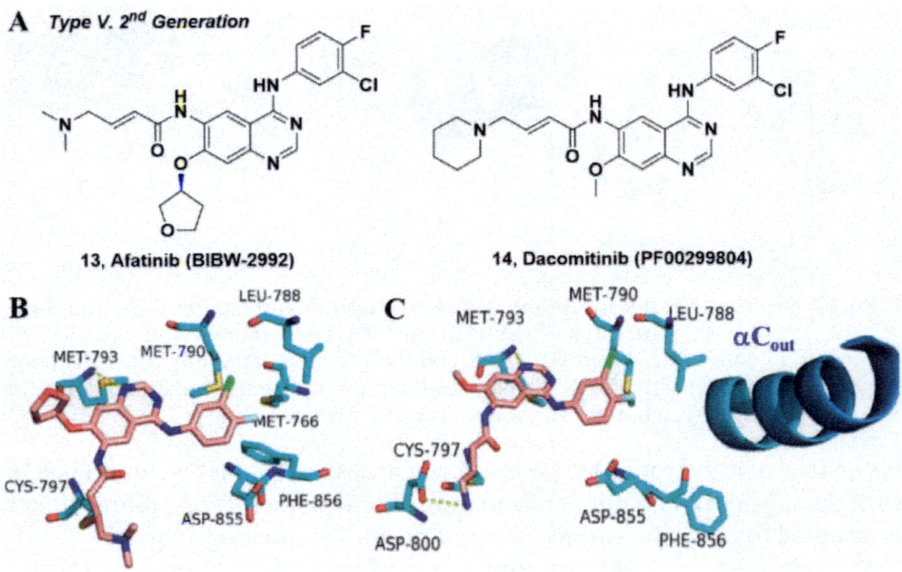

Figure 8.11 Type V EGFR irreversible inhibitors afatinib and dacomitinib (A) and X-ray crystal structures of afatinib (B) (PDB code 4G5P) and dacomitinib (C) (PDB code 4I24) with EGFR. The key residues are represented as cyan sticks and hydrogen bonds are shown as yellow dashes.

conformation (Figure 8.11C). However, the second-generation inhibitors suffer from an unacceptably low maximum tolerated dose (MTD) in patients due to their activity against WT EGFR and the normal physiological role of these kinases in skin and gastrointestinal tissues.[52,53]

Third-generation EGFR inhibitors with good selectivity for EGFR T790M over EGFR WT have been widely reported[17] and include WZ4002(**15**), osimertinib (**16**, AZD9291) and olmutinib (**17**, HM61713) (Figure 8.12A). Almost all of the third-generation EGFR inhibitors adopt a U-shaped conformation in the EGFR T790M active site to present an acrylamide group to Cys797 and selectively form a covalent complex. WZ4002,[54] a 4-aminopyrimidine-based compound, was first reported as a third-generation EGFR inhibitor. WZ4002 selectively inhibits Ba/F3-EGFR T790M/L858R and H1975 NSCLC cells with IC_{50} = 8 and 47 nM, respectively, and is less potent against Ba/F3-EGFR WT cells with an IC_{50} value of 113 nM. Moreover, compared with an EGFR WT model, it shows higher efficacy in mouse models harbouring either EGFR L858R/T790M or EGFR Del19/T790M. The X-ray crystal structure of WZ4002 and EGFR T790M demonstrates that WZ4002 binds EGFR T790M in a U-shaped conformation with the anilinopyrimidine core forming two hydrogen bonds with Met793 and the acrylamide group forming a covalent bond with Cys797 (Figure 8.12B). The chlorine interacts with the gatekeeper Met790, explaining its target selectivity. However, WZ4002 did not enter the clinic owing to an issue with patent rights.

Osimertinib, with an aminopyrimidine scaffold similar to that of WZ4002, selectively inhibits the proliferation of PC9 NSCLC cells that are driven by the Del19 activating mutation and of H1975 cells that express the T790M/L858R double mutant with IC_{50} values of 17 and 15 nM, respectively, while sparing EGFR WT-dependent LoVo cells (IC_{50} = 480 nM).[55,56] Osimertinib also induces significant dose-dependent tumour regression in both PC9 and H1975 tumour xenograft models dosed at 2.5 and 5 mg kg^{-1} per day with improved PK characteristics and reduced human ether-a-go-go related gene (hERG) affinity.[57] Systematic clinical trials confirmed that osimertinib has significant efficacy in patients with T790M-positive mutant advanced NSCLC (including those with CNS metastases) in whom disease had progressed during first-line EGFR inhibitor therapy.[58–60] To date, only osimertinib has been approved by the FDA for the treatment of patients with metastatic NSCLC whose tumours have EGFR exon 19 deletions or exon 21 (L858R) substitution mutations.[56,57]

Olmutinib, a thieno[3,2-*d*]pyrimidine-based EGFR L858R/T790M inhibitor,[61] was launched in Korea for the treatment of patients with locally advanced or metastatic EGFR T790M-positive NSCLC. Olmutinib exhibits potent *in vitro* and *in vivo* antiproliferative activity against H1975 and HCC827 cells with GI_{50} values of 9.2 and 10 nM, respectively. It displays an increased progression-free survival (PFS) for NSCLC patients with a disease control rate of about 91%. However, follow-up clinical practice indicates that the efficacy

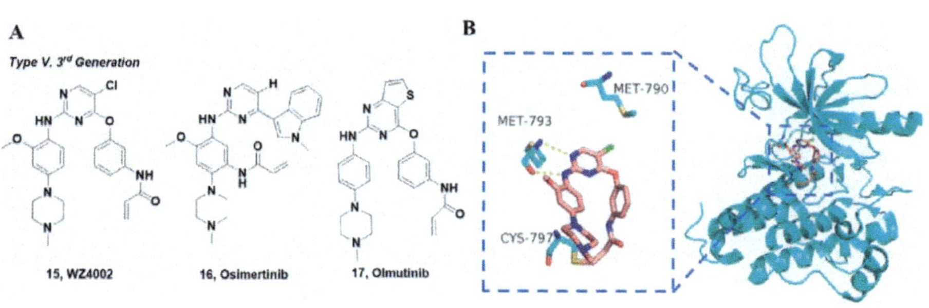

Figure 8.12 WZ4002, osimertinib and olmutinib (A) and X-ray crystal structure of WZ4002 with $EGFR^{T790M}$ (B) (PDB code 3IKA). Details of the binding interactions between WZ4002 and EGFR are shown in the inset on the left. The key residues are represented as cyan sticks and labelled and hydrogen bonds are shown as yellow dashes.

of osimertinib and olmutinib is mainly compromised by the dominant mutation C797S, which disturbs the covalent binding mode (Figure 8.10B).[62,63] Hence the development of fourth-generation EGFR inhibitors with the ability to overcome the C797S mutation is urgently needed.

As described above, the EGFR C797S mutation has become the most common mechanism of acquired resistance to third-generation EGFR inhibitors through disruption of covalent binding. Several type I reversible ATP-binding inhibitors and type III allosteric inhibitors overriding the EGFR C797S mutation have been reported,[17,64] but most of them are biochemically active. Only a few inhibitors (EAI045,[65] JBJ-04-125-02[66] and JND3229[67]) display promising activity *in vitro* and *in vivo*. These will be used as examples to describe their potency, selectivity and binding modes.

The first allosteric EGFR inhibitor, the thiazole amide-based EAI001 (**18**, Figure 8.13), was discovered by a high-throughput screening strategy designed to overcome EGFR T790M/C797S resistance.[68,69] EAI001 exhibits potent activity against EGFR L858R/T790M (IC_{50} = 24 nM) and high selectivity over EGFR WT (IC_{50} > 50 μM). The X-ray crystal structure of EAI001 with EGFR demonstrates that it binds to the allosteric site of EGFR T790M in a type III binding mode. The aminothiazole group makes direct contact with the gatekeeper Met790 while the NH group of the carboxamide acts as an H-bond donor with Asp855 (Figure 8.13B). The phenyl group engages in hydrophobic interactions with the DFG-in pocket and the 1-oxoisoindolinyl extends to the solvent-accessible region. Further optimization of EAI001 led to a more potent compound, EAI045 (**19**), with an IC_{50} value of 3 nM.

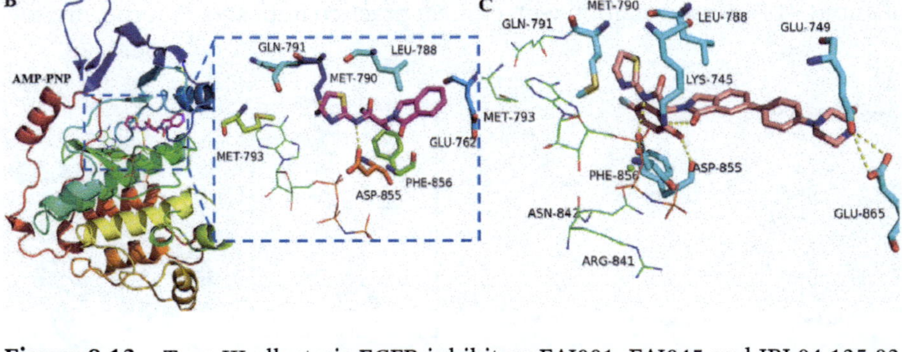

Figure 8.13 Type III allosteric EGFR inhibitors EAI001, EAI045 and JBJ-04-125-02 (A) and X-ray crystal structures of EAI001 (B) (PDB code 5D41) and JBJ-04-125-02 (C) (PDB code 6DUK). The key residues are represented as sticks and hydrogen bonds are shown as yellow dashes.

However, EAI045 showed weak antiproliferative activity in NSCLC cells harbouring the T790M mutant due to lack of access to the two subunits of the asymmetric EGFR kinase dimer. As a result, EAI045 has limited clinical efficacy as a single agent and had to be combined with cetuximab in mouse xenograft models harbouring the EGFR L858R/T790M mutation.[65] The same synergistic effect has also been seen in the tertiary C797S mutation models.

In a search for allosteric EGFR inhibitors more potent than EAI045 that would not require the co-administration of cetuximab, further medicinal chemistry optimization led to JBJ-04-125-02 (**20**) through incorporation of a phenylpiperazine group on the isoindolinone in EAI045.[66] The X-ray crystal structure of JBJ-04-125-02 with EGFR T790M demonstrates that, similarly to EAI001, it binds to the allosteric site of EGFR with the phenylpiperazine group oriented towards the solvent region by forming hydrogen bonds with Glu749 and Glu865 (Figure 8.13C). JBJ-04-125-02 exhibits sub-nanomolar potency against the EGFRL858R/T790M kinase with an IC_{50} value of 0.26 nM and it effectively inhibits cell proliferation and EGFR L858R/T790M/C797S signalling *in vitro* and *in vivo* as a single agent, whereas it is less effective against EGFR del19D. Further combination of JBJ-04-125-02 and osimertinib increased the efficacy *in vitro* and *in vivo*, suggesting that combinations of allosteric inhibitors with ATP-competitive EGFR SMKIs can suppress the emergence of resistance caused by an EGFR-driven mutation.

In 2018, our group reported the novel pyrimidopyrimidinone-based JND3229 (**21**, Figure 8.14A) as a type I EGFR C797S inhibitor with single-digit nM potency (IC_{50} = 5.8 nM).[67] JND3229 also exhibited good *in vitro* and *in vivo* mono-drug anticancer efficacy in a xenograft mouse model of BaF3/EGFR 19D/T790M/C797S cells. The X-ray crystal structure of JND3229 with the EGFR T790M/C797S complex demonstrates that JND3229 binds in the ATP-binding site with a reversible "U-shaped" configuration. The pyrido[2,3-*d*]pyrimidin-7-one undergoes bidentate hydrogen-bond interactions with Met793 and the carbonyl of pyrido[2,3-*d*]pyrimidin-7-one forms a hydrogen-bond network with Lys745 mediated by a water molecule. The 2-chlorophenyl group extends to the hydrophobic back pocket formed by Lys745, Glu762 and Leu788. The propionamide group is located in a solvent-exposed region and forms a water-mediated hydrogen bond with Leu718. The phenylpiperazine group also points towards the solvent region (Figure 8.14B). Although JND3229 exhibits *in vivo* mono-drug anticancer efficacy, it lacks selectivity over EGFR WT, which raises some concern about potential "on-target" toxicity. To improve the target selectivity, further optimization was conducted, which led to a new, selective EGFR L858R/T790M/C797S inhibitor, 8r-B (**22**, Figure 8.14A).[70] 8r-B exhibits an IC_{50} value of 27.5 nM against EGFR L858R/T790M/C797S and is significantly less potent against EGFR WT (IC_{50} > 1.0 μM). The X-ray crystal structure and molecular dynamics simulations demonstrated that the 5-methyl group in 8r-B is inserted into the "gatekeeper" residue Met790, contributing greatly to EGFR T790M selectivity over EGFR WT (Figure 8.14C). An *in vivo* study of 8r-B is in progress.

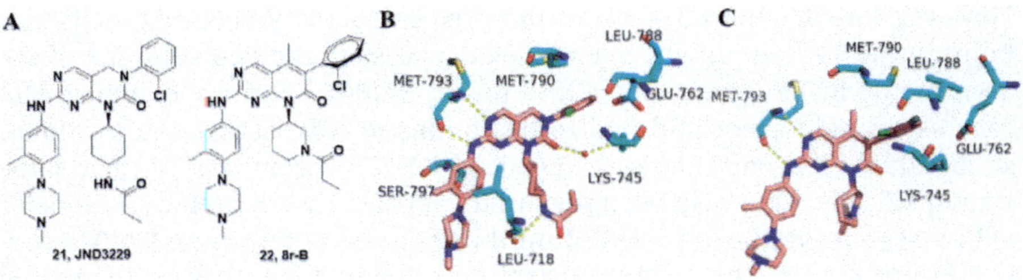

Figure 8.14 Type I EGFR reversible fourth-generation inhibitors JND3229 and 8r-B (A) and X-ray crystal structures of JND3229 (B) (PDB code 5ZTO) and 8r-B (C) (PDB code 6JRJ) with EGFR. The key residues are represented as cyan sticks and hydrogen bonds are shown as yellow dashes.

8.3.3 B-Raf

The PT/SK Raf are key components of Ras–Raf–MEK–Erk signalling. Among the three isoforms of the family (A-Raf, B-Raf and C-Raf), mutations of B-Raf are the most frequent in human cancers.[71,72] A mutation of Val600 to Glu600 in B-Raf (B-Raf V600E) accounts for more than 90% of the B-Raf oncogenic mutants, elevating the kinase activity about 600-fold compared with B-RAF WT.[73] As a result, B-Raf V600E became an effective target for the treatment of human cancers harbouring this mutation.[74] Currently, three small inhibitors of B-Raf, vemurafenib (**23**),[75] dabrafenib (**24**)[76] and encorafenib (**25**)[77] (Figure 8.15A), have been approved by the FDA for the treatment of metastatic and unresectable melanoma bearing the B-Raf V600E mutation. X-ray crystal analysis revealed that vemurafenib and dabrafenib bind to B-Raf with a DFG-in and αC-helix-out inactive conformation and are defined as type I½ inhibitors (Figure 8.15B and C).

Vemurafenib was discovered by structure-based drug design based on a 7-azaindole lead, PLX4720.[78,79] It exhibits potent activity against B-Raf V600E (IC_{50} = 31 nM) and has selectivity over WT B-Raf in biochemical and cellular assays. The X-ray crystal structure of vemurafenib with B-Raf V600E demonstrates that it binds to the ATP site of B-Raf V600E in an inactive αC_{out} conformation with DFG-in and is classified as a type I½ binding mode. The azaindole core forms bivalent hydrogen bonds with hinge residues Gln530 and Cys532 and the carbonyl group makes a hydrogen-bond network with Lys483 and Asp594 mediated by a water molecule. Moreover, the sulfonamide oxygen forms hydrogen bonds with Phe595 and Gly596. The propyl group occupies the hydrophobic DFG pocket by shifting Leu505 outwards in the αC-helix (Figure 8.15B). Vemurafenib exhibited good PK and safety profiles in preclinical and clinical studies. It was approved by the FDA in 2011 for the treatment of advanced B-Raf V600E mutant melanoma.

Dabrafenib, an aminopyrimidine-based derivative, exhibits strong inhibitory activity against B-Raf V600E (IC_{50} = 0.7 nM) and potent cellular activity

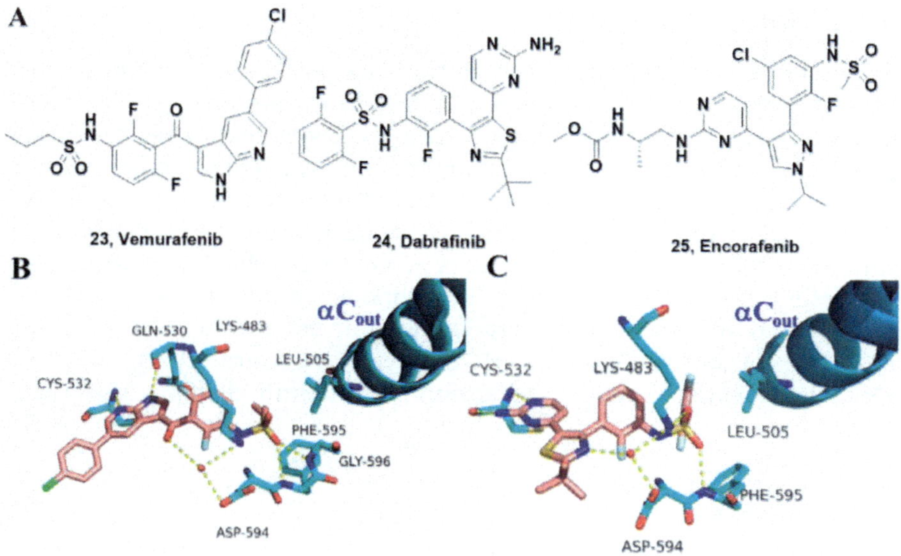

Figure 8.15 Type I½ B-Raf V600E inhibitors vemurafenib, dabrafenib and encorafenib (A) and X-ray crystal structures of vemurafenib (B) (PDB code 3OG7) and dabrafenib (C) (PDB code 4XV2) with B-Raf V600E. The key residues are represented as cyan sticks and hydrogen bonds are shown as yellow dashes.

against B-Raf V600E-driven melanoma.[80] Dabrafenib has a minimal effect on tumour cells with B-Raf WT and fails to harbour the activating B-Raf V600 mutation. The X-ray crystal structure of dabrafenib with B-Raf V600E demonstrates that similarly to vemurafenib, it also binds to the ATP site of B-Raf V600E in an inactive αC_{out} conformation with DFG-in and is classified as a type I½ inhibitor (Figure 8.15C).[81] Subsequent clinical trials showed that dabrafenib is safe for patients with solid tumours and delivers a significant response in patients with melanoma and other solid tumours.[82,83] Dabrafenib was approved as a single agent for the treatment of B-Raf V600E mutation-positive advanced melanomas or in combination with trametinib (an MEK1/2 inhibitor) for the treatment of melanomas or NSCLC harbouring a B-Raf V600E mutation.

Encorafenib is an aminopyrimidine-based derivative that was approved in combination of binimetinib (a MEK1/2 inhibitor) for the treatment of B-Raf V600E or B-Raf V600K mutation-positive, unresectable or metastatic melanoma.[77,84] However, there is no available information on the crystal structure of complexes involving encorafenib and B-Raf V600E. It is suspected that encorafenib binds to B-Raf V600E in a similar mode to vemurafenib and dabrafenib is a type I½ inhibitor thanks to the same terminal sulfonamide group.

8.3.4 MEK1/2

Activation of the MAPK pathway, comprising the Ras, Raf, MEK and Erk kinases, has been implicated in the pathogenesis and progression of various human malignant tumours. MEK1/2, as downstream effectors of Raf, play a critical role in the MAPK signalling pathway. Three highly selective and potent type III allosteric inhibitors of MEK1/2, trametinib (**26**),[85] cobimetinib (**27**)[86] and binimetinib (**28**),[87] have been approved in combination with Raf inhibitors for the treatment of melanoma or NSCLC (Figure 8.16A). Trametinib was the first allosteric small-molecule inhibitor of MEK1/2 to be approved by the FDA. It displays remarkable potency in SK-MEL-28 cells harbouring B-Raf V600E and in HCT116 cells harbouring KRAS G13D for ERK1/2 phosphorylation. Furthermore, trametinib displays significant

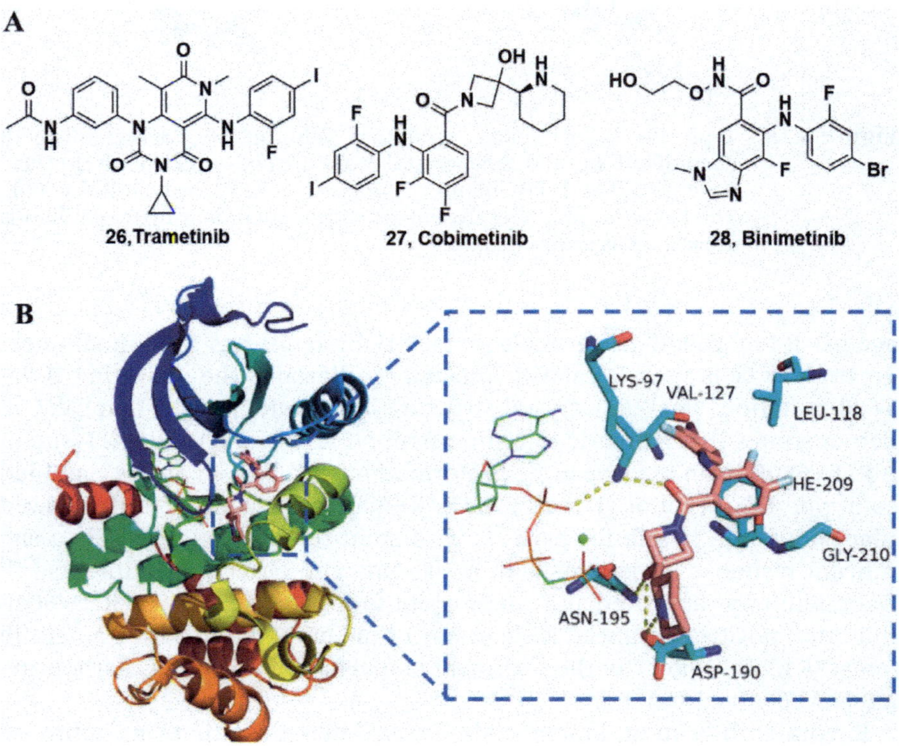

Figure 8.16 Type III allosteric MEK1/2 inhibitors trametinib, cobimetinib and binimetinib (A) and X-ray crystal structure of cobimetinib (B) (PDB code 4AN2). Details of the binding interactions are shown in the inset on the right. The key residues are represented as sticks and hydrogen bonds are shown as yellow dashes.

in vivo anti-tumour activity in KRas G12C, causing growth inhibition in a dose-dependent manner. However, there is no available information on the crystal structure of trametinib bound to MEK1/2. In 2013, trametinib was approved for the treatment of metastatic melanoma with either B-Raf V600E/V600K mutation. In 2014, the combination of dabrafenib and trametinib was approved to delay the emergence of resistance when using trametinib as a single-agent.[88,89]

A further two allosteric MEK1/2 inhibitors, cobimetinib and binimetinib, have also been approved in combination with B-Raf V600E inhibitors vemurafenib and encorafenib for the treatment of B-Raf V600E/V600K mutation-positive and advanced melanoma, respectively.[77,86] The X-ray crystal structure of MEK1–AMP–PCP with a cobimetinib ternary complex demonstrates that it binds to the allosteric site of MEK1 with the carbonyl of the amide forming a hydrogen bond with Lys97. The hydroxyl group and the nitrogen of the piperidine form a hydrogen-bond network with Asn195 and Asp190. The 2-fluoro-4-iodophenyl group binds to the hydrophobic pocket composed of Leu118, Val127 and Phe209.

8.4 Chemical Probes with New Binding Modes

8.4.1 Allosteric Target in the Extracellular Domain

Discoidin domain receptors (DDRs), including DDR1 and DDR2, are transmembrane receptor tyrosine kinases that specifically recognize collagens as extracellular ligands.[90] DDRs are characterized by the presence of six distinct protein domains: a discoidin (DS) domain, a DS-like domain, an extracellular juxtamembrane (EJXM) region, a transmembrane (TM) segment, an intracellular juxtamembrane (IJXM) region and an intracellular kinase domain (KD). The N-terminal DS, DS-like and EJXM domains are composed of an extracellular portion of the DDRs.[91] DDRs are involved in the regulation of cellular differentiation, proliferation and migration. Dysregulation of DDRs has been implicated in a variety of inflammatory diseases.[92,93] However, nearly all of the reported DDR inhibitors are ATP-competitive inhibitors and this reveals a lack of the selectivity of the subtype over other homologous kinases.[94–96] Just recently, the small-molecule WRG-28 (**29**) was characterized as a selective DDR2 inhibitor, by uniquely inhibiting DDR2–collagen binding interactions *via* allosteric targeting of the extracellular domain (ECD) of DDR2 (Figure 8.17).[15] It is postulated that WRG-28 binds to the allosteric site at the interface of the DS–DS-like domains, blocking the binding of the collagen to DS of DDR2 without affecting the conformation of the KD. WRG-28 displays good selectivity over DDR1 and exhibits potent inhibitory activity against acquired DDR2 mutations that are resistant to SMKIs.

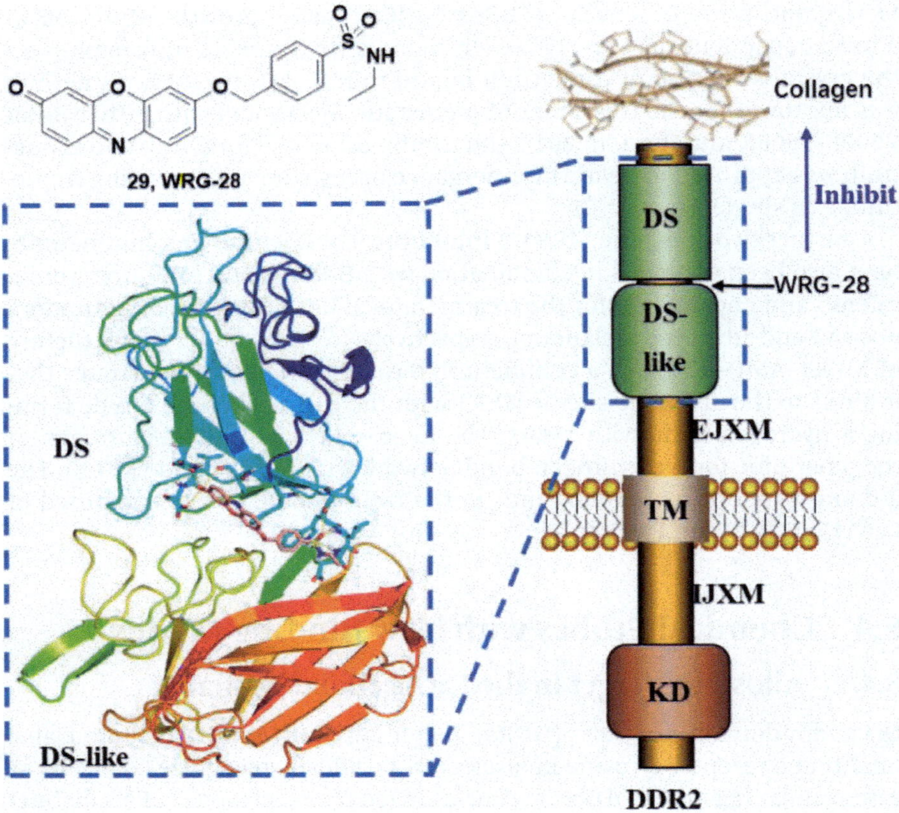

Figure 8.17 WRG-28 and overall view of the binding interactions of WRG-28 and DS–DS-like domains of DDR2.

8.4.2 PROTAC Degradation

As addressed in detail in Chapter 7, proteolysis-targeting chimeras (PROTACs) involve a heterobifunctional molecule containing two ligands, one for the target protein and the other for an E3 ligase, connected by a linker, and have been highlighted in the drug discovery area as a new therapeutic approach.[97,98] Compared with small-molecule inhibitors, PROTACs have increased selectivity due to the target protein inhibitor and E3 ligase ligand. Furthermore, a PROTAC requires only a binder, which does not necessarily need to be a functional inhibitor. It thus affords an additional opportunity to attack so-called undruggable targets, such as transcription factors or protein–protein interactions. Finally, a PROTAC approach has the possibility of overcoming the clinical resistance of SMKIs caused by a point mutation in the target protein. Targeted protein kinase degradation with PROTAC technology has been demonstrated *in vitro* for a wide range of kinases,[99] and a limited set of kinase degraders have been examined *in vivo*.[98] For example, TD-004 (**30**, Figure 8.18), an anaplastic lymphoma kinase (ALK) degrader designed with ceritinib linked to a ligand for

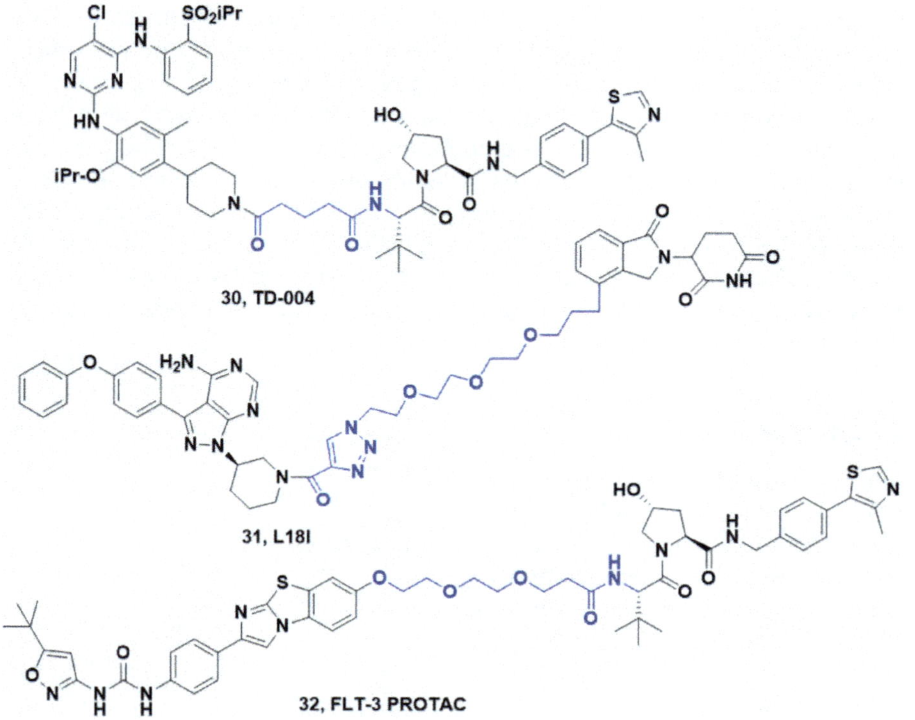

Figure 8.18 Chemical structures of AXL degrader TD-004 (30), BTK degrader L18I (31) and FLT3 degrader 32.

von Hippel–Lindau (VHL) E3 ligase, effectively induces ALK degradation in cellular assays and significantly reduces tumour growth in an ALK fusion-positive cell line H3122 xenograft model.[100] L18I (**31**), a Bruton's tyrosine kinase (BTK) degrader designed with an ibrutinib-linked lenalidomide, exhibits enhanced anti-proliferative effects on mantle cell lymphoma (MCL) cells harbouring the BTK C481S mutant *in vitro* and efficacy in mouse xenograft models inoculated with BTK C481S,[101] which indicates that it can overcome the clinical resistance to ibrutinib caused by the BTK C481S mutation. FLT-3-PROTAC (**32**), designed by linking quizartinib with a PEG linker, induces degradation of an FLT3 ITD mutant *in vitro* and *in vivo* with better selectivity than the warhead quizartinib.[102] These studies suggest that degradation of kinases may prove to be a useful means of therapeutic intervention.

8.5 Conclusion

Kinase-based drug discovery has seen dramatic progress in the past 20 years. SMKIs bring encouraging clinical benefits for cancer patients, but still face two critical challenges in both clinical and experimental oncology. First, because most SMKIs bind to the ATP-binding pocket of a kinase, they lack the relative selectivity with respect to the kinase family. Second, acquired drug resistance generally occurs after 1–2 years of treatment with SMKIs due to secondary or tertiary kinase point mutations. Allosteric kinase inhibitors offer a potentially complementary therapeutic strategy to ATP-competitive kinase inhibitors owing to their distinct sites of target binding with high selectivity and their ability to overcome acquired resistance. The development of allosteric kinase inhibitors targeting other less-conserved allosteric pockets, such as the EGFR dimerization interface (EDI) or the docking site of ERK FXF (DEF),[103] is being actively pursued. On the other hand, only a small subset of the human kinome has been studied, with SMKIs reported. Hence there is an urgent need to develop new chemical probes to explore the functions of the many unknown kinases.

Abbreviations

ALK	Anaplastic lymphoma kinase
ALL	Acute lymphoblastic leukaemia
AMP-PCP	Adenosine 5′-(β,γ-methylenetriphosphate)
BTK	Bruton's tyrosine kinase
CML	Chronic myelogenous leukaemia
DDR	Discoidin domain receptor
DEF	Docking site of ERK FXF
DFG	Asp–Phe–Gly
DS	Discoidin
ECD	Extracellular domain
EDI	EGFR dimerization interface
EGFR	Epidermal growth factor receptor
EJXM	Extracellular juxtamembrane
FDA	Food and Drug Administration
hERG	Human ether-a-go-go related gene
IJXM	Intracellular juxtamembrane
KD	Kinase domain
MCL	Mantle cell lymphoma
MTD	Maximum tolerated dose
NSCLC	Non-small cell lung cancer
Ph^+	Philadelphia chromosome-positive
PK	Pharmacokinetic
PROTAC	Proteolysis-targeting chimera
PS/TK	Protein serine/threonine kinase
PTK	Protein tyrosine kinase
SMKI	Small-molecule kinase inhibitor
TM	Transmembrane
VHL	Von Hippel–Lindau
WT	Wild-type

Acknowledgements

The authors are grateful for financial support from the National Natural Science Foundation of China (81922062 and 81425021) and Jinan University.

References

1. Z. Wang and P. A. Cole, *Methods Enzymol.*, 2014, **548**, 1–21.
2. G. Manning, D. B. Whyte, R. Martinez, T. Hunter and S. Sudarsanam, *Science*, 2002, **298**, 1912–1934.
3. R. Roskoski Jr, *Pharmacol. Res.*, 2015, **100**, 1–23.
4. P. Cohen, *Nat. Rev. Drug Discovery*, 2002, **1**, 309–315.
5. R. Roskoski Jr, *Pharmacol. Res.*, 2019, **144**, 19–50.
6. P. Wu, T. E. Nielsen and M. H. Clausen, *Drug discovery today*, 2016, **21**, 5–10.
7. D. R. Knighton, J. H. Zheng, L. F. Ten Eyck, N. H. Xuong, S. S. Taylor and J. M. Sowadski, *Science*, 1991, **253**, 414–420.
8. M. Tong and M. A. Seeliger, *ACS Chem. Biol.*, 2015, **10**, 190–200.
9. M. E. Noble, J. A. Endicott and L. N. Johnson, *Science*, 2004, **303**, 1800–1805.
10. F. Zuccotto, E. Ardini, E. Casale and M. Angiolini, *J. Med. Chem.*, 2010, **53**, 2681–2694.
11. R. Roskoski Jr, *Pharmacol. Res.*, 2016, **103**, 26–48.
12. L. K. Gavrin and E. Saiah, *MedChemComm*, 2013, **4**, 41–51.
13. A. A. Wylie, J. Schoepfer, W. Jahnke, S. W. Cowan-Jacob, A. Loo, P. Furet, A. L. Marzinzik, X. Pelle, J. Donovan, W. Zhu, S. Buonamici, A. Q. Hassan, F. Lombardo, V. Iyer, M. Palmer, G. Berellini, S. Dodd, S. Thohan, H. Bitter, S. Branford, D. M. Ross, T. P. Hughes, L. Petruzzelli, K. G. Vanasse, M. Warmuth, F. Hofmann, N. J. Keen and W. R. Sellers, *Nature*, 2017, **543**, 733–737.
14. T. K. Johnson and M. B. Soellner, *Bioconjugate Chem.*, 2016, **27**, 1745–1749.
15. W. R. Grither and G. D. Longmore, *Proc. Natl. Acad. Sci. U. S. A.*, 2018, **115**, E7786–E7794.
16. X. Y. Lu, Q. Cai and K. Ding, *Curr. Med. Chem.*, 2011, **18**, 2146–2157.
17. X. Lu, L. Yu, Z. Zhang, X. Ren, J. B. Smaill and K. Ding, *Med. Res. Rev.*, 2018, **38**, 1550–1581.
18. G. Q. Daley, R. A. Van Etten and D. Baltimore, *Science*, 1990, **247**, 824–830.
19. M. W. Renshaw, J. R. McWhirter and J. Y. Wang, *Mol. Cell. Biol*, 1995, **15**, 1286–1293.
20. R. Capdeville, E. Buchdunger, J. Zimmermann and A. Matter, *Nat. Rev. Drug Discovery*, 2002, **1**, 493–502.
21. T. Schindler, W. Bornmann, P. Pellicena, W. T. Miller, B. Clarkson and J. Kuriyan, *Science*, 2000, **289**, 1938–1942.

22. S. W. Cowan-Jacob, V. Guez, G. Fendrich, J. D. Griffin, D. Fabbro, P. Furet, J. Liebetanz, J. Mestan and P. W. Manley, *Mini-Rev. Med. Chem.*, 2004, **4**, 285–299.
23. M. E. Gorre, M. Mohammed, K. Ellwood, N. Hsu, R. Paquette, P. N. Rao and C. L. Sawyers, *Science*, 2001, **293**, 876–880.
24. T. O'Hare, C. A. Eide and M. W. Deininger, *Blood*, 2007, **110**, 2242–2249.
25. E. Weisberg, P. W. Manley, W. Breitenstein, J. Bruggen, S. W. Cowan-Jacob, A. Ray, B. Huntly, D. Fabbro, G. Fendrich, E. Hall-Meyers, A. L. Kung, J. Mestan, G. Q. Daley, L. Callahan, L. Catley, C. Cavazza, M. Azam, D. Neuberg, R. D. Wright, D. G. Gilliland and J. D. Griffin, *Cancer Cell*, 2005, **7**, 129–141.
26. M. Golemovic, S. Verstovsek, F. Giles, J. Cortes, T. Manshouri, P. W. Manley, J. Mestan, M. Dugan, L. Alland, J. D. Griffin, R. B. Arlinghaus, T. Sun, H. Kantarjian and M. Beran, *Clin. Cancer Res.*, 2005, **11**, 4941–4947.
27. T. O'Hare, D. K. Walters, E. P. Stoffregen, T. Jia, P. W. Manley, J. Mestan, S. W. Cowan-Jacob, F. Y. Lee, M. C. Heinrich, M. W. Deininger and B. J. Druker, *Cancer Res.*, 2005, **65**, 4500–4505.
28. J. S. Tokarski, J. A. Newitt, C. Y. Chang, J. D. Cheng, M. Wittekind, S. E. Kiefer, K. Kish, F. Y. Lee, R. Borzillerri, L. J. Lombardo, D. Xie, Y. Zhang and H. E. Klei, *Cancer Res.*, 2006, **66**, 5790–5797.
29. N. P. Shah, C. Tran, F. Y. Lee, P. Chen, D. Norris and C. L. Sawyers, *Science*, 2004, **305**, 399–401.
30. J. M. Golas, K. Arndt, C. Etienne, J. Lucas, D. Nardin, J. Gibbons, P. Frost, F. Ye, D. H. Boschelli and F. Boschelli, *Cancer Res.*, 2003, **63**, 375–381.
31. N. M. Levinson and S. G. Boxer, *PLoS One*, 2012, **7**, e29828.
32. E. Weisberg, P. W. Manley, S. W. Cowan-Jacob, A. Hochhaus and J. D. Griffin, *Nat. Rev. Cancer*, 2007, **7**, 345–356.
33. A. S. Corbin, E. Buchdunger, F. Pascal and B. J. Druker, *J. Biol. Chem.*, 2002, **277**, 32214–32219.
34. T. S. Lee, S. J. Potts, H. Kantarjian, J. Cortes, F. Giles and M. Albitar, *Cancer*, 2008, **112**, 1744–1753.
35. W. S. Huang, C. A. Metcalf, R. Sundaramoorthi, Y. Wang, D. Zou, R. M. Thomas, X. Zhu, L. Cai, D. Wen, S. Liu, J. Romero, J. Qi, I. Chen, G. Banda, S. P. Lentini, S. Das, Q. Xu, J. Keats, F. Wang, S. Wardwell, Y. Ning, J. T. Snodgrass, M. I. Broudy, K. Russian, T. Zhou, L. Commodore, N. I. Narasimhan, Q. K. Mohemmad, J. Iuliucci, V. M. Rivera, D. C. Dalgarno, T. K. Sawyer, T. Clackson and W. C. Shakespeare, *J. Med. Chem.*, 2010, **53**, 4701–4719.
36. T. O'Hare, W. C. Shakespeare, X. Zhu, C. A. Eide, V. M. Rivera, F. Wang, L. T. Adrian, T. Zhou, W. S. Huang, Q. Xu, C. A. Metcalf 3rd, J. W. Tyner, M. M. Loriaux, A. S. Corbin, S. Wardwell, Y. Ning, J. A. Keats, Y. Wang, R. Sundaramoorthi, M. Thomas, D. Zhou, J. Snodgrass, L. Commodore, T. K. Sawyer, D. C. Dalgarno, M. W. Deininger, B. J. Druker and T. Clackson, *Cancer Cell*, 2009, **16**, 401–412.

37. X. Ren, X. Pan, Z. Zhang, D. Wang, X. Lu, Y. Li, D. Wen, H. Long, J. Luo, Y. Feng, X. Zhuang, F. Zhang, J. Liu, F. Leng, X. Lang, Y. Bai, M. She, Z. Tu, J. Pan and K. Ding, *J. Med. Chem.*, 2013, **56**, 879–894.
38. F. J. Adrian, Q. Ding, T. Sim, A. Velentza, C. Sloan, Y. Liu, G. Zhang, W. Hur, S. Ding, P. Manley, J. Mestan, D. Fabbro and N. S. Gray, *Nat. Chem. Biol.*, 2006, **2**, 95–102.
39. Y. Choi, M. A. Seeliger, S. B. Panjarian, H. Kim, X. Deng, T. Sim, B. Couch, A. J. Koleske, T. E. Smithgall and N. S. Gray, *J. Biol. Chem.*, 2009, **284**, 29005–29014.
40. D. Fabbro, P. W. Manley, W. Jahnke, J. Liebetanz, A. Szyttenholm, G. Fendrich, A. Strauss, J. Zhang, N. S. Gray, F. Adrian, M. Warmuth, X. Pelle, R. Grotzfeld, F. Berst, A. Marzinzik, S. W. Cowan-Jacob, P. Furet and J. Mestan, *Biochim. Biophys. Acta*, 2010, **1804**, 454–462.
41. J. Schoepfer, W. Jahnke, G. Berellini, S. Buonamici, S. Cotesta, S. W. Cowan-Jacob, S. Dodd, P. Drueckes, D. Fabbro, T. Gabriel, J. M. Groell, R. M. Grotzfeld, A. Q. Hassan, C. Henry, V. Iyer, D. Jones, F. Lombardo, A. Loo, P. W. Manley, X. Pelle, G. Rummel, B. Salem, M. Warmuth, A. A. Wylie, T. Zoller, A. L. Marzinzik and P. Furet, *J. Med. Chem.*, 2018, **61**, 8120–8135.
42. W. Qiang, O. Antelope, M. S. Zabriskie, A. D. Pomicter, N. A. Vellore, P. Szankasi, D. Rea, J. M. Cayuela, T. W. Kelley, M. W. Deininger and T. O'Hare, *Leukemia*, 2017, **31**, 2844–2847.
43. T. J. Lynch, D. W. Bell, R. Sordella, S. Gurubhagavatula, R. A. Okimoto, B. W. Brannigan, P. L. Harris, S. M. Haserlat, J. G. Supko, F. G. Haluska, D. N. Louis, D. C. Christiani, J. Settleman and D. A. Haber, *N. Engl. J. Med.*, 2004, **350**, 2129–2139.
44. N. E. Hynes and H. A. Lane, *Nat. Rev. Cancer*, 2005, **5**, 341–354.
45. J. G. Paez, P. A. Janne, J. C. Lee, S. Tracy, H. Greulich, S. Gabriel, P. Herman, F. J. Kaye, N. Lindeman, T. J. Boggon, K. Naoki, H. Sasaki, Y. Fujii, M. J. Eck, W. R. Sellers, B. E. Johnson and M. Meyerson, *Science*, 2004, **304**, 1497–1500.
46. C. H. Yun, T. J. Boggon, Y. Li, M. S. Woo, H. Greulich, M. Meyerson and M. J. Eck, *Cancer Cell*, 2007, **11**, 217–227.
47. J. H. Park, Y. Liu, M. A. Lemmon and R. Radhakrishnan, *Biochem. J.*, 2012, **448**, 417–423.
48. J. Stamos, M. X. Sliwkowski and C. Eigenbrot, *J. Biol. Chem.*, 2002, **277**, 46265–46272.
49. D. W. Rusnak, K. Lackey, K. Affleck, E. R. Wood, K. J. Alligood, N. Rhodes, B. R. Keith, D. M. Murray, W. B. Knight, R. J. Mullin and T. M. Gilmer, *Mol. Cancer Ther.*, 2001, **1**, 85–94.
50. E. R. Wood, A. T. Truesdale, O. B. McDonald, D. Yuan, A. Hassell, S. H. Dickerson, B. Ellis, C. Pennisi, E. Horne, K. Lackey, K. J. Alligood, D. W. Rusnak, T. M. Gilmer and L. Shewchuk, *Cancer Res.*, 2004, **64**, 6652–6659.

51. C. H. Yun, K. E. Mengwasser, A. V. Toms, M. S. Woo, H. Greulich, K. K. Wong, M. Meyerson and M. J. Eck, *Proc. Natl. Acad. Sci. U. S. A.*, 2008, **105**, 2070–2075.
52. G. R. Oxnard, M. E. Arcila, J. Chmielecki, M. Ladanyi, V. A. Miller and W. Pao, *Clin. Cancer Res.*, 2011, **17**, 5530–5537.
53. G. Giaccone and Y. Wang, *Cancer Treat. Rev.*, 2011, **37**, 456–464.
54. W. Zhou, D. Ercan, L. Chen, C. H. Yun, D. Li, M. Capelletti, A. B. Cortot, L. Chirieac, R. E. Iacob, R. Padera, J. R. Engen, K. K. Wong, M. J. Eck, N. S. Gray and P. A. Janne, *Nature*, 2009, **462**, 1070–1074.
55. M. R. Finlay, M. Anderton, S. Ashton, P. Ballard, P. A. Bethel, M. R. Box, R. H. Bradbury, S. J. Brown, S. Butterworth, A. Campbell, C. Chorley, N. Colclough, D. A. Cross, G. S. Currie, M. Grist, L. Hassall, G. B. Hill, D. James, M. James, P. Kemmitt, T. Klinowska, G. Lamont, S. G. Lamont, N. Martin, H. L. McFarland, M. J. Mellor, J. P. Orme, D. Perkins, P. Perkins, G. Richmond, P. Smith, R. A. Ward, M. J. Waring, D. Whittaker, S. Wells and G. L. Wrigley, *J. Med. Chem.*, 2014, **57**, 8249–8267.
56. D. A. Cross, S. E. Ashton, S. Ghiorghiu, C. Eberlein, C. A. Nebhan, P. J. Spitzler, J. P. Orme, M. R. Finlay, R. A. Ward, M. J. Mellor, G. Hughes, A. Rahi, V. N. Jacobs, M. Red Brewer, E. Ichihara, J. Sun, H. Jin, P. Ballard, K. Al-Kadhimi, R. Rowlinson, T. Klinowska, G. H. Richmond, M. Cantarini, D. W. Kim, M. R. Ranson and W. Pao, *Cancer Discovery*, 2014, **4**, 1046–1061.
57. S. L. Greig, *Drugs*, 2016, **76**, 263–273.
58. P. A. Janne, J. C. Yang, D. W. Kim, D. Planchard, Y. Ohe, S. S. Ramalingam, M. J. Ahn, S. W. Kim, W. C. Su, L. Horn, D. Haggstrom, E. Felip, J. H. Kim, P. Frewer, M. Cantarini, K. H. Brown, P. A. Dickinson, S. Ghiorghiu and M. Ranson, *N. Engl. J. Med.*, 2015, **372**, 1689–1699.
59. G. Goss, C.-M. Tsai, F. A. Shepherd, L. Bazhenova, J. S. Lee, G.-C. Chang, L. Crino, M. Satouchi, Q. Chu, T. Hida, J.-Y. Han, O. Juan, F. Dunphy, M. Nishio, J.-H. Kang, M. Majem, H. Mann, M. Cantarini, S. Ghiorghiu and T. Mitsudomi, *Lancet Oncol.*, 2016, **17**, 1643–1652.
60. T. S. Mok, Y.-L. Wu, M.-J. Ahn, M. C. Garassino, H. R. Kim, S. S. Ramalingam, F. A. Shepherd, Y. He, H. Akamatsu, W. S. M. E. Theelen, C. K. Lee, M. Sebastian, A. Templeton, H. Mann, M. Marotti, S. Ghiorghiu and V. A. Papadimitrakopoulou, *N. Engl. J. Med.*, 2017, **376**, 629–640.
61. K.-O. Lee, M. Y. Cha, M. Kim, J. Y. Song, J.-H. Lee, Y. H. Kim, Y.-M. Lee, K. H. Suh and J. Son, *Cancer Res.*, 2014, **74**, LB-1–00.
62. K. S. Thress, C. P. Paweletz, E. Felip, B. C. Cho, D. Stetson, B. Dougherty, Z. Lai, A. Markovets, A. Vivancos, Y. Kuang, D. Ercan, S. E. Matthews, M. Cantarini, J. C. Barrett, P. A. Janne and G. R. Oxnard, *Nat. Med.*, 2015, **21**, 560–562.
63. D. Ercan, H. G. Choi, C. H. Yun, M. Capelletti, T. Xie, M. J. Eck, N. S. Gray and P. A. Janne, *Clin. Cancer Res.*, 2015, **21**, 3913–3923.
64. L. Chen, W. Fu, L. Zheng, Z. Liu and G. Liang, *J. Med. Chem.*, 2018, **61**, 4290–4300.
65. S. Wang, Y. Song and D. Liu, *Cancer Lett.*, 2017, **385**, 51–54.

66. C. To, J. Jang, T. Chen, E. Park, M. Mushajiang, D. J. H. De Clercq, M. Xu, S. Wang, M. D. Cameron, D. E. Heppner, B. H. Shin, T. W. Gero, A. Yang, S. E. Dahlberg, K. K. Wong, M. J. Eck, N. S. Gray and P. A. Janne, *Cancer Discovery*, 2019, **9**, 926–943.
67. X. Lu, T. Zhang, S. J. Zhu, Q. Xun, L. Tong, X. Hu, Y. Li, S. Chan, Y. Su, Y. Sun, Y. Chen, J. Ding, C. H. Yun, H. Xie and K. Ding, *ACS Med. Chem. Lett.*, 2018, **9**, 1123–1127.
68. Y. Jia, C. H. Yun, E. Park, D. Ercan, M. Manuia, J. Juarez, C. Xu, K. Rhee, T. Chen, H. Zhang, S. Palakurthi, J. Jang, G. Lelais, M. DiDonato, B. Bursulaya, P. Y. Michellys, R. Epple, T. H. Marsilje, M. McNeill, W. Lu, J. Harris, S. Bender, K. K. Wong, P. A. Janne and M. J. Eck, *Nature*, 2016, **534**, 129–132.
69. C. To, J. Jang, T. Chen, E. Park, M. Mushajiang, D. J. H. De Clercq, M. Xu, S. Wang, M. D. Cameron, D. E. Heppner, B. H. Shin, T. W. Gero, A. Yang, S. E. Dahlberg, K. K. Wong, M. J. Eck, N. S. Gray and P. A. Jänne, *Cancer Discovery*, 2016, **6**, 691.
70. J. Shen, T. Zhang, S. J. Zhu, M. Sun, L. Tong, M. Lai, R. Zhang, W. Xu, R. Wu, J. Ding, C. H. Yun, H. Xie, X. Lu and K. Ding, *J. Med. Chem.*, 2019, **62**, 7302–7308.
71. H. Davies, G. R. Bignell, C. Cox, P. Stephens, S. Edkins, S. Clegg, J. Teague, H. Woffendin, M. J. Garnett, W. Bottomley, N. Davis, E. Dicks, R. Ewing, Y. Floyd, K. Gray, S. Hall, R. Hawes, J. Hughes, V. Kosmidou, A. Menzies, C. Mould, A. Parker, C. Stevens, S. Watt, S. Hooper, R. Wilson, H. Jayatilake, B. A. Gusterson, C. Cooper, J. Shipley, D. Hargrave, K. Pritchard-Jones, N. Maitland, G. Chenevix-Trench, G. J. Riggins, D. D. Bigner, G. Palmieri, A. Cossu, A. Flanagan, A. Nicholson, J. W. Ho, S. Y. Leung, S. T. Yuen, B. L. Weber, H. F. Seigler, T. L. Darrow, H. Paterson, R. Marais, C. J. Marshall, R. Wooster, M. R. Stratton and P. A. Futreal, *Nature*, 2002, **417**, 949–954.
72. G. Maurer, B. Tarkowski and M. Baccarini, *Oncogene*, 2011, **30**, 3477–3488.
73. P. T. Wan, M. J. Garnett, S. M. Roe, S. Lee, D. Niculescu-Duvaz, V. M. Good, C. M. Jones, C. J. Marshall, C. J. Springer, D. Barford, R. Marais and P. Cancer Genome, *Cell*, 2004, **116**, 855–867.
74. M. Holderfield, M. M. Deuker, F. McCormick and M. McMahon, *Nat. Rev. Cancer*, 2014, **14**, 455–467.
75. G. Bollag, J. Tsai, J. Zhang, C. Zhang, P. Ibrahim, K. Nolop and P. Hirth, *Nat. Rev. Drug Discovery*, 2012, **11**, 873–886.
76. A. D. Ballantyne and K. P. Garnock-Jones, *Drugs*, 2013, **73**, 1367–1376.
77. M. Shirley, *Drugs*, 2018, **78**, 1277–1284.
78. J. Tsai, J. T. Lee, W. Wang, J. Zhang, H. Cho, S. Mamo, R. Bremer, S. Gillette, J. Kong, N. K. Haass, K. Sproesser, L. Li, K. S. Smalley, D. Fong, Y. L. Zhu, A. Marimuthu, H. Nguyen, B. Lam, J. Liu, I. Cheung, J. Rice, Y. Suzuki, C. Luu, C. Settachatgul, R. Shellooe, J. Cantwell, S. H. Kim, J. Schlessinger, K. Y. Zhang, B. L. West, B. Powell, G. Habets, C. Zhang, P. N. Ibrahim, P. Hirth, D. R. Artis, M. Herlyn and G. Bollag, *Proc. Natl. Acad. Sci. U. S. A.*, 2008, **105**, 3041–3046.

79. G. Bollag, P. Hirth, J. Tsai, J. Zhang, P. N. Ibrahim, H. Cho, W. Spevak, C. Zhang, Y. Zhang, G. Habets, E. A. Burton, B. Wong, G. Tsang, B. L. West, B. Powell, R. Shellooe, A. Marimuthu, H. Nguyen, K. Y. Zhang, D. R. Artis, J. Schlessinger, F. Su, B. Higgins, R. Iyer, K. D'Andrea, A. Koehler, M. Stumm, P. S. Lin, R. J. Lee, J. Grippo, I. Puzanov, K. B. Kim, A. Ribas, G. A. McArthur, J. A. Sosman, P. B. Chapman, K. T. Flaherty, X. Xu, K. L. Nathanson and K. Nolop, *Nature*, 2010, **467**, 596–599.
80. T. R. Rheault, J. C. Stellwagen, G. M. Adjabeng, K. R. Hornberger, K. G. Petrov, A. G. Waterson, S. H. Dickerson, R. A. Mook Jr, S. G. Laquerre, A. J. King, O. W. Rossanese, M. R. Arnone, K. N. Smitheman, L. S. Kane-Carson, C. Han, G. S. Moorthy, K. G. Moss and D. E. Uehling, *ACS Med. Chem. Lett.*, 2013, **4**, 358–362.
81. C. Zhang, W. Spevak, Y. Zhang, E. A. Burton, Y. Ma, G. Habets, J. Zhang, J. Lin, T. Ewing, B. Matusow, G. Tsang, A. Marimuthu, H. Cho, G. Wu, W. Wang, D. Fong, H. Nguyen, S. Shi, P. Womack, M. Nespi, R. Shellooe, H. Carias, B. Powell, E. Light, L. Sanftner, J. Walters, J. Tsai, B. L. West, G. Visor, H. Rezaei, P. S. Lin, K. Nolop, P. N. Ibrahim, P. Hirth and G. Bollag, *Nature*, 2015, **526**, 583–586.
82. G. S. Falchook, G. V. Long, R. Kurzrock, K. B. Kim, T. H. Arkenau, M. P. Brown, O. Hamid, J. R. Infante, M. Millward, A. C. Pavlick, S. J. O'Day, S. C. Blackman, C. M. Curtis, P. Lebowitz, B. Ma, D. Ouellet and R. F. Kefford, *Lancet*, 2012, **379**, 1893–1901.
83. G. V. Long, U. Trefzer, M. A. Davies, R. F. Kefford, P. A. Ascierto, P. B. Chapman, I. Puzanov, A. Hauschild, C. Robert, A. Algazi, L. Mortier, H. Tawbi, T. Wilhelm, L. Zimmer, J. Switzky, S. Swann, A. M. Martin, M. Guckert, V. Goodman, M. Streit, J. M. Kirkwood and D. Schadendorf, *Lancet Oncol.*, 2012, **13**, 1087–1095.
84. Z. Li, K. Jiang, X. Zhu, G. Lin, F. Song, Y. Zhao, Y. Piao, J. Liu, W. Cheng, X. Bi, P. Gong, Z. Song and S. Meng, *Cancer Lett.*, 2016, **370**, 332–344.
85. H. Abe, S. Kikuchi, K. Hayakawa, T. Iida, N. Nagahashi, K. Maeda, J. Sakamoto, N. Matsumoto, T. Miura, K. Matsumura, N. Seki, T. Inaba, H. Kawasaki, T. Yamaguchi, R. Kakefuda, T. Nanayama, H. Kurachi, Y. Hori, T. Yoshida, J. Kakegawa, Y. Watanabe, A. G. Gilmartin, M. C. Richter, K. G. Moss and S. G. Laquerre, *ACS Med. Chem. Lett.*, 2011, **2**, 320–324.
86. K. D. Rice, N. Aay, N. K. Anand, C. M. Blazey, O. J. Bowles, J. Bussenius, S. Costanzo, J. K. Curtis, S. C. Defina, L. Dubenko, S. Engst, A. A. Joshi, A. R. Kennedy, A. I. Kim, E. S. Koltun, J. C. Lougheed, J. C. Manalo, J. F. Martini, J. M. Nuss, C. J. Peto, T. H. Tsang, P. Yu and S. Johnston, *ACS Med. Chem. Lett.*, 2012, **3**, 416–421.
87. A. A. N. Rose, *Drugs Today*, 2019, **55**, 247–264.
88. G. V. Long, D. Stroyakovskiy, H. Gogas, E. Levchenko, F. de Braud, J. Larkin, C. Garbe, T. Jouary, A. Hauschild, J. J. Grob, V. Chiarion Sileni, C. Lebbe, M. Mandala, M. Millward, A. Arance, I. Bondarenko, J. B. Haanen, J. Hansson, J. Utikal, V. Ferraresi, N. Kovalenko, P. Mohr, V. Probachai, D. Schadendorf, P. Nathan, C. Robert, A. Ribas, D. J. DeMarini, J. G. Irani, M. Casey, D. Ouellet, A. M. Martin, N. Le, K. Patel and K. Flaherty, *N. Engl. J. Med.*, 2014, **371**, 1877–1888.

89. C. Robert, B. Karaszewska, J. Schachter, P. Rutkowski, A. Mackiewicz, D. Stroiakovski, M. Lichinitser, R. Dummer, F. Grange, L. Mortier, V. Chiarion-Sileni, K. Drucis, I. Krajsova, A. Hauschild, P. Lorigan, P. Wolter, G. V. Long, K. Flaherty, P. Nathan, A. Ribas, A. M. Martin, P. Sun, W. Crist, J. Legos, S. D. Rubin, S. M. Little and D. Schadendorf, *N. Engl. J. Med.*, 2015, **372**, 30–39.
90. H. Xu, N. Raynal, S. Stathopoulos, J. Myllyharju, R. W. Farndale and B. Leitinger, *Matrix Biol.*, 2011, **30**, 16–26.
91. Y. Li, X. Lu, X. Ren and K. Ding, *J. Med. Chem.*, 2015, **58**, 3287–3301.
92. C. M. Borza and A. Pozzi, *Matrix Biol.*, 2014, **34**, 185–192.
93. S. Kothiwale, C. M. Borza, E. W. Lowe Jr, A. Pozzi and J. Meiler, *Drug discovery today*, 2015, **20**, 255–261.
94. M. Gao, L. Duan, J. Luo, L. Zhang, X. Lu, Y. Zhang, Z. Zhang, Z. Tu, Y. Xu, X. Ren and K. Ding, *J. Med. Chem.*, 2013, **56**, 3281–3295.
95. D. Zhu, H. Huang, D. M. Pinkas, J. Luo, D. Ganguly, A. E. Fox, E. Arner, Q. Xiang, Z. C. Tu, A. N. Bullock, R. A. Brekken, K. Ding and X. Lu, *J. Med. Chem.*, 2019, **62**, 7431–7444.
96. Z. Wang, Y. Zhang, D. M. Pinkas, A. E. Fox, J. Luo, H. Huang, S. Cui, Q. Xiang, T. Xu, Q. Xun, D. Zhu, Z. Tu, X. Ren, R. A. Brekken, A. N. Bullock, G. Liang, K. Ding and X. Lu, *J. Med. Chem.*, 2018, **61**, 7977–7990.
97. M. Toure and C. M. Crews, *Angew. Chem., Int. Ed.*, 2016, **55**, 1966–1973.
98. G. F. Watt, P. Scott-Stevens and L. Gaohua, *Drug Discovery Today: Technol.*, 2019, **31**, 69–80.
99. L. H. Jones, *Cell Chem. Biol.*, 2018, **25**, 30–35.
100. C. H. Kang, D. H. Lee, C. O. Lee, J. Du Ha, C. H. Park and J. Y. Hwang, *Biochem. Biophys. Res. Commun.*, 2018, **505**, 542–547.
101. Y. Sun, N. Ding, Y. Song, Z. Yang, W. Liu, J. Zhu and Y. Rao, *Leukemia*, 2019, **33**, 2105–2110.
102. G. M. Burslem, J. Song, X. Chen, J. Hines and C. M. Crews, *J. Am. Chem. Soc.*, 2018, **140**, 16428–16432.
103. C. Yueh, J. Rettenmaier, B. Xia, D. R. Hall, A. Alekseenko, K. A. Porter, K. Barkovich, G. Keseru, A. Whitty, J. A. Wells, S. Vajda and D. Kozakov, *J. Med. Chem.*, 2019, **62**, 6512–6524.

CHAPTER 9

Chemical Probes to Control RNA Function

LIBERTY FRANÇOIS-MOUTAL[a,b], VICTOR MIRANDA[a,b], MICHAEL SANDINO[a,b] AND MAY KHANNA*[a,b]

[a]Department of Pharmacology, College of Medicine, University of Arizona, Tucson, AZ 85724, USA; [b]Center for Innovation in Brain Science, Tucson, AZ 85721, USA
*E-mail: maykhanna@email.arizona.edu

9.1 Introduction

Proteins have been the primary target for therapeutic modulation, with RNA being less represented in targeted drug discovery. RNA has not been considered a suitable drug target for several reasons. First, it was believed to be divided into three major groups, ribosomal RNA (rRNA), transfer RNA (tRNA) and messenger RNA (mRNA), all serving as an information relay between DNA and protein or as an aid for protein genesis. Second, there was an initial focus on the three-dimensional structures of tRNAs,[1–3] since they were easier to crystallize. Most other RNAs have traditionally been challenging to study, owing to their inherent flexibility. This is immediately clear when comparing the number of structures available in the Protein Data Bank (PDB) of RNA compared with protein (so far in 2020 there were 25 structures deposited for RNA and 4855 structures deposited for proteins). Finally, because the first compounds reported to target RNA were planar and charged, they were found to be associated with high toxicity, which remained affiliated to RNA-targeting strategies.

Chemical Biology No. 16
The Discovery and Utility of Chemical Probes in Target Discovery
Edited by Paul Brennan and Saleta Vazquez Rodriguez

Published by the Royal Society of Chemistry, www.rsc.org

Whereas rRNAs and tRNAs were detected in 1955–1958,[4,5] the remaining fraction of RNAs, considered complex and unstable, were presumed to be mRNA or junk intronic regions. The 1980s heralded a change in this thinking, with the isolation and detection of small RNAs that were markedly different from rRNA and tRNA.[6,7] A few other interesting discoveries were made that included autoimmune diseases showing targeting of small ribonucleoprotein particles[8] and the involvement of RNA in catalysis.[9,10] These changed the perception of RNA as playing only regulatory roles as it demonstrated that, like proteins, RNA could support catalytic functions that were critical in disease progression in some autoimmune diseases, for example. A growing list of new classes of RNAs, such as miRNA, snoRNA, snRNA and circular RNAs, have also been discovered in the last decade.[11] One further class of non-coding RNA that has been understudied and may become important in small-molecule RNA targeting is non-coding RNAs that are translated to peptides/small proteins. All these new RNAs have not been studied structurally as extensively as tRNA and rRNA. There is a critical need for high-resolution structures of RNAs to increase the diversity of available structures.

Structural characterization of RNA increased in the 1990s, with roughly 100 structures being solved by X-ray and NMR methods,[12] which increased our understanding of the complex diversity of structural elements within RNAs, such as hairpins, loops, triple helices and the presence of non-Watson–Crick base pairing. RNA exists as a dynamic ensemble of structural conformations, including well-defined secondary and tertiary folds, dictated by primary sequence, shaped by mono- and divalent ions.[13,14] Despite the increase in the number of available structures and the advent of high-resolution structures of the ribosome and the spliceosome obtained with X-ray and cryogenic electron microscopy (cryo-EM) methods, much is still unknown about the majority of the RNAs that are in the intronic regions of mRNAs. SHAPE (selective 2′-hydroxyl acylation analysed by primer extension) analysis has been used to define higher level secondary structural elements in RNA that may ultimately be incorporated into three-dimensional prediction. These are useful in developing unique probes for unique structural regions.

All of the above discoveries mentioned, along with technical progress, shifted the "central dogma" about RNA as a possible target. Two fundamental discoveries may have kickstarted the pursuit of targeting RNA as a therapeutic strategy: (i) the approval by the US Food and Drug Administration (FDA) in 1998 of fomivirsen (marketed as Vitravene),[15–17] the first antisense oligonucleotide strategy to treat cytomegalovirus chorioretinitis, and (ii) the discovery in 2002 of riboswitches,[18–20] regulatory elements of an mRNA able to bind specifically small molecules. It became clear that targeting RNA was a viable therapeutic strategy since antisense oligonucleotides (oligos) could modulate RNA levels and were translatable into therapies since RNAs were structurally complex and could bind to small molecules with high specificity.

In the past few years, several studies reported the use of small molecules targeting RNA for therapeutic use.[21–23] There is growing interest in RNA-targeting therapeutics, exemplified by the growth of pharmaceutical companies focusing solely on targeting RNA (*e.g.* Arrakis Therapeutics, Ribometrix, Expansion Therapeutics). In this chapter, we focus on why targeting RNA is of importance, what are known RNA-targeting molecules and difficulties of targeting RNA structures with consistency and specificity. We also describe current methods used for targeting RNA directly.

9.2 Importance of RNAs as Drug Discovery Targets

The continuing need for new drugs is a constant consequence of the urge to cure currently untreatable diseases, the increase in antibiotic resistance, the continuing emergence of new diseases and simply a need for better drugs with fewer side effects. Among the 20 000 proteins encoded by the human genome (which represents 3% of the genome), only 15% of them are considered disease related[24] and there has been an estimate that as little as 0.03% of the genome has been successfully targeted. Although recent efforts to address what seem to be "undruggable" proteins have emerged,[25,26] mainly due to increased structural and biological data available on new classes of proteins, there are still many failed attempts in drug discovery in both the academic and pharmaceutical settings.

One way to expand drug discovery efforts is to consider completely unique and new targets such as RNA. About 80% of the genome is transcribed into RNA, with most being non-coding RNAs.[27] Non-coding RNAs have been directly implicated in a large number of diseases, including cancers, neurodegenerative diseases, cardiovascular diseases and diabetes.[28–31]

Targeting RNA directly using small molecules offers a whole new range of therapeutic strategies. Here we outline possible strategies that target post-transcriptional regulation, functional disease regulation and structural regulation.

9.2.1 Post-transcriptional Regulation: RNA Splicing

RNA splicing, particularly alternative splicing (AS), produces significant proteomic expansion by controlling how mRNA removes introns. Based on extensive microarray data, it was estimated that ~95% of multi-exonic human genes give rise to transcripts with more than 100 000 distinct AS events.[32,33] AS events are cell, tissue and condition dependent,[32,33] making targeting AS events a promising targeting strategy since it can be both compartmentalized and specifically modulated in specific cells and tissues. There are different types of splicing events that yield different proteins, such as constitutive, mutually exclusive exons, cassette alternative exons, alternative 3′ and 5′ splice sites and intron retention, with the most prevalent pattern being cassette-type alternative exons (or exon skipping) in vertebrates and invertebrates.[34] For splicing to occur, there is an assembly of the spliceosome (a multi-RNA–protein complex) followed by splicing of pre-mRNA (by multiple enzymes that execute this through multiple splicing steps).[34] *Cis*-acting elements, in addition to *trans*-acting factors, are involved in guiding which exons end up in mature RNA.[34] The *cis*-acting elements, made up of RNAs, are known as exonic splicing enhancers (ESEs) and intronic splicing enhances (ISEs). Most relevant to direct targeting of RNA is the observation that enhancer activity can be abolished by a stable stem-loop structure as short as seven base pairs in RNA induced by competition and long-range RNA pairing.[35,36] This is one example of a complex structure modulating splicing, but many more will continue to emerge. This offers an attractive mechanism to target unique structures that modulate the intron borders or structures that interact with exonic or intronic enhancers to regulate splicing.

In a normal state, the average human gene contains approximately eight exons and seven introns with an average of three to four alternatively spliced variants.[37] In pathological conditions, defects in normal RNA splicing caused by mutations are thought to represent ~30% of pathological mutations.[38] Mis-splicing is associated with a broad spectrum of diseases, ranging from muscular dystrophies to neurodegenerative diseases.[39] Alternatively, induction of aberrant splicing can also be used to inhibit the expression of pathological protein or to skip exons containing disease-causing mutations. The spliceosome relies heavily on these RNA codes to instruct the proper splicing pattern. In a disease state, this code can be modified at the RNA level first, potentially through structural arrangements. This offers an attractive target strategy to modulate splicing by targeting structural RNAs near codes that dictate AS patterns particularly in disease states.

9.2.2 Post-transcriptional Regulation: RNA Modification

Chemical modification of RNA (epitranscriptomic modification) has a structural and functional influence on the transcript that can modify its stability, dynamics and binding to proteins and can add another layer of complexity in gene expression.[40] RNA modifications have been associated with complex folding, an aspect that is critical for direct RNA targeting.[41] However, modifications not only increase the intricacy of RNA folding but also have an effect on RNA–protein interactions.[41] Both of these features are desirable for directly targeting RNA since increased diversity in folding and RNA–protein interactions indicate that these modifications make the RNA structurally unique.

Epitranscriptomics profiling has associated patterns of RNA modification with disease states such as, but not limited to, cancer, type 2 diabetes, cardiac malfunction and intellectual impairment.[42] There are hundreds of known modifications of RNA ranging from simple methylation to more complex patterns.[43] The introduction of modifications to RNA can be through (i) RNA-guided mechanisms or (ii) direct conversion by enzymes without RNA guides. Guide RNAs such as Box H/ACA and Box C/D snoRNAs are used to define the sites that are then converted by enzymes.[44,45] These guides will introduce pseudouridine and methylation, respectively. The effects of pseudouridylation and methylation have been studied with various RNAs, but each can bring about different effects on stability.[46] Because pseudouridine, for example, contains an additional available NH group, it has been proposed to bring structural diversity into the overall three-dimensional structure of RNA. These are two well-defined modifications and the most prevalent; however, for more complex RNA modifications, the machinery responsible for RNA is still being deciphered and RNA mapping is far from being complete. The development of better tools to decipher RNA modifications, the guide RNAs, particularly in diseased states, and what are the writers, erasers and readers for these modifications (enzymes that can add and remove modifications) will be critical in the development of chemical probes that will offer new critically needed therapeutic avenues. Modified bases have not yet been used in better anchoring small molecules in the RNA therapeutics toolbox. This is partly due to the lack of structures that include modified RNAs. With the advent of better sequencing, SHAPE analysis and more information on three-dimensional structures with modified RNA, more opportunities to utilize these in the development of small molecules that can use modified RNAs as anchor points will emerge.

9.2.3 Post-transcriptional Regulation: RNA Stability

With the recent upsurge in characterizing long non-coding RNAs (lncRNAs) as a viable drug target, a study completed in 2012 by Clark *et al.* undertook a genome-wide analysis of lncRNA stability.[47] An *in vivo* study showed that the half-lives of lncRNAs could range from very short, <30 min, to as long as >48 h. On comparing the half-lives of lncRNA and mRNA, the median half-life of lncRNAs (3.5 h) was only about 1.6 h shorter than the median protein-coding RNA half-life (5.1 h). It is worth noting that these half-lives are comparable to those of proteins. The half-lives suggest that lncRNAs are a stable RNA subgroup.[47] With the high diversity of half-lives in lncRNA biomolecules, it is necessary to target an lncRNA molecule that has a longer half-life and is resistant to eukaryotic RNA degradation machineries. For example, MALAT1 (metastasis-associated lung adenocarcinoma transcript 1) has a relatively long half-life (ranging from 9 to >12 h) because of its triple helical expression and nuclear retention element (ENE) structure at its 3′ end that provides stability by engaging a downstream A-rich tract.[48,49] In addition to their stability, lncRNAs are structurally diverse and contain elements that form pockets for small-molecule binding. These RNAs are often between 100 and 500 nucleotides long with complex structures. Most of the structural studies have tended to be of RNAs that are 50–100 nucleotides long.

9.2.4 Disease Regulation: Pathological RNA

Tandem repeat sequences, trinucleotide repeats and G-quadruplexes are responsible for a large number of neurological and neuromuscular diseases.[50] The expansion of such repeats leading to polymorphic structures (such as G-quadruplexes), exceeding the physiological threshold, can lead to the accumulation of toxic RNA called RNA foci, sequestration of RNA-binding proteins and/or the expression of toxic peptides. One of the main genes responsible for the familial form of amyotrophic lateral sclerosis (ALS), an important expansion repeat, is c9orf72, $(G_4C_2)^n$.[51] Given that there is currently no effective treatment for ALS, targeting G_4C_2 expansion is a plausible option. Here, strategies could target the inhibition of pathological protein–RNA interactions or modulation of RNA degradation.

9.2.5 Disease Regulation: Viral, Bacterial and Fungal RNA

RNA targeting is a useful tool in targeting viruses. Despite viral genomes being highly variable, there are some sequences that are more closely conserved among viral families. In order for a virus to replicate, it must utilize eukaryotic cellular machinery from a host cell to reproduce its genome and produce viral proteins needed for survival and infection. The co-evolution of viruses and eukaryotic cells has yielded a variety of eukaryotic mechanisms to evade viral infection, in addition to viral mechanisms to overcome or evade defences by the host cell.[52]

Some alphaviruses alter ribosomal scanning, when a subgenomic RNA transcript is translated, *via* eIF2alpha phosphorylation inhibition or *via* a stable RNA stem-loop downstream of the start codon. The RNA stem-loop slows the ribosome sufficiently to allow Met-tRNAi to enter the initiation complex without an initiation factor present in the P site of the ribosome.[52] In a study by Ventoso on viral 26S transcripts, it was determined that the downstream loop (DLP), used for translation without the presence of an initiation factor, was relatively conserved amongst viral particles under the alphaviruses, specifically in Sindbis virus (SINV).[53] Disrupting these stable stem-loops from forming could improve host cellular immunity to viral RNA particles by translating only mRNA transcripts that specifically bind to initiation factors. One of the greatest obstacles in developing successful compounds for viral pathogens is that they utilize the host's cellular machinery to replicate. Hence targeting the mRNA transcripts with these conserved stem-loops could provide a much safer alternative by directly targeting the transcript and not the machinery used to produce viral proteins.

9.2.6 Structural Regulation: Protein–RNA Interactions

Targeting efforts have been predominantly focused on the protein aspect of protein–RNA interactions. It may be more challenging to target RNA when bound to the protein, particularly for those RNAs that are unwound and single stranded as they bind proteins.[54] However, there are two possible ways in which these types of complexes could be disrupted. The RNA portion that will interact with proteins can be targeted *prior* to the interaction, or RNAs that retain structure during the interaction with proteins can be targeted in their complexed form. Our laboratory focuses on RNA–protein interactions that involve unfolded RNA such as RNA–EXOSC3 (from the RNA exosome) and RNA–TDP-43 interactions; we have successfully targeted the protein portion of these complexes.[55,56] Those RNAs would be challenging to target since they are single-stranded RNA when they bind to the protein. However, we have measured the binding of TDP-43 to G-quadruplexes,[56] structures that have been targeted for therapeutic development.[57] There are also additional examples of RNA–protein complexes, such as nucleolin binding to protein and zinc finger domains binding to hairpin loop regions of RNA, that involve structural RNAs that can be targeted and that will likely disrupt the RNA–protein complex.[54] There is an urgent need for further high-resolution structural characterization of unique RNA structures that form complexes with proteins to accelerate drug discovery efforts.

9.2.7 Structural Regulation: Targeting miRNA

miRNAs are a novel type of small RNAs with unique structural elements that are 18–28 nt long with roles in post-transcriptional regulation and protein expression.[58] These RNAs are found in normal and pathological processes. One proposed mechanism of action is through base pairing with mRNA; however, miRNAs are known as multivalent binders since one miRNA will interact with multiple genes and regulate several different proteins. We propose two possible approaches to target miRNA: (i) *prior* to base pairing with mRNA or (ii) *after* base pairing with mRNA where they may form unusual structures. Because miRNAs influence multiple genes, targeting prior to base pairing will perturb several genes at once whereas if targeting is performed after miRNA binds to their cognate mRNA then targeting may be more specific to one gene. This offers new opportunities to combat pathogens, for example, by specifically focusing on miRNAs since these have been shown to be highly expressed in viruses. These types of RNAs have also been shown to be dysregulated in cancer[59] and in tumour-associated responses.[60] Directly targeting miRNAs would offer new therapeutics that could potentially be synergistic with current protein-centric targets.

9.2.8 Structural Regulation: Micropeptides from lncRNAs

There has been growing evidence of short proteins/peptides (~100 amino acids) produced from lncRNAs known as micropeptides (small proteins depending on the cut-off by genomics studies). Whereas proteomics identifies expressed peptides from protein databases, proteogenomics uses genomic or transcript sequences to identify peptides not previously annotated.[61] Recent applications of this approach utilize RNASeq data sets to identify peptides;[62] however, a more comprehensive approach is to create a peptide database using six-frame translations of genomic sequence. The application of proteogenomics identifies peptides expressed ubiquitously across the genome of vertebrates.[63] Initially, these peptides were considered to be an artefact of proteomic discovery; however, more rigorous peptide identification has not eliminated their occurrence. For example, a multi-tissue survey of chicken identified unique proteomic peptides at >100 times the false discovery rate for this data set.[63] Moreover, most of these novel peptides map to regions annotated as intergenic and may represent small open reading frames (ORFs) (<100 amino acids) that standard annotation pipelines discard. The identification of novel expressed peptides in intergenic regions occurred concurrently with the expansion of ncRNA types and, more recently, reports of ncRNAs that encode small peptides.[64] For example, chicken lncRNA–Six1 encodes a 7.26 kDa peptide that is required for *cis*-regulation of Six1.[65] The human orthologue of Six1 is also associated with an upstream lncRNA (SAL-RNA1), which is associated with senescence.[66]

Both proteogenomic peptides and micropeptides associated with ncRNAs are frequently shown to have relatively low expression and be expressed in a tissue-restricted manner, factors which contribute to their poor annotation. Improving our ability to predict and determine the function of both small ORFs and ncRNAs associated with micropeptides will fundamentally improve our understanding of lncRNAs and how small proteins regulate gene expression. While improving our understanding of this new class of protein-producing lncRNAs, this would be an excellent example of how targeting RNA could help on two fronts: these molecules would (i) help in understanding the roles of these RNAs and (ii) target lncRNAs in addition to micropeptide-producing RNAs – a new class of targetable RNA.

9.3 RNA-targeting Molecules

Several classes of molecules able to bind RNA directly were recently analysed and compiled into the RNA-targeted BIoactive ligaNd Database (R-BIND).[67,68] Aminoglycosides were excluded from that study based on their lack of specificity, but we describe them below as they are critical to define how Nature, specifically bacteria, has used small molecules to kill cells effectively through RNA targeting.

9.3.1 Bacterial/Fungal Elements

9.3.1.1 Riboswitches

The discovery of riboswitches in 2002[18–20] was evidence of RNA binding to small molecules in a specific manner and resulting in a structural and functional impact. Riboswitches are *cis*-acting elements able to regulate

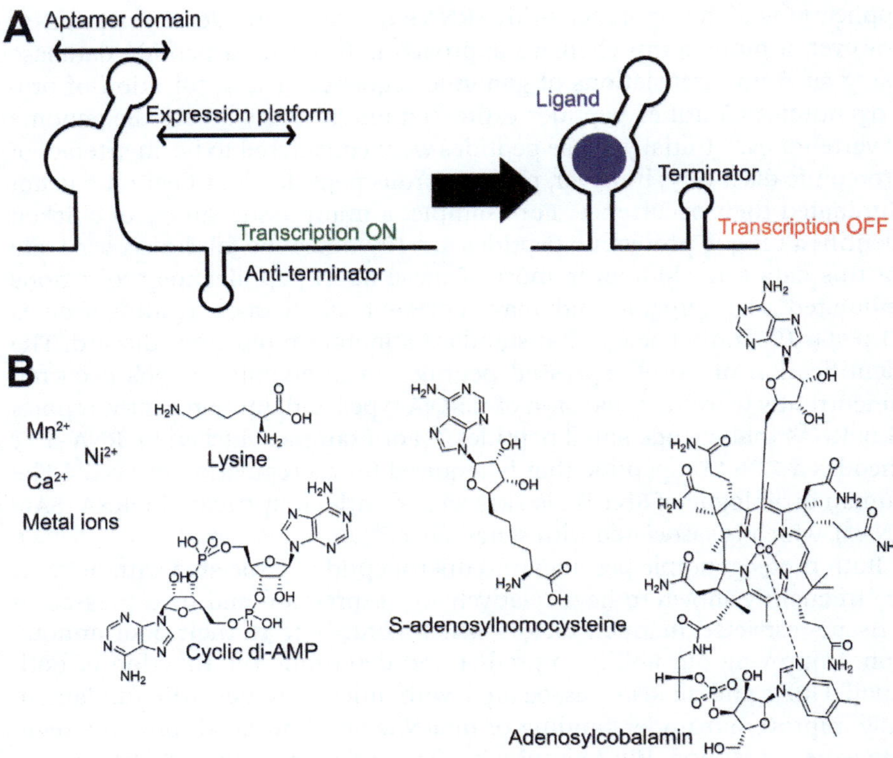

Figure 9.1 Riboswitch mode of action and ligands. (A) Example of riboswitch mechanism by premature transcription termination. (B) Examples of natural ligands for riboswitches.

protein expression *via* a structural change induced by the binding of a ligand (Figure 9.1A). There are more than 35 classes of riboswitches that bind different types of ligands, ranging from metal ions to complex metabolites such as adenosylcobalamin, an active form of vitamin B_{12}[69] (Figure 9.1B).

Riboswitches, found predominantly in bacteria, control the expression of essential genes for metabolite transport or biosynthesis, by modulating RNA at either the transcription or the translation level. In eukaryotes, riboswitches regulate gene expression at the splicing level. This makes riboswitches a prime target for therapeutic intervention. Indeed, several antibacterial and potentially antifungal molecules were later found to target riboswitches.[70] These molecules are usually natural compounds and analogues of riboswitch ligands.

Recently, Merck Research Laboratories of Kenilworth, NJ, USA reported the discovery and characterization of ribocil, a highly selective chemical modulator of bacterial riboflavin riboswitches, which was identified in a phenotypic screen and acts as a structurally distinct synthetic mimic of the natural ligand, to inhibit bacterial cell growth[21] (Figure 9.2).

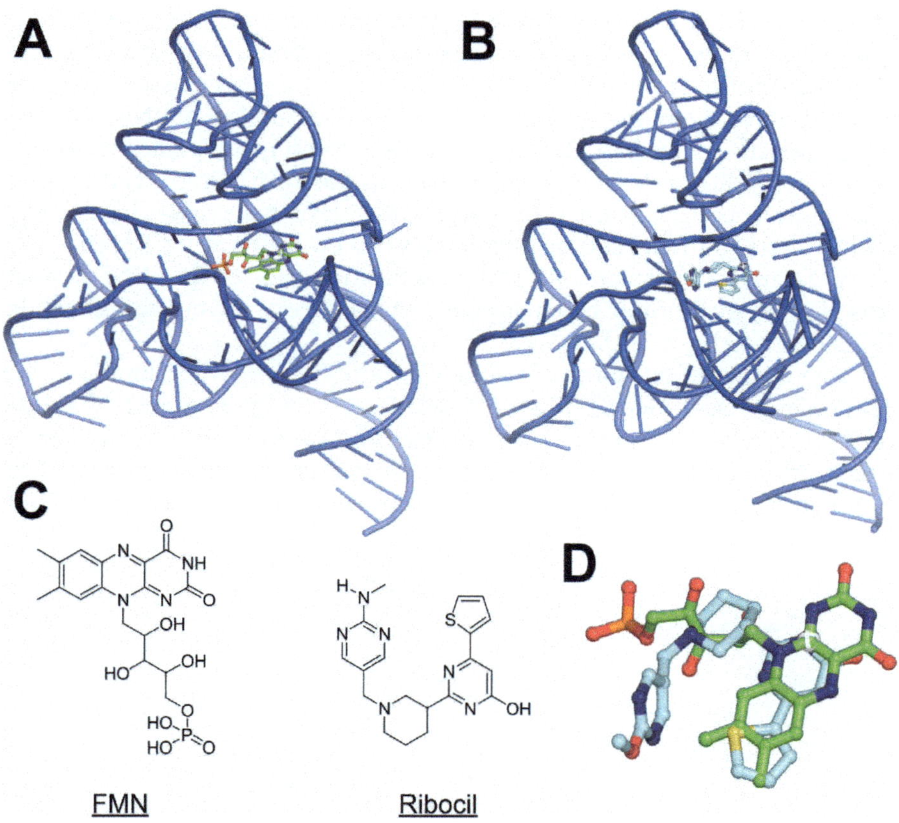

Figure 9.2 Successful targeting of riboswitch. FMN riboswitch (blue, ribbon representation) complexed with a natural ligand, FMN (A) (green, PDB code 2yie[112]) or with antibacterial ribocil (B) (cyan, PDB code 5kx9[113]). (C) Structures of FMN and ribocil. (D) Superposition of ribocil (cyan) and FMN (green) when complexed with FMN riboswitch.

9.3.1.2 Antibiotics Targeting rRNA

Aminoglycoside, macrolide, tetracycline, chloramphenicol, oxazolidinone and lincosamide antibiotics exert their antibacterial effects by binding to functionally critical points on ribosomal RNA (rRNA).

Aminoglycoside antibiotics were originally derived from bacterial species (Figure 9.3). These antibiotics include a glycosidic polycyclic structure the activity of which is modulated through the aminocyclitol ring that binds to the conserved region of prokaryotic 16S ribosomal RNA (Figure 9.4A). These antibiotics interfere with the identification of cognate tRNA, subsequently interrupting the translation of tRNA to mRNA.[71] Initially, antibiotic function is mediated by the positively charged aminoglycoside and the negatively charged phosphates that comprise the RNA backbone, which is part of the reason why they may bind non-specifically to other RNAs. Further binding is dependent on the number of hydrolysable groups extending from the aminosugar rings, giving each type of aminoglycoside antibiotic a slight variation on its complementary structure within the ribosome. The majority of aminoglycosides function by binding to the major groove of the codon–anticodon dsRNA complex. Corresponding conformational changes adversely displace important adenine residues out of their proofreading positions, stabilized by thermodynamic contributions from the aminoglycoside; RNA is misread, which leads to misfolded or improperly folded protein.[72,73] Because of a more extensive structural insight into aminoglycoside binding to RNA, this class of antibiotics has been used in numerous experimental validations (see section 9.4.4).

Other classes of antibiotics include chloramphenicol, lincosamides, macrolides and oxazolidinones that bind to 50S rRNA and tetracyclines that bind to 30S (Figures 9.3 and 9.4). The mechanism of action of these antibiotics works predominantly through inhibition of protein synthesis through RNA binding. All these antibiotics have systematically demonstrated that rRNA is

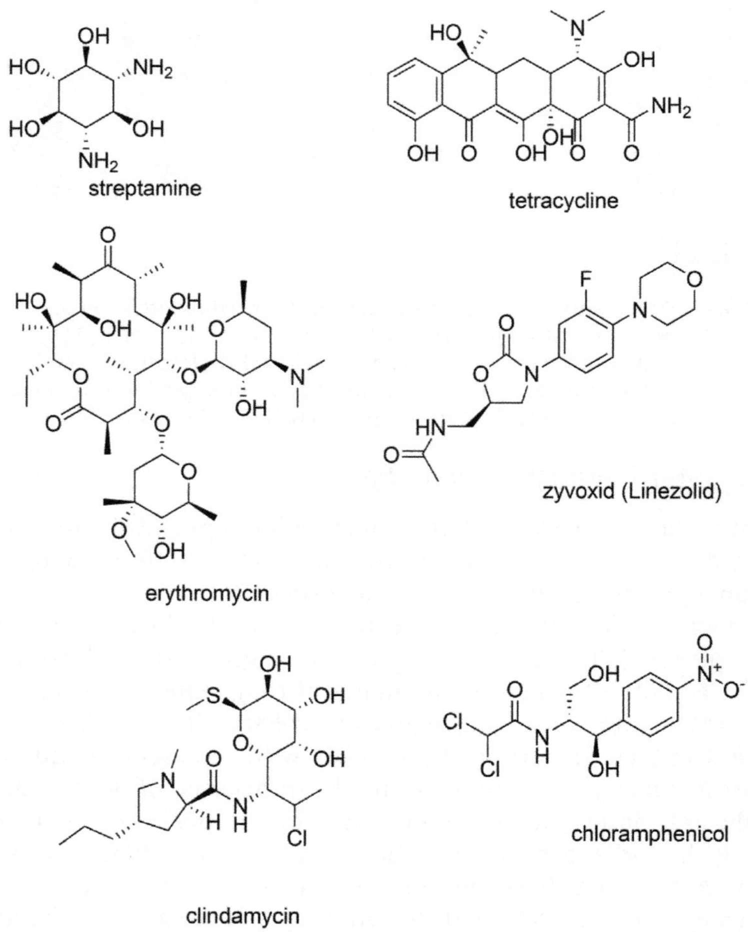

Figure 9.3 Classes of antibiotics targeting RNA.

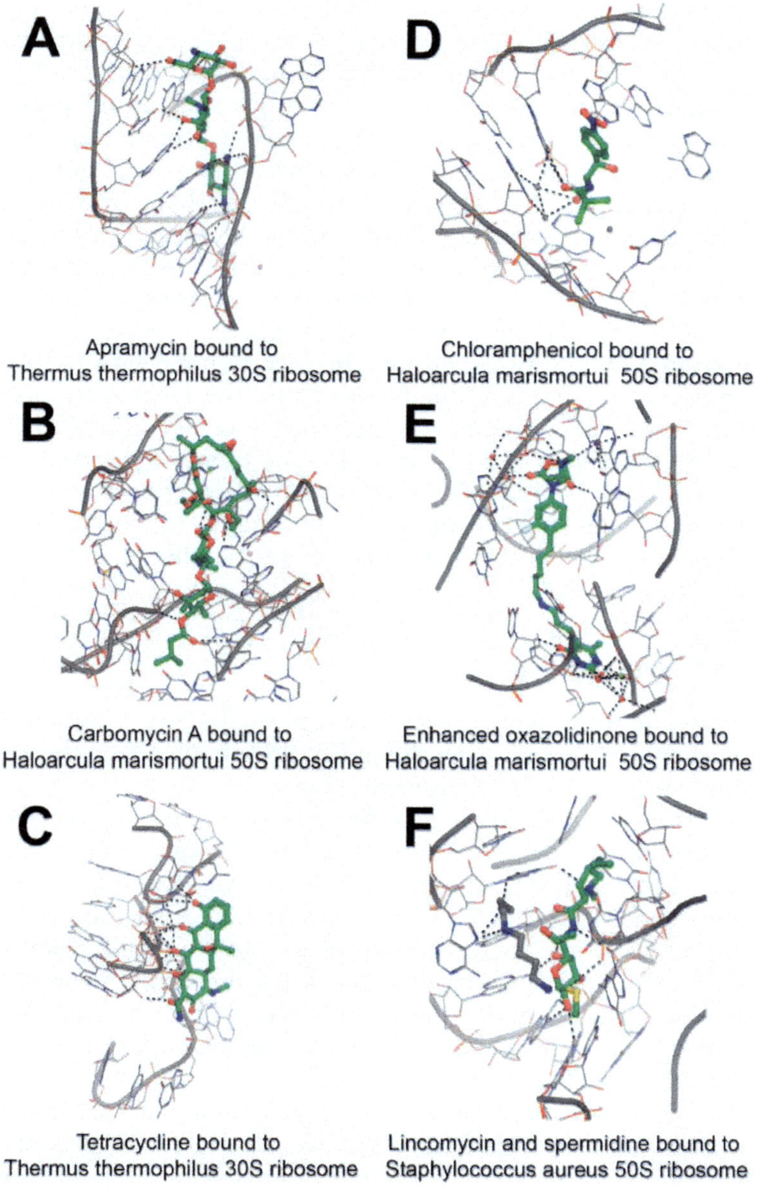

Figure 9.4 Binding of antibiotics to bacterial ribosome. (A) Apramycin (aminoglycoside) (PDB code 4AQY);[114] (B) carbomycin A (macrolide) (PDB code 1K8A);[115] (C) tetracycline (PDB code 1HNW);[116] (D) chloramphenicol (PDB code 1NJI);[117] (E) oxazolidinone (PDB code: 3CXC);[118] (F) lincomycin and spermidine (PDB code 5HKV).[119]

an excellent antibacterial target. However, there are caveats that temper their use as they are mostly non-specific and have multiple off-targets including inhibition of mitochondrial protein synthesis[74] or targeting eukaryotic mitochondrial ribosomes causing toxicity symptoms in patients with prolonged administration.[75]

Bacterial resistance can rapidly develop and off-site targeting often creates more concerns that outweigh benefits. The mechanisms of resistance are diverse, ranging from resistance mutations to increased cellular efflux.[76] Since the majority of these antibiotics are natural products, a medicinal chemistry approach was often used to create semi-synthetic structural analogues to modulate potency. With the advent of higher resolution co-structures characterized through cryo-EM, X-ray diffraction or nuclear magnetic resonance (NMR) methods, mechanisms of action at the structural level were better understood,[77,78] including mutations causing resistance. Research was expanded by revealing the specific residues and interactions known to occur at the peptidyl transferase centre.[79] Computational modelling could now be incorporated into the determination of antibiotic binding modes. Based on the mechanism of action and information on binding modes, antibiotics were modified to increase potency, although it appears that combination therapy of antibiotics with other compounds has been favoured over modifying antibiotics. Examples include the latest FDA-approved combination of pretomanid with bedaquiline and linezolid (targets 50S rRNA) for highly resistant tuberculosis (TB) of the lungs.[80]

9.3.2 Viral Targeting

Historically, the most common RNA target was the HIV-1 *trans*-activation response element (TAR) RNA, a well-studied RNA that binds to the viral protein Tat, a stimulator of viral gene expression.

In 1998, Hung *et al.* set out to target HIV-1 by disrupting a necessary translational frameshift to achieve viral replication.[81] Through high-throughput screening, a frameshifting stimulator (DB213) was found to bind HIV-1 frameshift signal (HIV-1 FS) RNA and decrease viral replication in cultured cells. Further NMR structural studies revealed the binding site of DB213 to HIV-1 FS, its binding affinity and the effect that the small molecule has on RNA stability. It was found that the small molecule bound to HIV-1 FS in the major groove located at the centre of the RNA duplex, with a moderate binding affinity.[82]

In 2002, Lind *et al.* targeted the 5′ bulged stem-loop of TAR RNA, an element essential for binding to Tat, using an *in silico* screen. The most favourable interactions from the docking experiments were then tested *in vitro* using gel shift assays, cellular assays and NMR spectroscopy. The results showed that their lead compounds interacted specifically with the 5′ bulge of TAR RNA.[83]

In 2007, Miller and co-workers used dynamic combinatorial chemistry to screen a compound library against HIV-1 frameshift regulatory stem-loop RNA. In order to develop their library for screening, they used resin-bound dynamic combinatorial chemistry that utilizes immobilized ligand candidates and a fluorescent target to identify possible binders. From their library of 11 325 unique agents, they identified a hit compound that was selective for the regulatory RNA stem-loop sequence with micromolar binding affinity.[84]

In a small-molecule microarray screening study, Schneekloth's group identified three thienopyridine derivatives that interact with HIV-1 TAR RNA hairpin using a melting temperature assay, differential scanning fluorimetry and a fluorescence titration curve. In a cellular assay using T-lymphocytes, it was shown that viral activity was decreased in cells that were treated with one of the derivatives.[85] Further characterization studies were performed on another derivative to identify its mechanistic mode of action by developing analogues and testing their affinities to HIV-1 TAR RNA hairpin. The lead analogue was found to bind to the 5′-UTR of HIV and displaced a Tat-derived peptide.[86]

Hargrove and co-workers developed a small library of derivatives of amiloride, an RNA-binding scaffold, with changes at the C5 and C6 positions to improve the binding affinity of amiloride to the loop and bulge of the HIV-1 TAR RNA. Using a pipeline including Tat peptide displacement assay, HMQC NMR spectroscopy and molecular docking, they found a hit compound whose inhibition activity was increased by more than 100-fold compared with the parent amiloride. This compound is stated to be one of the tightest non-aminoglycoside TAR ligands reported to date.[87]

9.3.3 Compounds Targeting Specific RNA Structures

9.3.3.1 Triple Helix of MALAT1

A recent study by Abulwerdi *et al.* revealed a selective compound for the unique stabilizing triple-helical structure of the lncRNA metastasis-associated lung adenocarcinoma transcript 1 (MALAT1), also known as an element for nuclear expression (ENE).[88] Because of its upregulation in cancer, it was hypothesized that targeting the stability element of the lncRNA could reduce the half-life of MALAT1 and influence its metastatic properties. A high-throughput screen using a high-throughput small-molecule microarray library and a follow-up screen with RT-qPCR (real-time quantitative polymerase chain reaction) to determine cellular levels of MALAT1 yielded two lead compounds (**5** and **16**, Figure 9.5). They applied molecular docking using AutoDock 4 to determine the binding location on the crystallized MALAT1–ENE triple helix (PDB code 4PLX)[48] using a flexible ligand and rigid receptor system, with each compound producing a different mechanism of binding to the triple helix. According to clusters formed from the molecular docking, **5** showed preferential binding to the major groove and **16** showed a preference for the surface of the minor groove. A dye displacement assay showed that the mode of action of **5** and **16** was not through RNA intercalation, further proving the feasibility of targeting specifically RNA triple helices.[88]

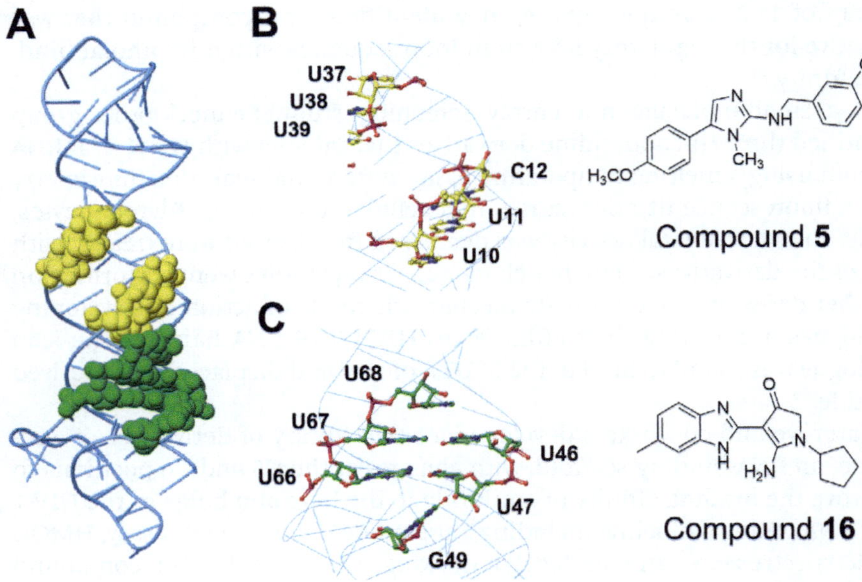

Figure 9.5 Targeting MALAT1 triple helix. (A) Cartoon tube diagram of crystallized MALAT1–ENE (PDB code 4PLX)[48] showing the top predicted binding sites through clustering of compounds **5** (shown in yellow) and **16** (shown in green) by molecular docking with AutoDock 4 software. Compound **5** binds in a major groove of the triple helix and **16** binds to the minor groove. (B) Compound **5** base interactions. (C) Compound **16** base interactions.[88]

9.3.3.2 *Three-way Junctions (3WJs)*

Three-way junctions are intramolecular constructs that occur in both DNA and RNA structures (Figure 9.6). Commonly found in ribozymes and riboswitches, and also rRNA ribonucleoprotein, these junctions serve as a unique structural motif that has the potential for small-molecule targeting.

The threefold symmetry of triptycene and its dimensions, which fit the central interface of a three-way junction, inspired the study of triptycene derivatives. Using UV thermal melting, circular dichroism and gel shift assays, Barros and Chenoweth demonstrated the ability of triptycene derivatives (Trip **1–3**, Figure 9.6) to stabilize DNA and also RNA 3WJs without impacting dsDNA, showing that triptycene compounds are selective for 3WJ structures and do not possess non-specific intercalative properties.[89] They further demonstrated the ability of Trip **1–3** to have anticancer properties, by having a high cytotoxic potency, similar to or higher than that of cisplatin.

9.3.3.3 *RNA Pseudoknots*

A study by Park *et al.* showed that pseudoknots could be another promising RNA motif for targeting using small molecules.[90] They targeted an mRNA pseudoknot that induces an essential programmed −1 ribosomal

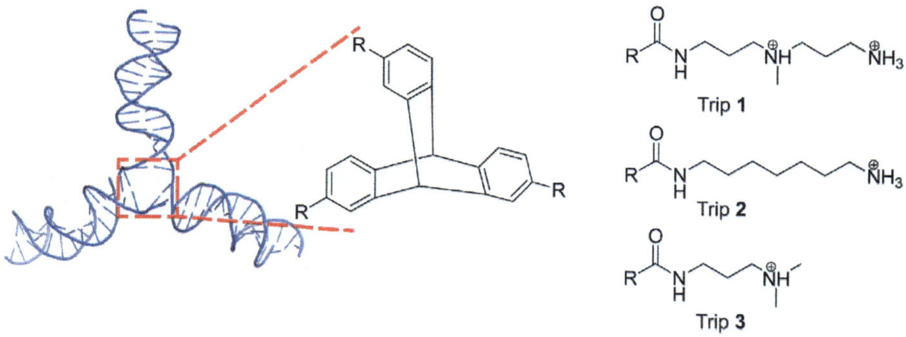

Figure 9.6 Targeting of three-way junction in DNA and RNA. Structure of an RNA three-way junction (PDB code 1F44[120]) along with the structure of triptycene derivatives (Trip **1–3**) utilized for targeting nucleic acid junctions. The red square is where triptycene is thought to bind.[89]

frameshift for the production of viral proteins in the SARS coronavirus (SARS-CoV). It was hypothesized that interfering with the frameshifting abilities of this viral transcript could decrease the efficiency at which the virus replicated.

Because the 3D structure of the SARS pseudoknot was not known, the structure was first predicted using PSEUDOVIEWER and further optimized using molecular dynamics with the Amber 8.0 program. A docking screen using this predicted 3D structure was completed using the Dock4.0 program with a grid centred around an essential residue, U21, implicated in maintaining frameshifting abilities. The screening library was compiled from about 80 000 compounds from LeadQuest (Tripos Inc.) and the top 58 compounds were chosen for further studies based on their docking scores and visual inspection. These compounds were then tested using an *in vitro* frameshifting assay with fluorescent reporter genes, an *in vitro* transcription and translation assay and cell-based −1 frameshifting assay.

9.3.3.4 RNA Repeat Elements

The most targeted repeat expansion RNA is probably the poly(CUG) RNA repeat element associated with myotonic dystrophy type 1 (DM1), a rare neuromuscular disease. Accumulation of toxic mRNA, containing a very large number of poly(CUG) repeats, leads to the formation of nuclear foci and sequestering of specific splicing regulators protein partners such as the MBNL (muscleblind) family and subsequent splicing defects.

In 2009, Zimmerman's group developed a compound through rational design that targets three different pathological entities of DM1.[91] Using a multi-target approach, a compound was found to (i) inhibit transcription of expanded CTG to expanded CUG, (ii) inhibit sequestering of MBNL1 by expanded CUG and (iii) yield RNase-like functionality to degrade expanded CUG transcripts. Using rational design, they developed this compound by starting with previously described ligands that selectively bind expanded CUG and inhibit sequestering of MBNL with a low K_I[91] through triaminotriazine moieties that bind to UU and TT mismatches in a Janus wedge-type binding mode, creating a base triplet with the mismatch. The use of an acridine and a bisamidinium group yielded a favourable interaction by A-form RNA intercalation and groove binding. Lastly, it had been shown that the more favourable active-site functional groups for RNase A small-molecule mimics were those that contained at least one ammonium ion and an imidazole or amino group. Thus, using the above information, Zimmerman and co-workers were able to design compounds that included the described moieties and screened using *in vitro* and *in vivo* assays.

9.3.4 Compounds Designed to Target Specific RNA Function

9.3.4.1 Splicing-targeting Compound: Branaplam

A truly exciting small molecule that has been developed to modulate splicing is LMI070/NVS-SM1 (branaplam) from Novartis[22] (Figure 9.7A). This molecule was developed to rescue mis-splicing of survival of motor neuron 2 (SMN2) that leads to spinal muscular atrophy (SMA), a motor neuron disease. In SMA, exon 7 of SMN2 is excluded during splicing, which induces the production of truncated and unstable SMN2 protein.

The authors used a phenotypic high-throughput screen to search for enhancers of exon 7 inclusion during SMN2 splicing and found several hits. One of them, NVS-SM1, was able to increase survival of an SMA mouse model by increasing the full-length protein expression. The compound displayed long-lasting effects and was orally active. This

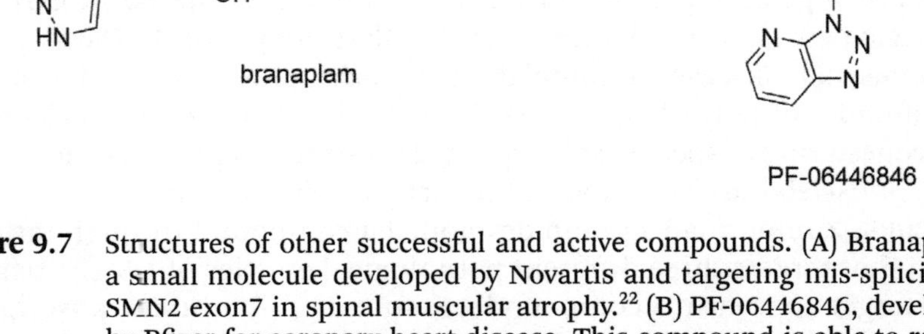

Figure 9.7 Structures of other successful and active compounds. (A) Branaplam, a small molecule developed by Novartis and targeting mis-splicing of SMN2 exon7 in spinal muscular atrophy.[22] (B) PF-06446846, developed by Pfizer for coronary heart disease. This compound is able to reduce the expression of PCSK9.[92]

sequence-selective small molecule was further shown to have a double mechanism of action: (i) stabilization of U1 snRNP–5′SS RNA complex, a key component of the spliceosome, and (ii) increase of the affinity of U1 snRNP to the splice site. Interestingly, NVS-SM1 affected the splicing of 35 off-target transcripts only and was considered as fairly selective. Branaplam is currently being assessed for safety and efficacy in a phase I/II clinical study.

9.3.4.2 Ribosomal Synthesis-targeting Compound: PF-06446846

Recently, Pfizer described a small molecule, called PF-06446846 (Figure 9.7B), that is able to reduce the expression of proprotein convertase subtilisin/kexin type 9 (PCSK9), a protein regulating the levels of plasma low-density lipoprotein cholesterol (LDL-C). Reducing the level of PCSK9 is likely to reduce the risk of coronary heart disease. PF-06446846 derives from a top hit of a high-throughput phenotypic screen, performed on CHO cells and looking for compounds able to inhibit accumulation of PCSK9. This compound was shown to selectively block PCSK9 translation *in vivo* by binding to 80S ribosome, leading to ribosomal stalling during elongation.[23,92]

9.4 Existing Strategies/Techniques to Target RNA

Although targeting RNA using antisense oligos has proved successful, this approach relies on primary sequence recognition by modified oligos, which is a disadvantage in terms of cell permeability and distribution. Even though those problems can be compensated for, small-molecule targeting is an alternative strategy owing to their more desirable properties, such as good penetration in intracellular regions, cytosol and nuclei and blood-brain barrier penetration. Moreover, compounds provide the possibility of targeting three-dimensional structural elements on RNA instead of relying on primary sequence.

Nevertheless, despite the excitement surrounding RNA, there are challenges to overcome when directly targeting RNA with small molecules. Specificity is perhaps the main concern and the greatest challenge, and small molecules targeting RNA should be tested for intercalating properties in addition to sequence and electrostatic complementarity. An additional disadvantage is that RNA folding is dynamic and flexible, which can lead to poorly structured and/or transient binding pockets.

For the most part, RNA-targeting small molecules for therapeutic development were found serendipitously, *via* phenotypic screens. Many are natural products, often non-specific, exhibiting toxicity and difficult to administer; charged compounds often do not cross cellular membranes. However, methods to target RNA specifically have been developed over the past few years and are being increasingly utilised.

9.4.1 Small-molecule Microarray (SMM)

The small-molecule microarray (SMM) technique, recently adapted for RNA structures, was derived from large-scale microarray-based screens developed to identify ligands that bind to undruggable protein targets. Originally, RNA-targeting libraries were focused on aminoglycosides and peptides. More recently, oligonucleotide-based libraries have been developed, such as the RIPtide microarray.[93,94] These approaches demonstrated the ability to find tight and specific binders for defined structures of RNA.

In order to find new RNA binders, Sztuba-Solinska *et al.* constructed an SMM using compounds with diverse nucleophilic functionalities and that comply with Lipinsky's rules.[85] This library did not include any previously known RNA or DNA binders. The array was then incubated with a fluorescently labelled nucleic acid, imaged and positive hits were determined based on the level of fluorescence of the spot.

9.4.2 Chemical Cross-linking Isolated by Pull-down: Chem-CLIP™

Chem-CLIP™ is a covalent small-molecule RNA cellular profiling approach[95,96] and takes its inspiration from the more traditional RNA-CLIP (RNA-cross-linking immunoprecipitation), where UV cross-linking and subsequent immunoprecipitation of the target followed ultimately by RNA-seq allows the mapping of specific protein–RNA interactions. Like RNA-CLIP, Chem-CLIP™ relies on the ability of a small molecule, biotinylated and linked to a cross-linking group, chlorambucil, to bind to RNA (Figure 9.8). A pull-down using streptavidin resin leads to isolation of the compound-containing complex(es) and the RNA can be further analysed by RT-qPCR or RNA-seq. This technique permits the detection in cells of on- and off-target

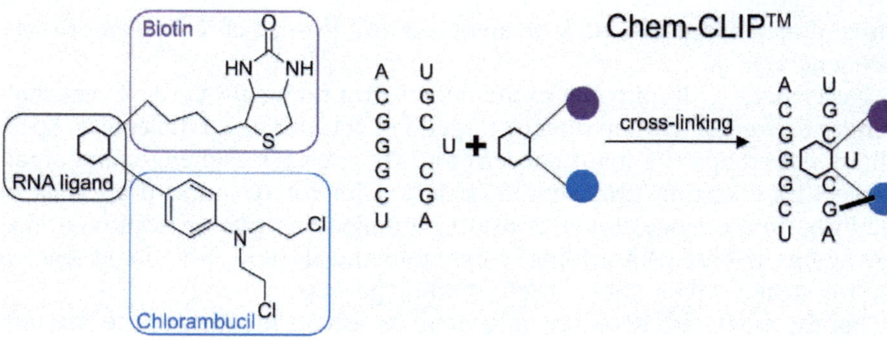

Figure 9.8 Concept of the Chem-CLIP™ technique. A known RNA ligand is linked to a biotin moiety for purification and a chlorambucil group for cross-linking properties. The compound is then incubated in cells, cross-linked and pulled down along with RNA binders. The pull-downed transcripts are then sequenced.

RNA binding to the small molecule of interest. There are different versions of Chem-CLIP™ such as C-Chem-CLIP™ and Chem-CLEAVE™ for target depletion and Chem-CLICK™ for target potency.[97–99]

9.4.3 Informa: Two-dimensional Combinatorial Screening (2DCS) Combined With Structure–Activity Relationships Through Sequencing (StARTS)

Developed by Disney's group, two-dimensional combinatorial screening (2DCS) is a library *versus* library type of screen, used to find small molecules that bind to specific types of RNA structures. A small-molecule microarray is generated by spotting a library on a functionalized agarose-coated microscope slide. The array is then incubated with a labelled library of RNA motifs under stringent conditions. The RNA motifs are small in length and include hairpins, internal loops and bulged loops.

The array is then imaged and the spots exhibiting a positive signal are excised mechanically from the array. The positive hits are then amplified by RT-PCR and the PCR products are ligated into a pGEM-T vector for transformation into DH5 alpha bacteria. After an X-gal screen, white colonies are amplified and sent for sequencing.[100]

In order to analyse data from 2DCS and define the RNA features that interact with a specific ligand, Disney and co-workers designed a computational method, called Structure–Activity Relationships Through Sequencing (StARTS).[101,102] For each small molecule, the corresponding RNA binders are analysed using the software RNA Privileged Space Predictor (RNA-PSP), also developed by Disney's group. The program compares the positive binding sequences for a specific structure (for example, a six-nucleotide loop) with the library containing all possible motifs for that RNA structure (4096 possible motifs for a six-nucleotide loop) and statistical *Z*-scores are calculated for each “trend” [a trend is considered as significant feature(s)], which are then ranked based on their *Z*-scores. For example, a kanamycin A derivative's most representative RNA feature was found to be a 5′UNNNC3′ loop (Figure 9.9).

A

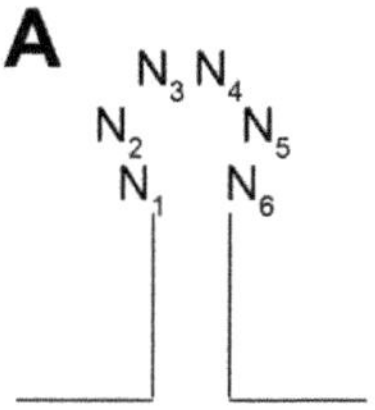

B

Ligand	Motif (5′-NNNNNN-3′)
kanamycin A derivative	5′UNNNC3′
tobramycin derivative	5′CNNA3′ and 5′ANNC3′
neamine derivative	5′RNNNNA3′
neomycin B derivative	5′UNNG3′

Figure 9.9 Two-dimensional combinatorial screening (2DCS). (A) Example of hairpin loop library used in 2DCS.[101] (B) Example of RNA motifs binding to aminoglycosides, found by 2DCS and analysed using StARTS. Data extracted from ref. 101.

Z-scores and the actual affinity of the ligand for selected RNA structures were correlated, based on the assumption that specific RNA features are driving specific ligand recognition. The sum of the *Z*-scores (ΣZ) of each feature displayed by a loop was plotted against the measured affinity for that loop with a good correlation (0.85) when fitted to an inverse first-order equation. This implied that (i) ΣZ is a good representation of RNA affinity and (ii) RNA selectivity for ligands can be predicted based on ΣZ values. A "fitness score" of a specific RNA was then created, by normalizing ΣZ of each ligand to the highest ΣZ value.

Inforna is a computational approach able to generate targetable motifs in a given RNA by comparing motifs found in the RNA with the Disney database.[103] The software then searches for corresponding small molecules for each targetable motif, based on their "fitness score" determined by StARTS. This approach was applied to human precursor miRNAs to try to identify leads binding to an miRNA and inhibit its maturation.[104] All known hairpin miRNA sequences were structurally modelled by energy minimization and loaded into Inforna. The hits were then further screened. The position of the targetable motif was restricted to a processing site in order to inhibit miRNA maturation and the chosen miRNA target had to be previously established as disease causing. Based on the fitness score, this method identified three compounds, which were targeting miR-96, miR-210 and miR-182. These three compounds were tested *in vivo* for their ability to reduce the expression level of their target miRNA and also other miRNAs. Only one compound was able to reduce miR-96 efficiently without affecting the other miRNAs and was shown to be more efficient than an oligonucleotide approach. This study showed a successful alternative approach to traditional high-throughput screens and computational methods in the design of small molecules binding to specific RNA.

9.4.4 *In Silico* Screening of RNA Structures

With the increasing number of three-dimensional RNA structures available, *in silico* docking offers an exciting opportunity to find new scaffolds that bind RNA. Predictions and ranking of the preferred orientation of each small molecule are measured for a target RNA molecule. RNA–ligand docking analyses have been sparsely used for structure-based drug design and virtual screening and are divided into two main approaches.

Existing methods and scoring functions, traditionally used for protein–ligand docking, were adapted to RNA targeting. One of the first examples of *in silico* docking on RNA was reported by Kuntz and co-workers.[105] Using DOCK3.5, they targeted the major groove of the RNA double helix, identified several aminoglycosides, including lividomycin, as the top hits and validated lividomycin binding in the targeted major groove of the RNA.

A few years later, an HIV–RNA loop was computationally screened using a four-step docking procedure: rigid docking first, followed by three steps of flexible docking using a pseudo-Brownian Monte Carlo minimization in torsion angle space with progressively more detailed conformational sampling on a progressively smaller list of top-ranking compounds.[106]

Moitessier *et al.* used a modified algorithm in AutoDock to model successfully the binding modes of aminoglycosides in the 16S rRNA major groove.[107] They were the first to account for the role of water molecules in the binding of compounds to RNA and this method considered the flexibility of both the ligand and the RNA. New algorithms and scoring functions were developed. Ribodock is one of the first examples of *de novo* approaches for targeting RNA *in silico*.[108] The method disfavoured non-complementary polar contacts and controlled non-specific charged interactions, which was successful in predicting RNA interactions. However, the method was tested on only 10 RNA structures. DrugScore^{RNA} is the first knowledge-based scoring function to predict RNA–ligand interactions accurately.[109] The authors used 670 RNA crystal structures as a training set to develop a function adapted to RNA and tested it on 31 RNA–ligand complexes. DrugScore^{RNA} succeeded in predicting good docking solutions on the first scoring rank in 42% of the cases. More recently, LigandRNA was developed, a knowledge-based scoring function for the prediction of RNA–small-molecule interactions that uses a grid-based algorithm. The authors used it in combination with the docking software Dock6 and found this method to be comparable to other docking-based techniques in the identification of near-native ligand poses.[110]

Overall, computational docking of compounds to RNA is more challenging compared with protein-based *in silico* methods since RNA is intrinsically more flexible and more highly charged than proteins. Moreover, most of the available structures of RNA in the PDB or RNADB are riboswitches, ribosomal RNA or aptamers, which do not constitute sufficient structural diversity for docking.

Although several laboratories have focused on improving docking methods such as inclusion of flexibility, inclusion of water and ligand rankings, there has been less emphasis on defining new unique regions within RNA and novel compounds. We believe that by using traditional methods created for proteins to define known binding sites of a particular target, pocket detection algorithms can be used to identify the most likely sites of RNAs for targeting. A previous computational analysis of binding pockets on RNA was performed by Warner *et al.* using the algorithm PocketFinder.[111] However, to determine if well-established tools used for proteins could be used for RNA, we screened the same RNAs, along with RNAs of other families, using Schrödinger's SiteMap tool (a software predominantly used for proteins) and found similar pockets in these RNA molecules (Figure 9.10). Our analysis shows that these RNA molecules have druggable pockets, with RNA structures of higher order producing up to five different binding sites. These pockets could be used to create a docking site for molecular docking. There are already established tools that can be immediately utilized to target RNA and begin, similarly to

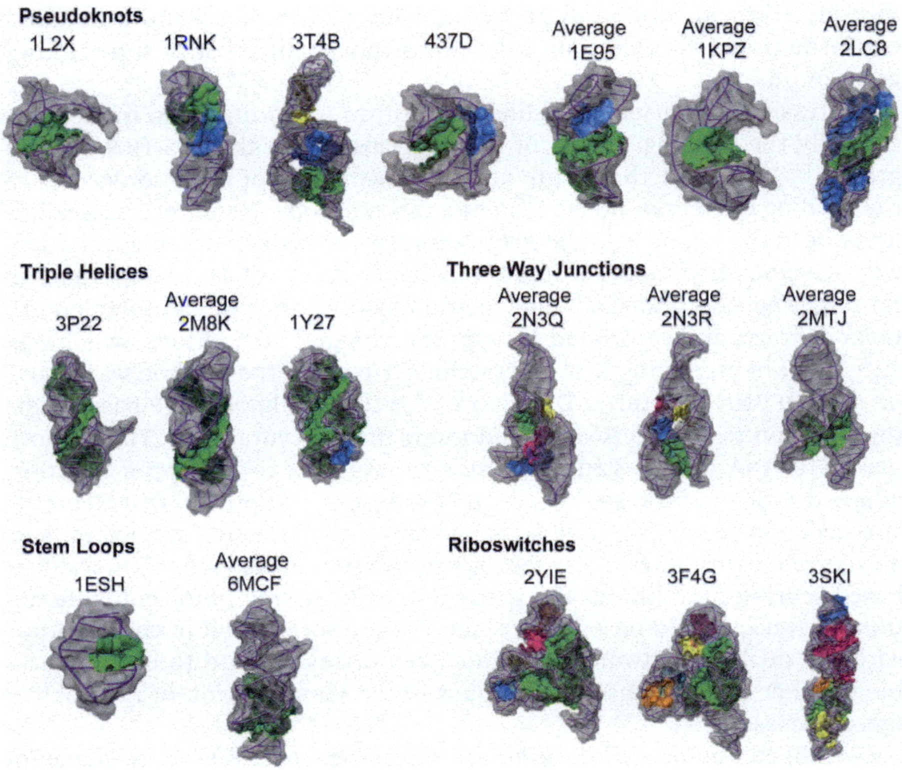

Figure 9.10 Mapping of possible pockets in different structures of RNA. Each RNA structure analysed was prepared using a traditional *in silico* workflow for protein molecular modelling using Schrödinger's Glide Suite and SiteMap tool. After downloading the structures from the PDB, they were processed using Glide's Protein Preparation Wizard and the structures were screened for possible binding sites. If the RNA structures were solved by NMR spectroscopy, an average structure was constructed from the top structures. The possible pockets were ranked by the scoring function SiteScore and coloured based on their respective rank: green, blue, yellow, pink and orange. The corresponding PDB code is given above each structure.

proteins, with weak binders that can then be improved with medicinal chemistry. It is unclear if the small-molecule library will need to be RNA focused on moving forward.

9.5 Concluding Remarks

RNA is a valid target and its use in therapeutics has been expanding rapidly, as exemplified by the increasing number of RNA-targeting molecules in clinical trials. The most unique and attractive idea of targeting RNA directly is that modulation occurs upstream of translation, hence each targeted mRNA could impact the production of many proteins. It could also help in fine tuning specific pathways implicating non-coding RNA.

The greatest challenge in targeting is to find unique and functional pockets on the RNA. Similarly to protein targets, there is a need for biological data on the RNA and the site being targeted. A key question is whether the small molecule will affect the structure/function of the RNA sufficiently to have a biological effect. Fortunately, in the case of RNA targeting, the level of transcripts being hit by the small molecule does not need to be as high as for proteins, but the specificity, similarly to proteins, needs to be high. Efforts aimed at RNA structure elucidation have identified complex pockets (riboswitches) but, compared with proteins, the variety of the building molecules is smaller (four bases *versus* 20 amino acids). In an elegant review by Weeks and co-workers, it was proposed that different RNA elements could account for bits.[111] For example, a simple base pair would be one bit, a bulge two bits and loops six to nine bits. They proposed that in order to achieve specificity, one would require at least 30 bits and more. It is immediately clear that complex folding would be necessary to achieve such levels for recognition and specificity. This is achieved by antibiotics, despite the level of non-specificity due to charged interactions. This will be the next level of challenge in developing better molecules that rely less heavily on charge interactions. There is a need for new and unique scaffolds for RNA targeting since protein-binding molecules have been used on RNA with very limited success and may have been one of the reasons why RNA was not considered for targeting.

There is unfortunately a lack of diversity in high-resolution three-dimensional RNA structures, which limits the use of computational methods. However, this hurdle is currently being overcome by medium-to high-throughput screening. However, these screening platforms have limitations. Inforna, for example, developed by Disney's group, is at present based solely on known data on RNA motif–ligand interactions, which potentially limits screening platforms to known motif–ligand combinations. Further development of *in silico* screening, higher resolution structures, better identification of modifications and a greater understanding of non-coding RNA will truly push the limits of RNA targeting and will expand the utility of this exciting new therapeutic strategy.

References

1. J. D. Robertus, J. E. Ladner, J. T. Finch, D. Rhodes, R. S. Brown, B. F. C. Clark and A. Klug, Structure of yeast phenylalanine TRNA at 3 Å resolution, *Nature*, 1974, **250**(5467), 546–551.
2. S. H. Kim, G. J. Quigley, F. L. Suddath, A. Mcpherson, D. Sneden, J. J. Kim, J. Weinzierl and A. Rich, Three-dimensional structure of yeast phenylalanine transfer RNA: folding of the polynucleotide chain, *Science*, 1973, 285–288.
3. S. H. Kim, F. L. Suddath, G. J. Quigley, A. McPherson, J. L. Sussman, A. H. J. Wang, N. C. Seeman and A. Rich, Three-dimensional tertiary structure of yeast phenylalanine transfer RNA, *Science*, 1974, 435–440.
4. G. E. A. Palade, Small particulate component of the cytoplasm, *J. Biophys. Biochem. Cytol.*, 1955, 59–68.

5. M. B. Hoagland, M. L. Stephenson, J. F. Scott, L. I. Hecht and P. C. Zamecnik, A soluble ribonucleic acid intermediate in protein synthesis, *J. Biol. Chem.*, 1958, 241–257.
6. G. W. Zieve, Two groups of small stable RNAs, *Cell*, 1981, 296–297.
7. H. Busch, R. Reddy, L. Rothblum and Y. C. Choi, SnRNAs, SnRNPs and RNA processing, *Annu. Rev. Biochem.*, 1982, 617–654.
8. M. R. Lerner and J. A. Steitz, Snurps and scyrps, *Cell*, 1981, 298–300.
9. C. Guerrier-Takada, K. Gardiner, T. Marsh, N. Pace and S. Altman, The RNA moiety of ribonuclease p is the catalytic subunit of the enzyme, *Cell*, 1983, 849–857.
10. K. Kruger, P. J. Grabowski, A. J. Zaug, J. Sands, D. E. Gottschling and T. R. Cech, Self-splicing RNA: autoexcision and autocyclization of the ribosomal RNA intervening sequence of tetrahymena, *Cell*, 1982, 147–157.
11. J. Sana, P. Faltejskova, M. Svoboda and O. Slaby, Novel classes of non-coding RNAs and cancer, *J. Transl. Med.*, 2012, 103.
12. O. C. Uhlenbeck, A. Pardi and J. Feigon, RNA structure comes of age, *Cell*, 1997, 833–840.
13. T. R. Sosnick and T. Pan, RNA folding: models and perspectives, *Curr. Opin. Struct. Biol.*, 2003, 309–316.
14. S. A. Woodson, Metal ions and RNA folding: a highly charged topic with a dynamic future, *Curr. Opin. Struct. Biol.*, 2005, 104–109.
15. L. Grillone and R. Lieberman, Fomivirsen, in *Kucers the Use of Antibiotics: A Clinical Review of Antibacterial, Antifungal, Antiparasitic, and Antiviral Drugs*, 7th edn, 2017.
16. M. D. de Smet, C. Meenken and G. J. Van Den Horn, Fomivirsen – a phosphorothioate oligonucleotide for the treatment of CMV retinitis, *Ocul. Immunol. Inflammation*, 1999, 189–198.
17. B. Roehr, Fomivirsen approved for CMV retinitis, *J. Int. Assoc. Provid. AIDS Care*, 1998, 14–16.
18. W. Winkler, A. Nahvi and R. R. Breaker, Thiamine derivatives bind messenger RNAs directly to regulate bacterial gene expression, *Nature*, 2002, 952–956.
19. A. S. Mironov, I. Gusarov, R. Rafikov, L. E. Lopez, K. Shatalin, R. A. Kreneva, D. A. Perumov and E. Nudler, Sensing small molecules by nascent RNA: A mechanism to control transcription in bacteria, *Cell*, 2002, 747–756.
20. A. Nahvi, N. Sudarsan, M. S. Ebert, X. Zou, K. L. Brown and R. R. Breaker, Genetic control by a metabolite binding MRNA, *Chem. Biol.*, 2002, 1043–1049.
21. J. A. Howe, H. Wang, T. O. Fischmann, C. J. Balibar, L. Xiao, A. M. Galgoci, J. C. Malinverni, T. Mayhood, A. Villafania and A. Nahvi, *et al.*, Selective small-molecule inhibition of an RNA structural element, *Nature*, 2015, 672–677.

22. J. Palacino, S. E. Swalley, C. Song, A. K. Cheung, L. Shu, X. Zhang, M. Van Hoosear, Y. Shin, D. N. Chin and C. G. Keller, *et al.*, SMN2 splice modulators enhance U1-Pre-MRNA association and rescue SMA mice, *Nat. Chem. Biol.*, 2015, 511–517.
23. D. N. Petersen, J. Hawkins, W. Ruangsiriluk, K. A. Stevens, B. A. Maguire, T. N. O'Connell, B. N. Rocke, M. Boehm, R. B. Ruggeri and T. Rolph, *et al.*, A small-molecule anti-secretagogue of PCSK9 targets the 80S ribosome to inhibit PCSK9 protein translation, *Cell Chem. Biol.*, 2016, 1362–1371.
24. M. J. Hangauer, I. W. Vaughn and M. T. McManus, Pervasive transcription of the human genome produces thousands of previously unidentified long intergenic noncoding RNAs, *PLoS Genet.*, 2013, e1003569.
25. C. V. Dang, E. P. Reddy, K. M. Shokat and L. Soucek, Drugging the "undruggable" cancer targets, *Nat. Rev. Cancer*, 2017, 502–508.
26. A. Russo, C. Aiello, P. Grieco and D. Marasco, Targeting "undruggable" proteins: design of synthetic cyclopeptides, *Curr. Med. Chem.*, 2016, 748–762.
27. E. Pennisi, ENCODE project writes eulogy for junk DNA, *Science*, 2012, 1159–1161.
28. R. Johnson, Long non-coding RNAs in huntington's disease neurodegeneration, *Neurobiol. Dis.*, 2012, 245–254.
29. M. Esteller, Non-coding RNAs in human disease, *Nat. Rev. Genet.*, 2011, 861–874.
30. S. Uchida and S. Dimmeler, Long noncoding RNAs in cardiovascular diseases, *Circ. Res.*, 2015, 737–750.
31. E. Lekka and J. Hall, Noncoding RNAs in disease, *FEBS Lett.*, 2018, 2884–2900.
32. Q. Pan, O. Shai, L. J. Lee, B. J. Frey and B. J. Blencowe, Deep surveying of alternative splicing complexity in the human transcriptome by high-throughput sequencing, *Nat. Genet.*, 2008, 1413–1415.
33. E. T. Wang, R. Sandberg, S. Luo, I. Khrebtukova, L. Zhang, C. Mayr, S. F. Kingsmore, G. P. Schroth and C. B. Burge, Alternative isoform regulation in human tissue transcriptomes, *Nature*, 2008, 470–476.
34. Y. Wang, J. Liu, B. Huang, Y.-M. Xu, J. Li, L.-F. Huang, J. Lin, J. Zhang, Q.-H. Min and W.-M. Yang, *et al.*, Mechanism of alternative splicing and its regulation, *Biomed. Rep.*, 2015, 152–158.
35. W. Liu, Y. Zhou, Z. Hu, T. Sun, A. Denise, X. D. Fu and Y. Zhang, Regulation of splicing enhancer activities by RNA secondary structures, *FEBS Lett.*, 2010, 4401–4407.
36. Y. Jin, Y. Yang and P. Zhang, New insights into RNA secondary structure in the alternative splicing of pre-MRNAs, *RNA Biol.*, 2011, 450–457.
37. Y. Lee and D. C. Rio, Mechanisms and regulation of alternative Pre-MRNA splicing, *Annu. Rev. Biochem.*, 2015, 291–323.

38. T. Sterne-Weiler, J. Howard, M. Mort, D. N. Cooper and J. R. Sanford, Loss of exon identity is a common mechanism of human inherited disease, *Genome Res.*, 2011, 1563–1571.
39. M. M. Scotti and M. S. Swanson, RNA mis-splicing in disease, *Nat. Rev. Genet.*, 2016, 19–32.
40. E. Peer, G. Rechavi and D. Dominissini, Epitranscriptomics: regulation of MRNA metabolism through modifications, *Curr. Opin. Chem. Biol.*, 2017, 93–98.
41. C. J. T. Lewis, T. Pan and A. Kalsotra, RNA modifications and structures cooperate to guide RNA-protein interactions, *Nat. Rev. Mol. Cell Biol.*, 2017, 202–210.
42. N. Jonkhout, J. Tran, M. A. Smith, N. Schonrock, J. S. Mattick and E. M. Novoa, The RNA modification landscape in human disease, *RNA*, 2017, 1754–1769.
43. X. Zhang, A. E. Cozen, Y. Liu, Q. Chen and T. M. Lowe, Small RNA modifications: integral to function and disease, *Trends Mol. Med.*, 2016, 1025–1034.
44. G. Yu, Y. Zhao and H. Li, The multistructural forms of box C/D ribonucleoprotein particles, *RNA*, 2018, 1625–1633.
45. M. D. De Zoysa and Y. T. Yu, Posttranscriptional RNA pseudouridylation, in *Enzymes*, 2017, pp. 151–167.
46. H. M. P. Chui, M. Meroueh, S. A. Scaringe and C. S. Chow, Synthesis of a 3-methyluridine phosphoramidite to investigate the role of methylation in a ribosomal RNA hairpin, *Bioorg. Med. Chem.*, 2002, 325–332.
47. M. B. Clark, R. L. Johnston, M. Inostroza-Ponta, A. H. Fox, E. Fortini, P. Moscato, M. E. Dinger and J. S. Mattick, Genome-wide analysis of long noncoding RNA stability, *Genome Res.*, 2012, 885–898.
48. J. A. Brown, D. Bulkley, J. Wang, M. L. Valenstein, T. A. Yario, T. A. Steitz and J. A. Steitz, Structural insights into the stabilization of MALAT1 noncoding RNA by a bipartite triple helix, *Nat. Struct. Mol. Biol.*, 2014, **21**, 633–640.
49. H. Tani, Y. Nakamura, K. Ijiri and N. Akimitsu, Stability of MALAT-1, a nuclear long non-coding RNA in mammalian cells, varies in various cancer cells, *Drug Discoveries Ther.*, 2010, 235–239.
50. N. Zhang and T. Ashizawa, RNA toxicity and foci formation in microsatellite expansion diseases, *Curr. Opin. Genet. Dev.*, 2017, 17–29.
51. V. Kumar, G. M. Hasan and M. I. Hassan, Unraveling the Role of RNA mediated toxicity of C9orf72 repeats in C9-FTD/ALS, *Front. Neurosci.*, 2017, **11**, 711.
52. Z. A. Jaafar and J. S. Kieft, Viral RNA structure-based strategies to manipulate translation, *Nat. Rev. Microbiol.*, 2019, 110–123.
53. I. Ventoso, Adaptive changes in alphavirus MRNA translation allowed colonization of vertebrate hosts, *J. Virol.*, 2012, 9484–9494.

54. A. Re, T. Joshi, E. Kulberkyte, Q. Morris and C. T. Workman, RNA-protein interactions: an overview, *Methods Mol. Biol.*, 2014, 491–521.
55. L. François-Moutal, S. Jahanbakhsh, A. D. L. Nelson, D. Ray, D. D. Scott, M. R. Hennefarth, A. Moutal, S. Perez-Miller, A. J. Ambrose and A. Al-Shamari, *et al.*, A chemical biology approach to model pontocerebellar hypoplasia type 1B (PCH1B), *ACS Chem. Biol.*, 2018, **13**, 3000–3010.
56. L. Francois-Moutal, D. D. Scott, R. Felemban, V. G. Miranda, M. R. Sayegh, S. Perez-Miller, R. Khanna, V. Gokhale, D. C. Zarnescu and M. Khanna, A small molecule targeting TDP-43's RNA recognition motifs reduces locomotor defects in a drosophila model of ALS, *ACS Chem. Biol.*, 2019, 2006–2013.
57. S. Balasubramanian, L. H. Hurley and S. Neidle, Targeting G-quadruplexes in gene promoters: A novel anticancer strategy?, *Nat. Rev. Drug Discovery*, 2011, 261–275.
58. T. Desvignes, P. Batzel, E. Berezikov, K. Eilbeck, J. T. Eppig, M. S. McAndrews, A. Singer and J. H. Postlethwait, MiRNA nomenclature: A view incorporating genetic origins, biosynthetic pathways, and sequence variants, *Trends Genet.*, 2015, 613–626.
59. M. V. Iorio and C. M. Croce, MicroRNA dysregulation in cancer: diagnostics, monitoring and therapeutics. A comprehensive review, *EMBO Mol. Med.*, 2012, 143–159.
60. L. Paladini, L. Fabris, G. Bottai, C. Raschioni, G. A. Calin and L. Santarpia, Targeting microRNAs as key modulators of tumor immune response, *J. Exp. Clin. Cancer Res.*, 2016, 103.
61. J. D. Jaffe, H. C. Berg and G. M. Church, Proteogenomic mapping as a complementary method to perform genome annotation, *Proteomics*, 2004, 59–77.
62. K. V. Ruggles and D. Fenyö, Next generation sequencing data and proteogenomics, in *Advances in Experimental Medicine and Biology*, 2016, pp. 11–19.
63. F. M. McCarthy, A. M. Cooksey, N. Wang, S. M. Bridges, G. T. Pharr and S. C. Burgess, Modeling a whole organ using proteomics: The avian bursa of fabricius, *Proteomics*, 2006, 2759–2771.
64. J. Ruiz-Orera, X. Messeguer, J. A. Subirana and M. M. Alba, Long noncoding RNAs as a source of new peptides, *Elife*, 2014, e03523.
65. B. Cai, Z. Li, M. Ma, Z. Wang, P. Han, B. A. Abdalla, Q. Nie and X. Zhang, LncRNA-Six1 encodes a micropeptide to activate Six1 in Cis and is involved in cell proliferation and muscle growth, *Front. Physiol.*, 2017, 230.
66. K. Abdelmohsen, A. Panda, M. J. Kang, J. Xu, R. Selimyan, J. H. Yoon, J. L. Martindale, S. De, W. H. Wood and K. G. Becker, *et al.*, Senescence-associated LncRNAs: senescence-associated long noncoding RNAs, *Aging Cell*, 2013, 890–900.

67. B. S. Morgan, J. E. Forte, R. N. Culver, Y. Zhang and A. E. Hargrove, Discovery of key physicochemical, structural, and spatial properties of RNA-targeted bioactive ligands, *Angew. Chem., Int. Ed.*, 2017, 13498–13502.
68. B. S. Morgan, J. E. Forte and A. E. Hargrove, Survey and summary insights into the development of chemical probes for RNA, *Nucleic Acids Res.*, 2018, 8025–8037.
69. P. J. Mccown, K. A. Corbino, S. Stav, M. E. Sherlock and R. R. Breaker, Riboswitch diversity and distribution, *RNA*, 2017, 995–1011.
70. P. Machtel, K. Bąkowska-Żywicka and M. Żywicki, Emerging applications of riboswitches – from antibacterial targets to molecular tools, *J. Appl. Genet.*, 2016, 531–541.
71. S. B. Vakulenko and S. Mobashery, Versatility of aminoglycosides and prospects for their future, *Clin. Microbiol. Rev.*, 2003, 430–450.
72. G. Yang, J. Trylska, Y. Tor and J. A. McCammon, Binding of aminoglycosidic antibiotics to the oligonucleotide a-site model and 30s ribosomal subunit: poisson-boltzmann model, thermal denaturation, and fluorescence studies, *J. Med. Chem.*, 2006, 5478–5490.
73. M. Chittapragada, S. Roberts and Y. W. Ham, Aminoglycosides: molecular insights on the recognition of rna and aminoglycoside mimics, *Perspect. Med. Chem.*, 2009, 21–37.
74. K. L. Leach, S. M. Swaney, J. R. Colca, W. G. McDonald, J. R. Blinn, L. M. M. Thomasco, R. C. Gadwood, D. Shinabarger, L. Xiong and A. S. Mankin, The site of action of oxazolidinone antibiotics in living bacteria and in human mitochondria, *Mol. Cell*, 2007, 393–402.
75. J. D. Mortison, M. Schenone, J. A. Myers, Z. Zhang, L. Chen, C. Ciarlo, E. Comer, S. K. Natchiar, S. A. Carr and B. P. Klaholz, *et al.*, Tetracyclines modify translation by targeting key human RRNA substructures, *Cell Chem. Biol.*, 2018, 1506–1518.e13.
76. K. Poole, Efflux-mediated antimicrobial resistance, in *Antibiotic Discovery and Development*, 2014.
77. J. H. Cate, M. M. Yusupov, G. Z. Yusupova, T. N. Earnest and H. F. Noller, X-ray crystal structures of 70s ribosome functional complexes, *Science*, 1999, 2095–2104.
78. W. M. Clemons, J. L. C. May, B. T. Wimberly, J. P. McCutcheon, M. S. Capel and V. Ramakrishnan, Structure of a bacterial 30S ribosomal subunit at 5.5 Å resolution, *Nature*, 1999, 833–840.
79. M. Selmer, C. M. Dunham, F. V. Murphy, A. Weixlbaumer, S. Petry, A. C. Kelley, J. R. Weir and V. Ramakrishnan, Structure of the 70S ribosome complexed with MRNA and TRNA, *Science*, 2006, 1935–1942.
80. J. Xu, S. Y. Li, D. V. Almeida, R. Tasneen, K. Barnes-Boyle, P. J. Converse, A. M. Upton, K. Mdluli, N. Fotouhi and E. L. Nuermberger, Contribution of pretomanid to novel regimens containing bedaquiline with either linezolid or moxifloxacin and pyrazinamide in murine models of tuberculosis, *Antimicrob. Agents Chemother.*, 2019, e00021-19.

81. M. Hung, P. Patel, S. Davis and S. R. Green, Importance of ribosomal frameshifting for human immunodeficiency virus type 1 particle assembly and replication, *J. Virol.*, 1998, 4819–4824.
82. R. J. Marcheschi, M. Tonelli, A. Kumar and S. E. Butcher, Structure of the HIV-1 frameshift site RNA bound to a small molecule inhibitor of viral replication, *ACS Chem. Biol.*, 2011, 857–864.
83. K. E. Lind, Z. Du, K. Fujinaga, B. M. Peterlin and T. L. James, Structure-based computational database screening, *in vitro* assay, and NMR assessment of compounds that target TAR RNA, *Chem. Biol.*, 2002, 185–193.
84. B. R. McNaughton, P. C. Gareiss and B. L. Miller, Identification of a selective small-molecule ligand for HIV-1 frameshift-inducing stem-loop RNA from an 11,325 member resin bound dynamic combinatorial library, *J. Am. Chem. Soc.*, 2007, 11306–11307.
85. J. Sztuba-Solinska, S. R. Shenoy, P. Gareiss, L. R. H. Krumpe, S. F. J. Le Grice, B. R. O'Keefe and J. S. Schneekloth, Identification of biologically active, HIV TAR RNA-binding small molecules using small molecule microarrays, *J. Am. Chem. Soc.*, 2014, 8402–8410.
86. F. A. Abulwerdi, M. D. Shortridge, J. Sztuba-Solinska, R. Wilson, S. F. J. Le Grice, G. Varani and J. S. Schneekloth, Development of small molecules with a noncanonical binding mode to HIV-1 trans activation response (TAR) RNA, *J. Med. Chem.*, 2016, 11148–11160.
87. N. N. Patwardhan, L. R. Ganser, G. J. Kapral, C. S. Eubanks, J. Lee, B. Sathyamoorthy, H. M. Al-Hashimi and A. E. Hargrove, Amiloride as a new RNA-binding scaffold with activity against HIV-1 TAR, *Medchemcomm*, 2017, 1022–1036.
88. F. A. Abulwerdi, W. Xu, A. A. Ageeli, M. J. Yonkunas, G. Arun, H. Nam, J. S. Schneekloth, T. K. Dayie, D. Spector and N. Baird, *et al.*, Selective small-molecule targeting of a triple helix encoded by the long noncoding RNA, MALAT1, *ACS Chem. Biol.*, 2019, 223–235.
89. S. A. Barros and D. M. Chenoweth, Recognition of nucleic acid junctions using triptycene-based molecules, *Angew. Chem., Int. Ed.*, 2014, 13746–13750.
90. S. J. Park, Y. G. Kim and H. J. Park, Identification of RNA pseudoknot-binding ligand that inhibits the - 1 ribosomal frameshifting of SARS-coronavirus by structure-based virtual screening, *J. Am. Chem. Soc.*, 2011, 10094–10100.
91. J. F. Arambula, S. R. Ramisetty, A. M. Baranger and S. C. Zimmerman, A simple ligand that selectively targets CUG trinucleotide repeats and inhibits MBNL protein binding, *Proc. Natl. Acad. Sci. U. S. A.*, 2009, 16068–16073.
92. N. G. Lintner, K. F. McClure, D. Petersen, A. T. Londregan, D. W. Piotrowski, L. Wei, J. Xiao, M. Bolt, P. M. Loria and B. Maguire, *et al.*, Selective stalling of human translation through small-molecule engagement of the ribosome nascent chain, *PLoS Biol.*, 2017, e2001882.

93. L. Gude, S. S. Berkovitch, W. L. Santos, P. S. Kutchukian, A. R. Pawloski, R. Kuimelis, G. McGall and G. L. Verdine, Mapping targetable sites on human telomerase RNA pseudoknot/template domain using 2′-OMe RNA-interacting polynucleotide (RIPtide) microarrays, *J. Biol. Chem.*, 2012, 18843–18853.
94. R. Kierzek, D. H. Turner and E. Kierzek, Microarrays for identifying binding sites and probing structure of RNAs, *Nucleic Acids Res.*, 2015, 1–12.
95. S. P. Velagapudi, Y. Luo, T. Tran, H. S. Haniff, Y. Nakai, M. Fallahi, G. J. Martinez, J. L. Childs-DIsney and M. D. Disney, Defining RNA-small molecule affinity landscapes enables design of a small molecule inhibitor of an oncogenic noncoding RNA, *ACS Cent. Sci.*, 2017, 205–216.
96. S. P. Velagapudi, M. D. Cameron, C. L. Haga, L. H. Rosenberg, M. Lafitte, D. R. Duckett, D. G. Phinney and M. D. Disney, Design of a small molecule against an oncogenic noncoding RNA, *Proc. Natl. Acad. Sci. U. S. A.*, 2016, 5898–5903.
97. S. G. Rzuczek, L. A. Colgan, Y. Nakai, M. D. Cameron, D. Furling, R. Yasuda and M. D. Disney, Precise small-molecule recognition of a toxic CUG RNA repeat expansion, *Nat. Chem. Biol.*, 2017, 188–193.
98. S. P. Velagapudi, Y. Li and M. D. Disney, A cross-linking approach to map small molecule-RNA binding sites in cells, *Bioorg. Med. Chem. Lett.*, 2019, 1532–1536.
99. L. Guan and M. D. Disney, Small-molecule-mediated cleavage of RNA in living cells, *Angew. Chem., Int. Ed.*, 2013, 1462–1465.
100. O. Aminova and M. D. Disney, A microarray-based method to perform nucleic acid selections, *Methods Mol. Biol.*, 2010, 209–224.
101. D. J. Paul, S. J. Seedhouse and M. D. Disney, Two-dimensional combinatorial screening and the RNA privileged space predictor program efficiently identify aminoglycoside RNA hairpin loop interactions, *Nucleic Acids Res.*, 2009, 5894–5907.
102. S. P. Velagapudi, S. J. Seedhouse and M. D. Disney, Structure-activity relationships through sequencing (StARTS) defines optimal and suboptimal RNA motif targets for small molecules, *Angew. Chem., Int. Ed.*, 2010, 3816–3818.
103. M. D. Disney, A. M. Winkelsas, S. P. Velagapudi, M. Southern, M. Fallahi and J. L. Childs-Disney, Inforna 2.0: A platform for the sequence-based design of small molecules targeting structured RNAs, *ACS Chem. Biol.*, 2016, 1720–1728.
104. S. P. Velagapudi, S. M. Gallo and M. D. Disney, Sequence-based design of bioactive small molecules that target precursor microRNAs, *Nat. Chem. Biol.*, 2014, 291–297.
105. Q. Chen, R. H. Shafer and I. D. Kuntz, Structure-based discovery of ligands targeted to the RNA double helix, *Biochemistry*, 1997, 11402–11407.

106. A. V. Filikov, V. Mohan, T. A. Vickers, R. H. Griffey, P. D. Cook, R. A. Abagyan and T. L. James, Identification of ligands for RNA targets *via* structure-based virtual screening: HIV-1 TAR, *J. Comput.-Aided Mol. Des.*, 2000, 593–610.
107. N. Moitessier, E. Westhof and S. Hanessian, Docking of aminoglycosides to hydrated and flexible RNA, *J. Med. Chem.*, 2006, 1023–1033.
108. S. D. Morley and M. Afshar, Validation of an empirical RNA-ligand scoring function for fast flexible docking using RiboDock®, *J. Comput.-Aided Mol. Des.*, 2004, 189–208.
109. P. Pfeffer and H. Gohlke, DrugScoreRNA - knowledge-based scoring function to predict RNA - ligand interactions, *J. Chem. Inf. Model.*, 2007, 1868–1876.
110. A. Philips, K. Milanowska, G. Łach and J. M. Bujnicki, LigandRNA: computational predictor of RNA-ligand interactions, *RNA*, 2013, 1605–1616.
111. K. D. Warner, C. E. Hajdin and K. M. Weeks, Principles for targeting RNA with drug-like small molecules, *Nat. Rev. Drug Discovery*, 2018, 1605–1616.
112. Q. Vicens, E. Mondragón and R. T. Batey, Molecular sensing by the aptamer domain of the FMN riboswitch: A general model for ligand binding by conformational selection, *Nucleic Acids Res.*, 2011, 8586–8598.
113. J. A. Howe, L. Xiao, T. O. Fischmann, H. Wang, H. Tang, A. Villafania, R. Zhang, C. M. Barbieri and T. Roemer, Atomic resolution mechanistic studies of ribocil: A highly selective unnatural ligand mimic of the E. Coli FMN riboswitch, *RNA Biol.*, 2016, 946–954.
114. T. Matt, C. L. Ng, K. Lang, S. H. Sha, R. Akbergenov, D. Shcherbakov, M. Meyer, S. Duscha, J. Xie and S. R. Dubbaka, *et al.*, Dissociation of antibacterial activity and aminoglycoside ototoxicity in the 4-monosubstituted 2-deoxystreptamine apramycin, *Proc. Natl. Acad. Sci. U. S. A.*, 2012, 10984–10989.
115. J. L. Hansen, J. A. Ippolito, N. Ban, P. Nissen, P. B. Moore and T. A. Steitz, The structures of four macrolide antibiotics bound to the large ribosomal subunit, *Mol. Cell*, 2002, 117–128.
116. D. E. Brodersen, W. M. Clemons, A. P. Carter, R. J. Morgan-Warren, B. T. Wimberly and V. Ramakrishnan, The structural basis for the action of the antibiotics tetracycline, pactamycin, and hygromycin B, on the 30S ribosomal subunit, *Cell*, 2000, 1143–1154.
117. J. L. Hansen, P. B. Moore and T. A. Steitz, Structures of five antibiotics bound at the peptidyl transferase center of the large ribosomal subunit, *J. Mol. Biol.*, 2003, 1061–1075.
118. J. Zhou, A. Bhattacharjee, S. Chen, Y. Chen, E. Duffy, J. Farmer, J. Goldberg, R. Hanselmann, J. A. Ippolito and R. Lou, *et al.*, Design at the atomic level: Design of biaryloxazolidinones as potent orally active antibiotics, *Bioorg. Med. Chem. Lett.*, 2008, 6175–6178.

119. D. Matzov, Z. Eyal, R. I. Benhamou, M. Shalev-Benami, Y. Halfon, M. Krupkin, E. Zimmerman, H. Rozenberg, A. Bashan and M. Fridman, *et al.*, Structural Insights of Lincosamides Targeting the Ribosome of Staphylococcus Aureus, *Nucleic Acids Res.*, 2017, 10284–10292.
120. K. C. Woods, S. S. Martin, V. C. Chu and E. P. Baldwin, Quasi-equivalence in site-specific recombinase structure and function: crystal structure and activity of trimeric Cre recombinase bound to a three-way lox DNA junction, *J. Mol. Biol.*, 2001, 49–69.

CHAPTER 10

Assays to Characterize the Cellular Pharmacology of a Chemical Probe

JAKUB STEFANIAK[a,b], SÉBASTIEN R. G. GALAN[c] AND KILIAN V. M. HUBER*[a,b]

[a]Structural Genomics Consortium, Nuffield Department of Medicine, University of Oxford, Oxford OX3 7DQ, UK; [b]Target Discovery Institute, Nuffield Department of Medicine, University of Oxford, Oxford OX3 7FZ, UK; [c]Department of Chemistry, Chemistry Research Laboratory, University of Oxford, Oxford OX1 3TA, UK
*E-mail: kilian.huber@sgc.ox.ac.uk

10.1 Introduction

To support the successful intended application of chemical probes for the identification and elucidation of cellular signalling pathways and consequently drug target discovery and validation, these tool compounds need to meet certain key requirements.[1] Commonly applied criteria involve a clear demonstration of direct target engagement in a cellular model, preferably at submicromolar cellular concentrations, and 30-fold selectivity over structurally related proteins to limit the confounding effects of compound promiscuity and polypharmacology. To address this issue at least partially, it is desirable for a chemical probe to have a structurally related analogue

Chemical Biology No. 16
The Discovery and Utility of Chemical Probes in Target Discovery
Edited by Paul Brennan and Saleta Vazquez Rodriguez

Published by the Royal Society of Chemistry, www.rsc.org

that is inactive against the cognate target and therefore can help account for potential compound scaffold-related off-targets. There is a plethora of cellular assays that can be used to explore biology using chemical probes and, equally important, help determine the selectivity of such tool compounds. This chapter conceptually distinguishes assays that require modification of either the chemical probe, protein or pathway of interest *versus* 'label-free' approaches that do not alter the chemical structure of the compound or the respective cellular context.

10.2 Assays Requiring Chemical Modification of the Probe

Chemical probes require rigorous testing to ensure that observed cellular effects can be matched with *in vitro* data in addition to, *e.g.*, target knockout/knockdown studies. To this end, chemical biology has developed a number of tools to study the interactions of small molecules inside cells, in terms of both targeted approaches and more global proteome and signalling changes. The first branch of these techniques is concerned with investigating the binding profile of the molecule in a cellular setting and can be divided into target engagement methods (whether the chemical probe binds the protein it is supposed to) and target identification methods (to explore the proteins bound by the molecule on a global level), with the former usually relying on molecular biology-based readouts, whereas the latter are often based on protein mass spectrometry (MS). Furthermore, all of these techniques can be classified into those that require modification ('labelling'), usually in the form of chemical alterations of the probe or genetic manipulation of the cells, and 'label-free' approaches (Table 10.1).

10.2.1 Affinity Chromatography-based Techniques

Affinity-capture or pulldown methods were one of the earliest methods of assaying small-molecule–target interactions.[2] At their most basic, they involve the immobilization of the compound on a resin or other solid support and incubation with a cell lysate, which will result in enrichment for not only the expected target protein(s), but also potential off-target and indirect interactions. A competitor in the form of an unbound molecule can help to distinguish specific from non-specific binders. Visualization of the results can be achieved *via* western blotting (WB) with specific antibodies for individual proteins or MS to survey targets on a more global level (Figure 10.1).

A prominent example of affinity proteomics being used to detect off-target interactions of a small molecule was the profiling of dasatinib, originally developed as an inhibitor of the fusion protein Bcr-Abl.[3,4] In order to

Table 10.1 Commonly used biophysical methods to assay interactions between ligands and proteins within a cellular context.

Technique	Biophysical principle	Labelling
BRET/FRET	Energy transfer across a small distance (<10 nm)	Yes, requires a luciferase and/or fluorescent tag on both the protein and the molecule
CETSA	Thermal stabilization of protein by ligand	No
Affinity chromatography	Immobilized ligand precipitates the targets with which it interacts	Yes, requires the modification of the small molecule to immobilize it
Fluorescence anisotropy	Differential anisotropy exhibited by bound *versus* unbound fluorescent ligands	Yes, requires the attachment of an appropriate fluorophore to the small molecule
Enzyme complementation	Recovery of protein activity *via* reassembly of fragment	Yes, requires the expression of the fragmented protein reporter on the target
FRAP	Rates of fluorescence recovery following laser photobleaching represent the diffusion and binding of the fluorescent target to its partner	Yes, requires a fluorescent tag on the protein

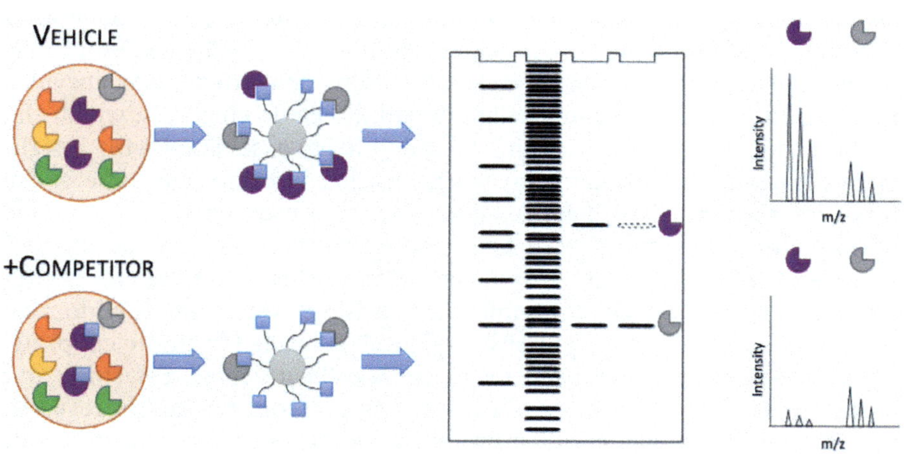

Figure 10.1 Affinity chromatography methods for target deconvolution. Immobilization of the compound on a resin permits the capture of cognate targets, which can then be visualized by WB, to validate target engagement, or assayed by MS, for target identification. Preincubation with the unconjugated compound allows differentiation between true targets and non-specific interactors.

immobilize the compound, the authors synthesized an analogue, dubbed c-dasatinib, with an amine handle to facilitate coupling. The compound was subsequently immobilized on Sepharose beads and lysate from K562 cells (a chronic myeloid leukaemia cell line susceptible to dasatinib treatment) was incubated with it. Following washing and elution, the flow-through was analysed by sodium dodecyl sulfate polyacrylamide gel electrophoresis (SDS-PAGE), followed by band excision and MS analysis. In addition to the cognate target, other targets were identified, including Bruton's tyrosine kinase (BTK). Subsequent validation in cellular models confirmed the hit, which was followed by structural analysis that revealed specific BTK mutants that show susceptibility to dasatinib, in addition to cell assays to examine the compound's mode of action, in particular its effects on inflammation and histamine release.

An example of the application of affinity proteomics for the identification of off-targets is Avandia (rosiglitazone), which is a thiazolidinedione (TZD)-based peroxisome proliferator-activated receptor gamma (PPARγ) agonist that was initially approved for the treatment of diabetes.[5] In 2007, a large meta-analysis of 42 clinical trials involving rosiglitazone revealed that the use of the drug carries an increased risk of myocardial infarction and, albeit borderline insignificant, an increased risk of death from cardiovascular events,[6] which eventually led to the US Food and Drug Administration (FDA) issuing a restriction on the drug until the expiration of the patent in 2012. The off-target profile of its scaffold was analysed by immobilizing the TZD core on an epoxy resin followed by incubation with extracts from rat heart tissue. Subsequent washing of the resin and elution of the bound proteins for subsequent MS analysis suggested that the TZD scaffold may be binding a diverse range of proteins, including dehydrogenases, enzymes involved in glucose metabolism and small-molecule channels, which the authors suggested could explain the side effects associated with rosiglitazone treatment,[7] although it is worth bearing in mind that none of the hits from the assay have been validated. Another exciting example of how affinity chromatography-based target deconvolution has opened up new areas is the discovery of the protein cereblon (CRBN) as a target for thalidomide. Here, the authors immobilized a carboxyl-containing thalidomide derivative on ferrite–glycidyl methacrylate (FG) beads *via* amide coupling. This was followed by incubation with HeLa cell extracts and washing and elution with free thalidomide. Subsequent proteomic analysis revealed CRBN and DNA damage-binding protein 1 (DDB1) as cellular interactors of thalidomide.[8]

Whereas amide and epoxide coupling reactions can be used to immobilize compounds on a solid phase (Table 10.2), the development of bioorthogonal 'click' reactions that can be performed *in situ*, *e.g.* the copper-catalysed azide–alkyne cycloaddition (CuAAC) reaction, has led to widespread use in biological chemistry.[9,10] This includes the use of alkyne-containing affinity probes, which serve as capture compounds that are incubated with the cells and subsequently conjugated *via* the

Table 10.2 Commonly used strategies for compound immobilization on beads.

Conjugation method	Characteristics
Streptavidin	The covalent attachment of streptavidin allows enrichment for biotin-containing probes and their targets
N-Hydroxysuccinimide (NHS)	Readily reacts with free amines to form amides
Epoxide	Readily reacts with free amines to form β-hydroxyamines
Carboxyl groups	Less reactive on their own; require activation with carbodiimide compounds 1-ethyl-3-(3-dimethyla minopropyl)carbodiimide (EDC) and dicyclohexylcarbodiimide (DCC)
Iodoacetyl/maleimide	Readily reacts with thiol groups

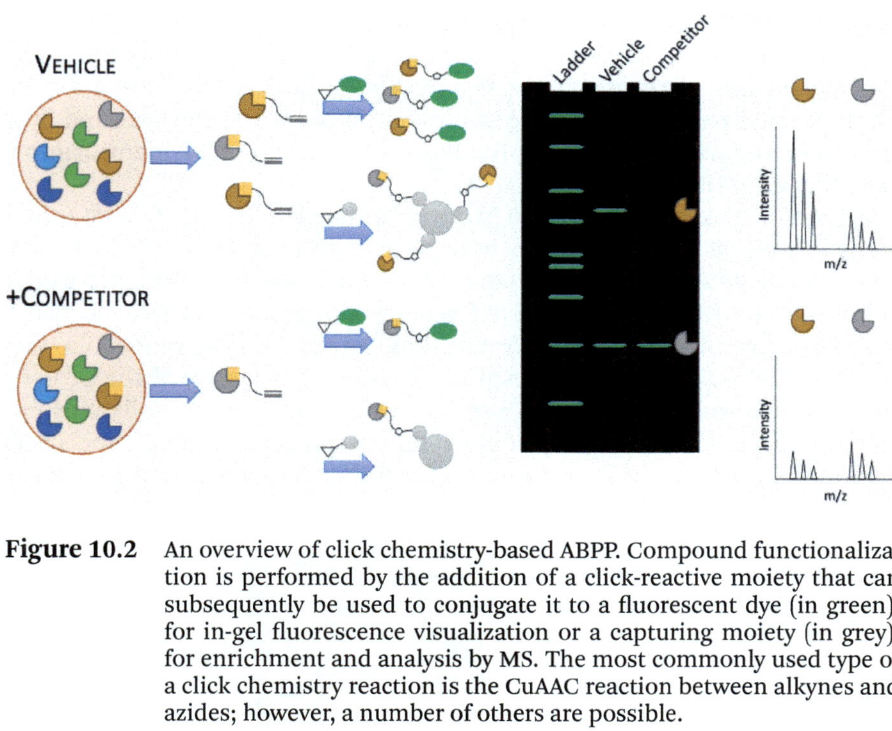

Figure 10.2 An overview of click chemistry-based ABPP. Compound functionalization is performed by the addition of a click-reactive moiety that can subsequently be used to conjugate it to a fluorescent dye (in green), for in-gel fluorescence visualization or a capturing moiety (in grey), for enrichment and analysis by MS. The most commonly used type of a click chemistry reaction is the CuAAC reaction between alkynes and azides; however, a number of others are possible.

click reaction to resins following lysis (Figure 10.2). The advantage of this approach is that it accounts for cell permeability effects. In a recent study of compounds targeting ubiquitin-specific proteases (USPs), an irreversible USP inhibitor was profiled employing an alkyne-containing derivative that was incubated with U2OS cells. Subsequently, the cells were lysed and the probes clicked to the capture reagent azido-TAMRA-biotin (AzTb), which permeits not only visualization of protein targets by SDS-PAGE in-gel fluorescence but also the capture of the probe on a NeutrAvidin resin.

MS analysis revealed that the probe binds members of the USP family, and also some oxygenases, dehydrogenases and glutathione transferases, likely due to its covalent 'warhead'.[11]

Recently, a click-based approach revealed broad proteome labelling of C646, a putative p300 histone acetyltransferase (HAT) inhibitor. Similarly, an affinity probe derived from the putative murine double minute X (MDMX) protein inhibitor SJ-172550 also indicated indiscriminate labelling of proteins in U2OS cells in a concentration-dependent manner.[12]

In the case of C646 and SJ-172550, the tool compounds themselves bear reactive chemical moieties that permit covalent conjugation to protein targets, which allows for extensive washing during sample preparation, and even protein-denaturing reagents can be tolerated. If the compound of interest is a non-covalent binder, activity-based protein profiling (ABPP) using pan-inhibitor probes can be attempted. In a typical ABPP experiment, the affinity probe(s) contain a core scaffold or chemotype [the 'bait', *e.g.* a purine mimetic for adenosine triphosphate (ATP)-binding proteins] plus a capturing warhead[13] (such as a reactive sulfonyl fluoride or a diazirine photoreactive moiety capable of covalent modification) connected by a linker to a reporter tag, such as an alkyne, which allows for identification or purification (Figure 10.3).

Shi *et al.*[14] used this method to investigate the cellular targets of staurosporine, a natural product that has been prolifically reported in the literature for years as a potent and promiscuous kinase inhibitor. They designed a cell-permeable staurosporine derivative featuring a diazirine warhead and also a clickable alkyne handle. Incubation of HepG2 cells with the probe followed by ultraviolet (UV) irradiation resulted in efficient labelling of the protein targets of staurosporine, which was further exploited by initially clicking rhodamine azide to the probe (for in-gel fluorescence scanning readout) and then biotin azide for pulldowns and MS analysis, which in turn revealed a

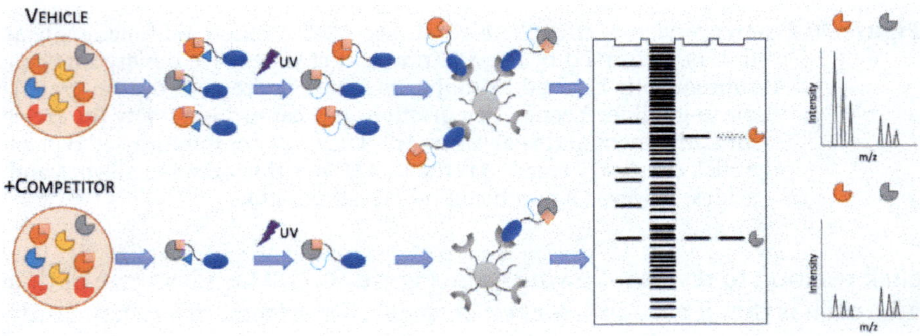

Figure 10.3 The use of photoreactive probes in ABPP. The functionalization of a compound is performed by the addition of a photoreactive moiety (represented by the blue triangle) for capture and an enrichment tag (represented by the blue oval). The most commonly used photoreactive group is diazirine, but many others can be used, such as (nitro) phenyl azides, benzophenone and psoralen.

total of 43 kinases putatively engaged by the small molecule.[14] This approach has also been successfully applied to clinical compounds and drugs such as dasatinib, a dual Src/Abl kinase inhibitor.[15] In order to facilitate the development of ABPP-compatible probes, various groups have developed new synthetic routes towards cell-permeable clickable linkers containing diazirine warheads[16] or even live-cell imaging tags.[17]

One reason why compounds fail in clinical trials is often off-target toxicity/adverse effects, which is why a strong case can be made for chemoproteomic target identification approaches as a de-risking strategy. In 2016, a clinical trial of BIA 10-2474, a putative fatty acid amide hydrolase (FAAH) inhibitor, came to an abrupt halt after five participants developed irreversible neurological damage and necrosis leading to one fatality.[18] A subsequent ABPP study elegantly uncovered several potential off-target interactors of the molecule and its metabolites.[19,20] The authors found that not only does the compound itself inhibit a variety of serine hydrolases involved in lipolysis, but also its metabolites irreversibly block neuronal aldehyde dehydrogenases. Lipidomic studies further showed major disruptions in the lipid profile of cells treated with it, which could account for the observed neurotoxic effects. In this respect, metabolomics and lipidomics, which specifically examine the cellular profile of metabolites and lipids, respectively, are useful tools to complement chemoproteomics.

Affinity capture techniques can be used very effectively to look across target classes using specific reactivity. To this end, acetamide-containing cysteine-reactive probes scan compounds that are covalent inhibitors targeting thiol groups. In a typical experiment, cells are incubated with either a cysteine-reactive probe or the vehicle, which is followed by the addition of a click-compatible non-discriminating alkylating agent, usually iodoacetamide (IA). Clicking the IA–alkyne to an affinity tag containing a tobacco etch virus (TEV) protease recognition motif and also a tandem mass tag allows subsequent purification, digestion and quantitative MS analysis to identify the precise location of IA-modified cysteines.[21] In a more recent example, isoTOP-ABPP was directly involved in the design of a novel proteolysis-targeting chimera (PROTAC) derived from the cytotoxic natural product nimbolide. The compound was found to elicit its cytotoxic effects by covalently binding a cysteine on the E3 ubiquitin ligase ring finger 114 (RNF114), thereby disrupting the proteasomal degradation of proteins, such as the tumour suppressor p21. The authors were then able to devise a bromodomain-containing protein (BRD4)-degrading PROTAC using nimbolide.[22]

10.2.2 Fluorescence Anisotropy Methods

The use of fluorescence anisotropy to monitor cellular target engagement of a small molecule is an extension of the conventional fluorescence polarization assay.[23] The biophysical phenomenon that underpins it is based on slower rates of rotation of a fluorophore when bound to a large macromolecule – fluorescently tagged chemical probes will tumble more slowly when complexed

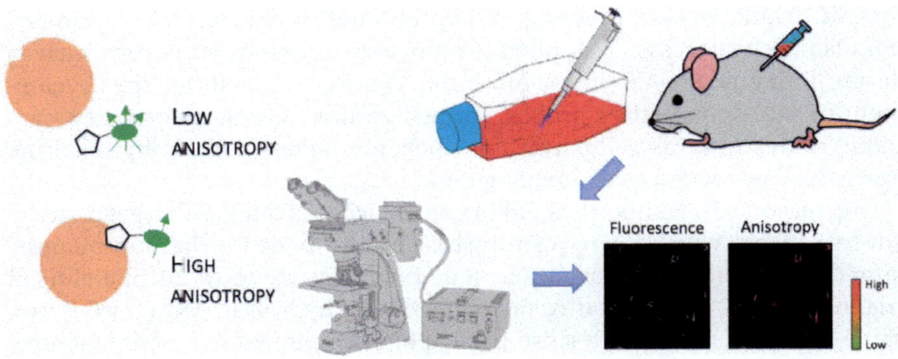

Figure 10.4 Anisotropy can be exploited to study target engagement in cells and live animals, by attaching a fluorescent tag to the molecule.

with proteins. As a result, one can monitor the anisotropy of emission following plane-polarized excitation. Free molecules (*i.e.* not bound to the target) will emit in an isotropic fashion, whereas those bound to a molecule will likely exhibit high anisotropy of emission.[24] Therefore, by covalently attaching a fluorescent dye to a molecule, one can directly monitor its cellular target engagement (Figure 10.4).

In order to adapt the technique to a cellular environment, the dye should meet a number of criteria, among them being cell permeability and non-polarity such that it does not interfere with the compound's normal mode of action. In addition, it should have a high quantum yield and a spectrum that does not shift substantially in relation to solvent polarity or pH. Initial studies by Dubach *et al.* used boron dipyrromethene (BODIPY)-labelled AZD2281 (olaparib), which is a poly-ADP ribose polymerase (PARP) inhibitor.[25] The advantage of the technique is that it can be applied to study compound engagement in tissue or even live animals using window chambers. The approach has more recently also been used to examine the cellular target engagement of ibrutinib, a BTK inhibitor.[26]

10.3 Assays Requiring Genetic Modification of the Target Protein

10.3.1 Reporter Gene Assays

An alternative way to visualize and therefore study cellular target engagement is to label the proteins of interest with exogenous reporter proteins. This in-cell labelling of proteins is based on reporter gene assay methodology (Figure 10.5). Reporter gene assays are extensively used to interrogate the regulation of a specific gene in phenotypic screening. Conceptually, the reporter is fused to the regulatory element of a gene of interest, replacing, or becoming attached to, the corresponding coding region (Figure 10.5A). The construct is then transfected into the cell type of choice. The activation of the gene of interest triggers the translation of the reporter protein alone or as a fusion to the protein of interest. In this latter case, the reporter protein allows the direct or indirect visualization of the protein target (Figure 10.5B). For example, fluorescence can be measured directly when the reporter gene expresses a fluorescent protein. To permit multiplexing or to overcome potential assay interference issues, alternatives such as green (GFP), yellow (YFP) and red fluorescent protein (RFP) may be used. Alternatively, the readout is the result of the indirect transformation of a substrate by the exogenous enzyme encoded by the reporter gene, *e.g.* a luciferase enzyme and bioluminescence. Luciferases catalyse the oxidation of luciferin to produce oxyluciferin in an excited state and relaxation to the ground state generates the light signal for detection. Commonly employed luciferase enzymes are firefly (FLuc), renilla (RLuc) and the more recently developed nano (NLuc) luciferases. Other exogenous enzymes with specific substrates have also been investigated, such as chloramphenicol acetyltransferase (CAT), β-glucuronidase (GUS), neomycin phosphotransferase, β-galactosidase (β-gal), β-lactamase (LacZ) and secreted alkaline phosphatase (SEAP).

Such systems therefore allow the detection and quantification of the effect of a chemical probe on a specific biological pathway *via* expression of a specific gene. Small molecules have been evaluated in wingless-related integration site (Wnt), transforming growth factor beta (TGFβ) and cyclic adenosine monophosphate (cAMP) signalling pathways using luciferase reporter genes, as exemplified in a review by Wang *et al.*[27] Another example

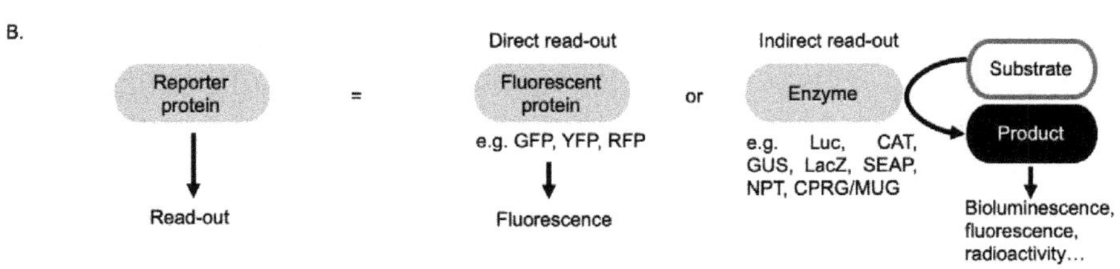

Figure 10.5 Principle of reporter gene assay. (A) In a transcriptional fusion, the reporter gene is inserted in place of the coding region. The activation of the gene of interest therefore triggers the expression of the reporter protein. In a translational fusion, the reporter gene is added to the coding region. The activation of the gene of interest therefore triggers the expression of both the reporter protein and the protein of interest. (B) Reporter proteins can be fluorescent proteins and therefore provide a direct readout or enzymes that catalyse a reaction that indirectly produces a signal which can be quantified.

is the development of GFP reporter assays in epigenetic drug discovery. Here, the expression of GFP is directly correlated with the epigenetically regulated promoter activity, *e.g.* the regulation of p21 and cytomegalovirus (CMV) promoters by histone deacetylases (HDACs). HDAC modulation by chemical probes can, as a result, be monitored.[28–30] As discussed in more detail below, the reporter gene methodology may also be used to detect specific protein binding of chemical probes and consequently their cellular target engagement. Examples include enzyme complementation, resonance energy transfer assay (RET) and fluorescence recovery after photobleaching (FRAP) assays.

10.3.2 Complementation Assays

Complementation assays rely on the recovery of the activity of a specific reporter protein *via* induced reassembly of inactive protein fragments. In these assays, each reporter fragment is linked to a different partner. Upon binding of the partners, the reporter protein is reconstituted, allowing measurement of its activity. Initially pioneered by the yeast two-hybrid (Y2H) approach, multiple additional reporter systems have been developed.

Cellular enzyme fragment complementation (EFC) assays developed by DiscoverX, namely InCELL Hunter™ and In CELL Pulse™,[31] use β-galactosidase (β-gal) as the reporter protein. To this end, β-gal is split into an enzyme donor enhanced ProLabel (ePL) and enzyme acceptor (EA) fragment. In cellular target engagement validation, the protein of interest is fused to the ePL. Upon treatment with a chemical probe, the protein of interest is either stabilized or destabilized, which is monitored by the restoration of the β-gal activity after addition of EA (Figure 10.6A). This approach has been successfully employed for determining target engagement of protein arginine methyltransferase 3 (PRMT3),[32] mitogen-activated protein kinase kinase 1 (MEK1)[33] and BRD4[34] inhibitors.

Whereas protein-fragment complementation assays (PCAs) use the same principle as EFC assays, PCA mainly refers to a method that identifies and quantifies protein–protein interactions.[35,36] In PCA, each gene of interest, designated 'bait' and 'prey', is fused to the C-terminal (CPR) and the N-terminal (NPR) reporter gene fragments, respectively. These two constructs are co-expressed in a cell type of choice. When binding occurs between the 'bait' and the 'prey', the reporter protein is reconstituted and its activity restored (Figure 10.6B). In addition to the reporter proteins described in Section 10.3.1, dihydrofolate reductase (DHFR), ubiquitin and TEV protease are also commonly used. This technique allows evaluation of the direct or indirect effects of compounds on protein–protein interactions.[37,38] The complementary bimolecular fluorescence complementation (BiFC) assay[39–41] follows the same principle but uses fluorescent proteins as reporter.

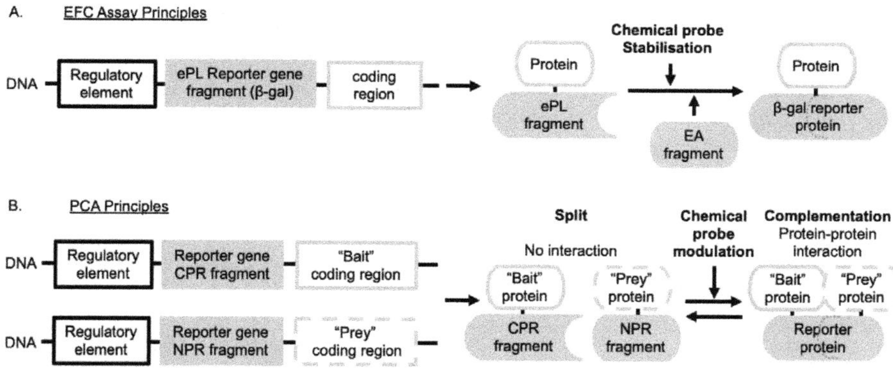

Figure 10.6 Principles of complementation assays. (A) In an enzyme fragment complementation (EFC) assay, the protein of interest is linked to a small fragment of β-galactosidase (β-gal), in this case referred to as the enhanced ProLabel (ePL). Upon compound treatment, the level of the protein of interest is either stabilized or destabilized. Addition of the large β-gal enzyme acceptor (AE) reconstitutes the functional β-gal reporter protein, which then generates the signal. (B) In a protein complementation assay (PCA), the two proteins of interest ('bait' and 'prey') are each linked to respective complementary fragments of a reporter protein called C-terminal (CPR) and N-terminal (NPR). If the bait and prey proteins interact, the reporter is reconstituted and thus provides an interaction-dependent signal.

10.3.3 Resonance Energy Transfer (RET) Assays

Fluorescence or Förster resonance energy transfer (FRET), based on the transfer of the excitation energy of a donor fluorophore to a close acceptor chromophore, is a phenomenon that has been extensively applied to the study of biological systems. Essentially, the donor fluorescence is quenched by the acceptor, which then emits light at a longer wavelength. This transfer is possible only if the donor and acceptor are close, typically <10 nm, in a proper orientation, and their corresponding emission and absorption spectra are overlapping. For protein–protein interaction studies and analogous to the previously described reporter gene methodology, two proteins are labelled with a fluorescent donor and an acceptor, respectively (Figure 10.7). To evaluate compound target engagement of a specific individual protein, both protein and ligand are labelled with a fluorophore. This can be combined with fluorescence lifetime imaging microscopy (FLIM),[42] which then allows for both spatial and temporal monitoring of target engagement. For example, FRET–FLIM assays successfully determined target engagement of acyl protein thioesterase 1 (APT1) by palmostatin B,[43] nucleophosmin (NPM) and exportin 1 (XPO1) by centrocountins[44] and 3′,5′-cyclic GMP phosphodiesterase delta (PDEδ) by deltarasin.[45]

In bioluminescent resonance transfer (BRET) assays, the donor is bioluminescent, typically a luciferase enzyme.[46] BRET does not suffer from issues inherent to the optical excitation of the donor needed in FRET, *i.e.*

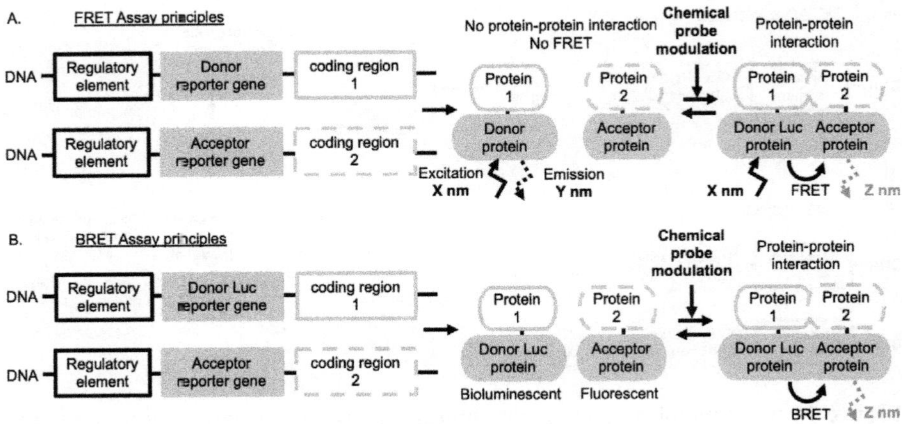

Figure 10.7 Principles of resonance energy transfer. (A) In fluorescence resonance energy transfer (FRET) assay, the two proteins of interest 1 and 2 are linked to a donor and an acceptor fluorescent protein. The interaction between proteins 1 and 2 brings the donor and acceptor fluorescent proteins closer. At a distance of <10 nm, upon excitation of the donor protein the subsequent emission energy is transferred to the acceptor protein (FRET), which triggers its emission. (B) In bioluminescence resonance energy transfer (BRET) assay, the donor protein is bioluminescent, *e.g.* a luciferase (Luc).

background fluorescence, excitation of both donor and acceptor and photobleaching. BRET, similarly to FRET, has been employed for both protein–protein interaction studies[47] and target engagement of chemical probes in cells.[48,49] More recently, the development of nanoluciferase (NLuc) and its use as part of the NanoBRET platform together with the HaloTag (HT) system as fluorescent acceptor has widened the applications of BRET assays. For instance, the action of I-BET151, an inhibitor of the bromodomain protein BRD4 and chromatin interaction, was evaluated in living cells expressing NLuc–BRD4 and histone H3.3–HT fusion proteins.[50] The binding of analogues of BI2536, a known dual BRD4 bromodomain/polo-like kinase 1 (PLK1) kinase inhibitor, to members of the bromodomain and extra-terminal domain (BET) family was also investigated using NLuc–BET and the known BET bromodomain probe JQ1 labelled with a fluorescent acceptor.[51] The study of ligand–receptor binding, such as G-protein coupled receptors and receptor tyrosine kinase receptors (GPCRs and RTKRs), previously challenging with early BRET versions due to the large size of the FLuc and RLuc interfering with the receptor protein, was accessible in living cells with N-terminal NLuc-tagged receptors and bound fluorescent ligands.[52,53] A real-time NanoBRET-based binding assay was also performed to study the interaction of unlabelled ligands with N-terminal NLuc-tagged histamine H3 and H4 receptors (H3R and H4R) against fluorophore-labelled histamine and clobenpropit.[54]

Finally, RET and complementation principles can be combined. BiFC–FRET and BRET–BiFC allow the visualization and study of protein complexes such as GPCR oligomerization.[41,55,56] The BRET signal resulting from the combination of bimolecular luminescence complementation (BiLC) of dopamine D2 receptor (D2R)–adenosine receptor A2a (A2aR) heterodimers with a YFP-tagged β-arrestin in the presence of D2R agonists quinpirole or UNC9994 demonstrated that the formation of D2R–A2aR heterodimers was necessary to promote β-arrestin2 recruitment.[57] The NanoBIT technology takes advantage of the NanoBRET technology with an NLuc-derived split luciferase reporter.[53] While NanoBIT technology is also used for the visualization of protein complexes, small-molecule–protein interactions have also been investigated, *e.g.* the stimulation of β-arrestin2 recruitment to cannabinoid receptors by synthetic cannabinoids and their metabolites.[58]

10.3.4 Fluorescence Recovery After Photobleaching (FRAP)

Fluorescence recovery after photobleaching (FRAP) measures the binding of a compound to proteins labelled with a fluorescent protein such as GFP.[59] Cells expressing the fusion protein are exposed to a high-intensity laser pulse that induces bleaching. The recovery of fluorescence measured by confocal microscopy is then plotted against time. The resulting curves therefore represent the diffusion and binding of the fluorescent target to its partner, *e.g.* the interaction of fluorescent bromodomains with chromatin. In this case, an inhibitor of this interaction accelerates the recovery of fluorescence.

10.4 Label-free Target Deconvolution Assays

10.4.1 Image-based Assays for Compound Characterization

Although systems-level approaches such as proteomics and ribonucleic acid (RNA) sequencing can complement target-specific assays and permit broad profiling of drugs and tool compounds in cellular systems, the precise tracking of compound action over time and *via* multiple markers, particularly in relevant model systems such as patient samples and primary cells, has remained a challenge. Recent advances in high-content microscopy and, importantly, image data processing are closing in on this gap and bear potential for immediate application in clinical settings.[60]

In principle, image-based readouts have been widely applied for both compound and genomic screening purposes. Early examples of using multivariate image features to profile compounds and predict mechanisms of action used 4′,6-diamidino-2-phenylindole (DAPI) staining in combination with cytoskeletal and other markers.[61]

10.4.2 Signature Profiling

In order to derive a mechanism of action for a given compound with unknown target(s), several groups have put forward approaches that generate a compound-specific signature using a particular readout or set of features and then compare this fingerprint with a panel of reference compounds (Figure 10.8). A major programme utilizing this 'guilt-by-association' principle is the National Cancer Institute's Developmental Therapeutics Program, for which thousands of compounds have been screened in cell proliferation assays against a defined panel of cell lines representing different tumour types. A 'COMPARE' algorithm then allows the comparison of compound profiles ('fingerprints') *versus* all previously tested compounds with known mechanism of action. Noteworthy examples of the successful application of this methodology include the identification of kenpaullone as an inhibitor of cyclin-dependent kinases (CDKs)[62] and induction of deoxyribonucleic acid (DNA) damage as a key mechanism for 'survivin suppressants' such as YM155.[63] More recently, the concept has been expanded to different readouts such as gene expression (*e.g.* the 'Connectivity Map'[64,65]), protein expression,[66] various image-based readouts[61,67,68] and cytokine release.[69] In contrast to most of these approaches, which mainly infer pharmacological activity, BioMAP (DiscoverX) represents a panel of pharmacological assays with direct readouts.

10.4.3 Cellular Thermal Shift Assay (CETSA)

The cellular thermal shift assay (CETSA) is based on the thermal shift assay (TSA), which has been used for high-throughput screens in drug discovery and structural biology projects for many years.[70] Both assays exploit the biophysical principle of thermal stabilization of a macromolecule due to the free energy of binding of a small molecule. In a conventional TSA, this is exploited by the addition of a dye that fluoresces as it interacts with the hydrophobic interior of the protein, thus monitoring protein unfolding. CETSA, on the other hand, relies on the removal of unfolded protein aggregates by

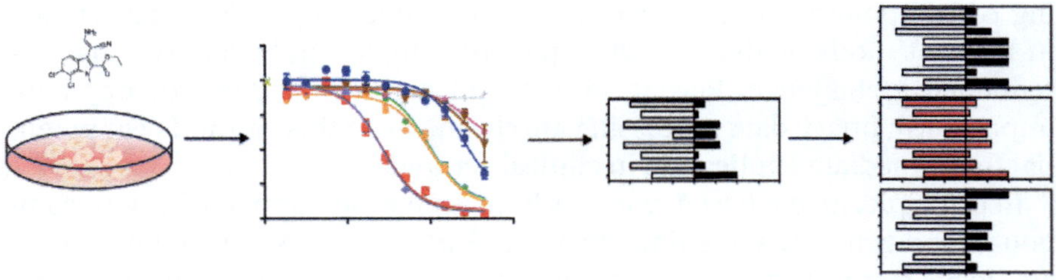

Figure 10.8 Mono- or multiparametric signature profiling of small molecules *versus* a set of reference fingerprints generated for compounds with known targets can reveal similarities and thus point to potential mechanisms.

centrifugation; the folded proteins that remain in solution are separated and those that were engaged by the ligand will be stabilized at higher temperatures, with WB being the commonly used method for target engagement validation.[71]

In a CETSA–WB assay, cells (or cell lysate) are incubated with a small molecule for a period of time that permits compound permeation and binding to its target(s). The cells are then aliquoted into tubes and the aliquots are treated with a temperature gradient. Following lysis, centrifugation removes aggregated and unfolded proteins and the samples are loaded onto an SDS-PAGE system for later analysis by immunoblotting. An increase in signal for the treated group at higher temperatures is indicative of thermal stabilization and thus target engagement (Figure 10.9).

Initially, the proof-of-concept CETSA experiments revolved largely around validating existing probes present in the literature, such as p38α kinase inhibitors[71] or dihydrofolate reductase (DHFR) antagonists in both live cells and lysate,[72] exhibiting strong thermal stabilization effects. Working with lysates instead of intact cells permits the measurement of target engagement to natural endogenous proteins for compounds that are not cell permeable. This was demonstrated, for example, with murine double minute 2 (MDM2) family-targeting probes based on stapled peptides, many of which exhibited strong thermal shifts in lysate but lack thereof in live cells, likely attributable to poor cell permeability.[73] A similar approach has also been used to distinguish between active and inactive nicotinamide phosphoribosyltransferase (NAMPT) inhibitors.[74]

Since then, the use of CETSA to validate small-molecule target engagement in early stages of medicinal chemistry or drug discovery pipelines has increased, as represented by a sharp increase in the number of publications

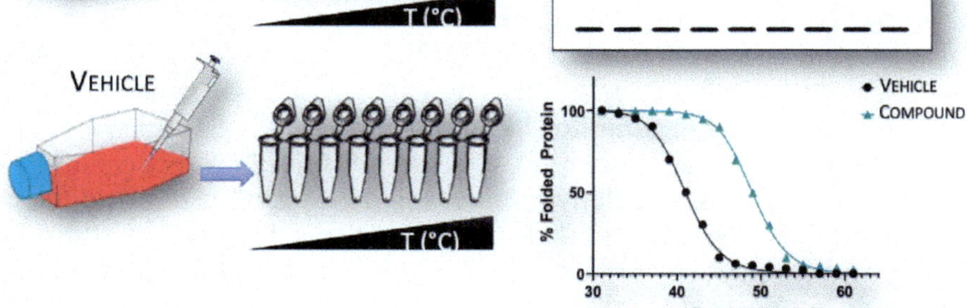

Figure 10.9 Cellular target engagement assay to study target engagement with a WB readout.

showing the applicability of the approach for a number of different target proteins.[75] In a recent study, the compounds ABT-199 (venetoclax) and RG7388 (idasanutlin), currently being tested in combination in clinical trials, were profiled for their off-target interactions by affinity chromatography and revealed protein hits. These hits, such as the DNA repair protein PMS1 (Postmeiotic Segregation Increased 1) were then validated by live cell CETSA immunoblotting.[76]

Because CETSA, unlike conventional assays using recombinant proteins, takes into account the effects of the target's environment on compound binding, it is in principle possible to harness the technology to explore mechanisms of resistance. For instance, anaplastic lymphoma kinase (ALK)-positive cells that exhibit resistance to crizotinib were shown to reduce strongly the thermal stabilization of ALK by the drug in responsive cell lines.[77]

Conventionally, when examining the target engagement of a compound using CETSA, the readout method of choice is SDS-PAGE in conjunction with WB. However, although the technique is relatively simple and allows for fast data acquisition and quantification *via* densitometry, it has a number of drawbacks, including limited availability of high-quality antibodies leading to spurious WB readouts,[78] and also issues with densitometry itself.[79] In addition, CETSA sample preparation relies on the removal of thermally denatured proteins *via* centrifugation following lysis, which can create additional problems if the detergent used for cell lysis has the capacity to resolubilize unfolded aggregates. These problems prompted the development of readout methods that not only circumvent the need for WB but also discriminate between the folded and the unfolded versions of the target protein *in situ* without the need for separation.

10.4.4 Thermal Profiling for Target Deconvolution

Although useful for assaying cellular target engagement, the CETSA immunoblot readout is naturally limited to only testing targets of a particular small molecule. The combination of the CETSA principle with quantitative protein MS using isobaric mass tags[80] [thermal (proteome) profiling, TPP or MS-CETSA], has paved the way towards unbiased small-molecule target engagement profiling on the global level (Figure 10.10). Initial studies investigated kinase inhibitors allowing for 'label-free' profiling of over 3000 proteins in intact cells. These studies revealed, *e.g.*, ferrochelatase as an off-target of clinical kinase inhibitor[81] and methotrexate being a thermal stabilizer of thymidylate synthase (TS), likely as a result of its conversion to a more potent polyglutamate derivative, raising the possibility of tracking not only small molecules, but also their metabolites, inside cells.[82] In addition, despite not being a direct target of the Abelson murine leukaemia viral oncogene homologue 1 (ABL1) inhibitor dasatinib, the v-crk avian sarcoma virus CT10 oncogene homologue-like protein (CRKL) protein was found to be thermally destabilized by it. As CRKL interacts with phosphorylated proteins, it

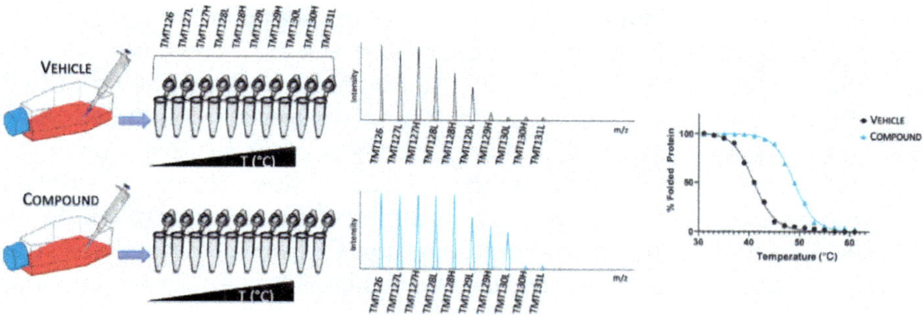

Figure 10.10 Thermal profiling combines CETSA with quantitative mass spectrometry for target identification.

is possible that the negative thermal shift is due to dasatinib blocking the phosphorylation of CRKL interactors by ABL1. As a result, intact cell CETSA also has potential to be used for profiling small molecules in the context of broader signalling pathways. More recently, CETSA was also used to identify parasitic purine nucleoside phosphorylase as the target of the antimalarial agents quinine and mefloquine, showing that the assay can be extended to projects beyond oncology.[83]

In a variation of the above-described thermal profiling approach, the experiment can also be performed in a 2D format where both temperature and compound concentration are varied and taken into account for analysis. This methodology was recently used to examine the off-targets of the pan-HDAC inhibitor panobinostat;[84] in order to distinguish between direct effects of the drug and indirect effects (such as a concentration-dependent increase in expression), the authors performed the experiment on both HepG2 liver cancer cells and cell extracts and looked for the overlap in concentration-dependent thermal stability changes. Four proteins were identified in this way; two of them were canonical panobinostat targets, namely histone deacetylases (HDAC1 and HDAC2), whereas the other two, TTC38 and phenylalanine hydroxylase (PAH), were previously unidentified. Subsequent validation including drug affinity chromatography confirmed both proteins to be direct interactors of panobinostat.

10.4.5 Alternative Strategies for CETSA Output Readout

One of these approaches involves combining CETSA with the aforementioned complementation assay (see Section 10.3.2). The cells are transfected with a constitutively expressed gene encoding a target of choice fused with a split reporter, thus thermal stabilization of the target by the probe will result in a subsequent signal following enzyme complementation. Although this assay bypasses the common issues associated with WB-CETSA, careful optimization must be performed to ensure that a number of factors do not confound the results: protein stabilization by the tag, instead of protein, stabilization by the compound of choice and concomitant unfolding of both protein and tag, the compound of choice does not stabilize the tag instead of the target, and that the unfolding of the protein also results in the concomitant unfolding of the tag. An elegant example of this approach was demonstrated with a galactosidase tag (ProLabel®) to examine binders of the histone lysinemethyltransferase SMYD3,[85] and also with a split luciferase (NanoLuc®) to examine mutant isocitrate dehydrogenase inhibitors.[86] Although expanding the approach to potentially many more even 'non-CETSAble' targets, this strategy is no longer strictly 'label free' as it necessitates transfection of cells with a fusion protein.

A separate approach involves the use of enzyme-linked immunosorbent assay (ELISA)-compatible antibody pairs that specifically recognize only the native folded target protein. By coupling two AlphaScreen® proximity assay beads[87] to two ELISA-specific antibodies against complementary epitopes on the target, one can in principle elegantly differentiate between the native and denatured/aggregated forms of the protein. In a typical experiment, following the heat pulse and lysis, the bead and antibody mix is added to the lysate. After incubation, the stability of the target protein is measured using a plate reader (Figure 10.11). The additional advantage of this approach is its applicability to high-throughput screening, which allows CETSA to augment and potentially even guide drug discovery at the initial HTS. On the other hand, this method requires extensive optimization, from antibody selection and buffer conditions to the duration of incubation and even selection of appropriate microplates and other hardware. There are publications detailing the use of AlphaLISA–CETSA to screen for cell-active compounds.[86,88]

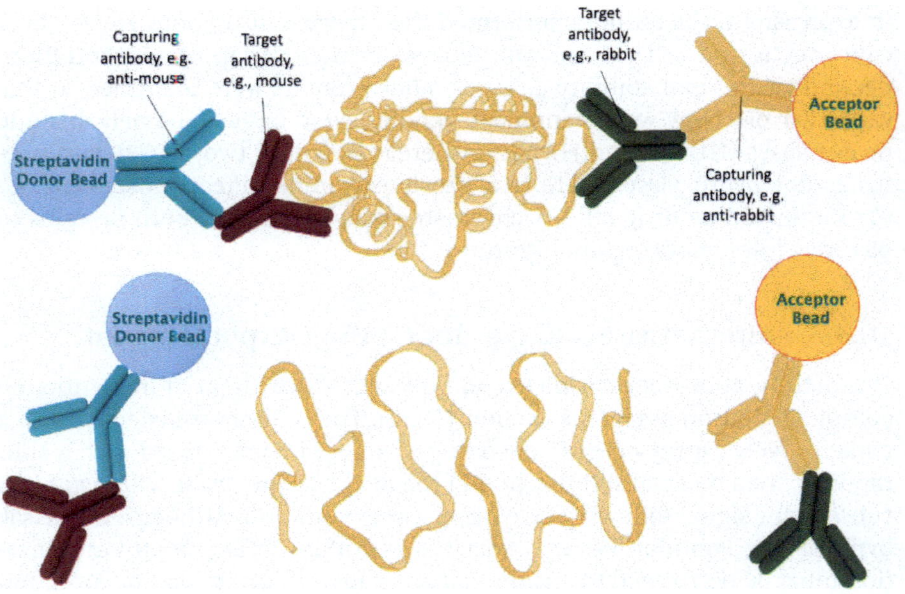

Figure 10.11 Adaptation of the AlphaScreen technology as a CETSA readout.

10.5 Concluding Remarks

Chemical probes are versatile tools for exploring new targets and signalling pathways in live cells, patient samples and, where applicable, animal model systems. As such, the number of potential assays is almost infinite and, ideally, these compounds should be tested in as many as possible to permit insights into the likely context-dependent biology of the respective target. Here, we have focused on methods and concepts that are widely applicable and allow broad phenotypic profiling in various different settings, taking advantage of the multiparametric readouts offered by recent systems biology approaches. Many of these methods can also be used to determine target selectivity and engagement in cells, but there are also techniques that have been optimized for a particular context and may offer advantages under certain circumstances. Therefore, strengths and weaknesses need to be assessed carefully depending on the scientific context, and the use of complementary methods should be considered during project planning. In combination, the power of selective chemical probes and systems pharmacology methods including novel label-free and image-based techniques provides unprecedented potential to discover new biology, develop new drugs and inform therapy.

Abbreviations

A2AR	Adenosine A2 receptor
ABL1	Abelson murine leukaemia viral oncogene homologue 1
ABPP	Activity-based protein profiling
ADP	Adenosine diphosphate
ALK	Anaplastic lymphoma kinase
APT1	Acyl protein thioesterase 1
ATP	Adenosine triphosphate
AzTb	Azido-TAMRA-biotin
BET	Bromodomain and extra-terminal domain
BiFC	Bimolecular fluorescence complementation
BODIPY	Boron dipyrromethene
BRD4	Bromodomain-containing protein 4
BRET	Bioluminescent resonance transfer
BTK	Bruton's tyrosine kinase
cAMP	Cyclic adenosine monophosphate
CAT	Chloramphenicol acetyltransferase
CDK	Cyclin-dependent kinase
CETSA	Cellular thermal shift assay
CMV	Cytomegalovirus
CPR	C-terminal positively charged region/fragment
CRBN	Cereblon
CRKL	v-crk avian sarcoma virus CT10 oncogene homologue-like protein
CuAAC	Copper-catalysed azide–alkyne cycloaddition

D2R	Dopamine D2 receptor
DAPI	4′,6-Diamidino-2-phenylindole
DBB1	DNA damage-binding protein 1
DHFR	Dihydrofolate reductase
DNA	Deoxyribonucleic acid
EA	Enzyme acceptor
EFC	Enzyme fragment complementation
ELISA	Enzyme-linked immunosorbent assay
ePL	Enhanced ProLabel
FAAH	Fatty acid amide hydrolase
FDA	Food and Drug Administration
FG	Ferrite–glycidyl methacrylate
FLIM	Fluorescence lifetime imaging microscopy
FLuc	Firefly luciferase
FRAP	Fluorescence recovery after photobleaching
FRET	Fluorescence or Förster resonance energy transfer
β-gal	β-Galactosidase
GFP	Green fluorescent protein
GPCR	G-protein coupled receptor
GUS	β-Glucuronidase
H3R	Histamine H3 receptor
H4R	Histamine H4 receptor
HAT	Histone acetyltransferase
HDAC	Histone deacetylase
HT	Halo-tag
HTS	High-throughput screen
IA	Iodoacetamide
LacZ	β-Lactamase
MDM2	Murine double minute 2
MDMX	Murine double minute X
MEK1	Mitogen-activated protein kinase kinase 1
MS	Mass spectrometry
NAMPT	Nicotinamide phosphoribosyltransferase
NLuc	NanoLuc® luciferase
NPM	Nucleophosmin
NPR	N-terminal positively charged region/fragment
PAH	Phenylalanine hydroxylase
PARP	Poly-ADP ribose polymerase
PCA	Protein-fragment complementation assay
PDEδ	3′,5′-Cyclic GMP phosphodiesterase delta
PLK1	Polo-like kinase 1
PMS1	Postmeiotic Segregation Increased 1
PPARγ	Peroxisome proliferator-activated receptor gamma
PRMT3	Protein arginine methyltransferase 3
PROTAC	proteolysis-targeting chimera

RET	Resonance energy transfer
RFP	Red fluorescent protein
RLuc	Renilla luciferase
RNA	Ribonucleic acid
RNF114	Ring finger 114
RTKR	Receptor tyrosine kinase receptor
SDS-PAGE	Sodium dodecyl sulfate polyacrylamide gel electrophoresis
SEAP	Secreted alkaline phosphatase
TAMRA	5-Carboxytetramethylrhodamine
TEV	Tobacco etch virus
TGFβ	Transforming growth factor beta
TPP	Thermal proteome profiling
TS	Thymidylate synthase
TSA	Thermal shift assay
TZD	Thiazolidinone
USP	Ubiquitin-specific protease
UV	Ultraviolet
WB	Western blotting
Wnt	Wingless-related integration site
XPO1	Exportin 1
Y2H	Yeast 2-hybrid
YFP	Yellow fluorescent protein

References

1. C. H. Arrowsmith, J. E. Audia, C. Austin, J. Baell, J. Bennett, J. Blagg, C. Bountra, P. E. Brennan, P. J. Brown, M. E. Bunnage, C. Buser-Doepner, R. M. Campbell, A. J. Carter, P. Cohen, R. A. Copeland, B. Cravatt, J. L. Dahlin, D. Dhanak, A. M. Edwards, S. V. Frye, N. Gray, C. E. Grimshaw, D. Hepworth, T. Howe, K. V. M. Huber, J. Jin, S. Knapp, J. D. Kotz, R. G. Kruger, D. Lowe, M. M. Mader, B. Marsden, A. Mueller-Fahrnow, S. Müller, R. C. O'Hagan, J. P. Overington, D. R. Owen, S. H. Rosenberg, B. Roth, R. Ross, M. Schapira, S. L. Schreiber, B. Shoichet, M. Sundström, G. Superti-Furga, J. Taunton, L. Toledo-Sherman, C. Walpole, M. A. Walters, T. M. Willson, P. Workman, R. N. Young and W. J. Zuercher, *Nat. Chem. Biol.*, 2015, **11**, 536–541.
2. R. E. Moellering and B. F. Cravatt, How Chemoproteomics Can Enable Drug Discovery and Development, *Chem. Biol.*, 2012, **19**, 11–22.
3. O. Hantschel, U. Rix, U. Schmidt, T. Burckstummer, M. Kneidinger, G. Schutze, J. Colinge, K. L. Bennett, W. Ellmeier, P. Valent and G. Superti-Furga, The Btk tyrosine kinase is a major target of the Bcr-Abl inhibitor dasatinib, *Proc. Natl. Acad. Sci. U. S. A.*, 2007, **104**, 13283–13288.
4. J. Li, U. Rix, B. Fang, Y. Bai, A. Edwards, J. Colinge, K. L. Bennett, J. Gao, L. Song, S. Eschrich, G. Superti-Furga, J. Koomen and E. B. Haura,

A chemical and phosphoproteomic characterization of dasatinib action in lung cancer, *Nat. Chem. Biol.*, 2010, **6**, 291–299.

5. V. Bermúdez, F. Finol, N. Parra, M. Parra, A. Pérez, L. Peñaranda, D. Vílchez, J. Rojas, N. Arráiz and M. Velasco, PPAR-γ Agonists and Their Role in Type 2 Diabetes Mellitus Management, *Am. J. Ther.*, 2010, **17**, 274–283.
6. S. E. Nissen and K. Wolski, Effect of Rosiglitazone on the Risk of Myocardial Infarction and Death from Cardiovascular Causes, *N. Engl. J. Med.*, 2007, **356**, 2457–2471.
7. B. R. Hoffmann, M. F. El-Mansy, D. S. Sem and A. S. Greene, Chemical Proteomics-Based Analysis of Off-Target Binding Profiles for Rosiglitazone and Pioglitazone: Clues for Assessing Potential for Cardiotoxicity, *J. Med. Chem.*, 2012, **55**, 8260–8271.
8. T. Ito, H. Ando, T. Suzuki, T. Ogura, K. Hotta, Y. Imamura, Y. Yamaguchi and H. Handa, Identification of a Primary Target of Thalidomide Teratogenicity, *Science*, 2010, **327**, 1345–1350.
9. L. Liang and D. Astruc, The copper(I)-catalyzed alkyne-azide cycloaddition (CuAAC) "click" reaction and its applications. An overview, *Coord. Chem. Rev.*, 2011, **255**, 2933–2945.
10. S. I. Presolski, V. P. Hong and M. G. Finn, Copper-Catalyzed Azide-Alkyne Click Chemistry for Bioconjugation, *Curr. Protoc. Chem. Biol.*, 2011, **3**, 153–162.
11. J. A. Ward, L. McLellan, M. Stockley, K. R. Gibson, G. A. Whitlock, C. Knights, J. A. Harrigan, X. Jacq and E. W. Tate, Quantitative Chemical Proteomic Profiling of Ubiquitin Specific Proteases in Intact Cancer Cells, *ACS Chem. Biol.*, 2016, **11**, 3268–3272.
12. J. Stefaniak, A. M. Lewis, D. Conole, S. R. G. Galan, C. J. R. Bataille, G. M. Wynne, M. P. Castaldi, T. Lundbäck, A. J. Russell and K. V. M. Huber, Chemical Instability and Promiscuity of Arylmethylidenepyrazolinone-Based MDMX Inhibitors, *ACS Chem. Biol.*, 2018, 2849–2854.
13. M. Gehringer and S. A. Laufer, Emerging and Re-emerging warheads for targeted covalent inhibitors: applications in medicinal chemistry and chemical biology, *J. Med. Chem.*, 2019, **62**, 5673–5724.
14. H. Shi, X. Cheng, S. K. Sze and S. Q. Yao, Proteome profiling reveals potential cellular targets of staurosporine using a clickable cell-permeable probe, *Chem. Commun.*, 2011, **47**, 11306.
15. H. Shi, C.-J. Zhang, G. Y. J. Chen and S. Q. Yao, Cell-based proteome profiling of potential dasatinib targets by use of affinity-based probes, *J. Am. Chem. Soc.*, 2012, **134**, 3001–3014.
16. Z. Li, P. Hao, L. Li, C. Y. J. Tan, X. Cheng, G. Y. J. Chen, S. K. Sze, H.-M. Shen and S. Q. Yao, Design and synthesis of minimalist terminal alkyne-containing diazirine photo-crosslinkers and their incorporation into kinase inhibitors for cell- and tissue-based proteome profiling, *Angew. Chem., Int. Ed.*, 2013, **52**, 8551–8556.

17. C. Guo, Y. Chang, X. Wang, C. Zhang, P. Hao, K. Ding and Z. Li, Minimalist linkers suitable for irreversible inhibitors in simultaneous proteome profiling, live-cell imaging and drug screening, *Chem. Commun.*, 2019, **55**, 834–837.
18. R. Kaur, P. Sidhu and S. Singh, What failed BIA 10-2474 Phase I clinical trial? Global speculations and recommendations for future Phase I trials, *J. Pharmacol. Pharmacother.*, 2016, **7**, 120.
19. A. C. M. van Esbroeck, A. P. A. Janssen, A. B. Cognetta, D. Ogasawara, G. Shpak, M. van der Kroeg, V. Kantae, M. P. Baggelaar, F. M. S. de Vrij, H. Deng, M. Allarà, F. Fezza, Z. Lin, T. van der Wel, M. Soethoudt, E. D. Mock, H. den Dulk, I. L. Baak, B. I. Florea, G. Hendriks, L. De Petrocellis, H. S. Overkleeft, T. Hankemeier, C. I. De Zeeuw, V. Di Marzo, M. Maccarrone, B. F. Cravatt, S. A. Kushner and M. van der Stelt, Activity-based protein profiling reveals off-target proteins of the FAAH inhibitor BIA 10-2474, *Science*, 2017, **356**, 1084–1087.
20. Z. Huang, D. Ogasawara, U. I. Seneviratne, A. B. Cognetta, C. W. am Ende, D. M. Nason, K. Lapham, J. Litchfield, D. S. Johnson and B. F. Cravatt, Global portrait of protein targets of metabolites of the neurotoxic compound bia 10–2474, *ACS Chem. Biol.*, 2019, **14**, 192–197.
21. E. Weerapana, C. Wang, G. M. Simon, F. Richter, S. Khare, M. B. D. Dillon, D. A. Bachovchin, K. Mowen, D. Baker and B. F. Cravatt, Quantitative reactivity profiling predicts functional cysteines in proteomes, *Nature*, 2010, **468**, 790–795.
22. J. N. Spradlin, X. Hu, C. C. Ward, S. M. Brittain, M. D. Jones, L. Ou, M. To, A. Proudfoot, E. Ornelas, M. Woldegiorgis, J. A. Olzmann, D. E. Bussiere, J. R. Thomas, J. A. Tallarico, J. M. McKenna, M. Schirle, T. J. Maimone and D. K. Nomura, Harnessing the anti-cancer natural product nimbolide for targeted protein degradation, *Nat. Chem. Biol.*, 2019, **15**, 747–755.
23. W. A. Lea and A. Simeonov, Fluorescence polarization assays in small molecule screening, *Expert Opin. Drug Discovery*, 2011, **6**, 17–32.
24. H. Zhang, Q. Wu and M. Y. Berezin, Fluorescence anisotropy (polarization): from drug screening to precision medicine, *Expert Opin. Drug Discovery*, 2015, **10**, 1145–1161.
25. J. M. Dubach, C. Vinegoni, R. Mazitschek, P. Fumene Feruglio, L. A. Cameron and R. Weissleder, In vivo imaging of specific drug–target binding at subcellular resolution, *Nat. Commun.*, 2014, **5**, 3946.
26. J. M. Dubach, E. Kim, K. Yang, M. Cuccarese, R. J. Giedt, L. G. Meimetis, C. Vinegoni and R. Weissleder, Quantitating drug-target engagement in single cells *in vitro* and *in vivo*, *Nat. Chem. Biol.*, 2017, **13**, 168–173.
27. Y. Wang, A. Cornett, F. J. King, Y. Mao, F. Nigsch, C. G. Paris, G. McAllister and J. L. Jenkins, Evidence-based and quantitative prioritization of tool compounds in phenotypic drug discovery, *Cell Chem. Biol.*, 2016, **23**, 862–874.

28. E. D. Martinez, A. B. Dull, J. A. Beutler and G. L. Hager, High-content fluorescence-based screening for epigenetic modulators, *Methods Enzymol.*, 2006, **414**, 21–36.
29. P. Atadja, Development of the pan-DAC inhibitor panobinostat (LBH589): successes and challenges, *Cancer Lett.*, 2009, **280**, 233–241.
30. D. Son, C. S. Kim, K. R. Lee and H.-J. Park, Identification of new quinic acid derivatives as histone deacetylase inhibitors by fluorescence-based cellular assay, *Bioorg. Med. Chem. Lett.*, 2016, **26**, 2365–2369.
31. R. M. Eglen, Enzyme fragment complementation: a flexible high throughput screening assay technology, *Assay Drug Dev. Technol.*, 2002, **1**, 97–104.
32. H. Ü. Kaniskan, M. M. Szewczyk, Z. Yu, M. S. Eram, X. Yang, K. Schmidt, X. Luo, M. Dai, F. He, I. Zang, Y. Lin, S. Kennedy, F. Li, E. Dobrovetsky, A. Dong, D. Smil, S.-J. Min, M. Landon, J. Lin-Jones, X.-P. Huang, B. L. Roth, M. Schapira, P. Atadja, D. Barsyte-Lovejoy, C. H. Arrowsmith, P. J. Brown, K. Zhao, J. Jin and M. Vedadi, A potent, selective and cell-active allosteric inhibitor of protein arginine methyltransferase 3 (PRMT3), *Angew. Chem., Int. Ed. Engl.*, 2015, **54**, 5166–5170.
33. D. S. Auld, C. A. Davis, M. Jimenez, S. Knight and J. P. Orme, Examining ligand-based stabilization of proteins in cells with mek1 kinase inhibitors, *Assay Drug Dev. Technol.*, 2015, **13**, 266–276.
34. J. Schulze, D. Moosmayer, J. Weiske, A. Fernández-Montalván, C. Herbst, M. Jung, B. Haendler and B. Bader, Cell-based protein stabilization assays for the detection of interactions between small-molecule inhibitors and BRD4, *J. Biomol. Screening*, 2015, **20**, 180–189.
35. S. W. Michnick, P. H. Ear, E. N. Manderson, I. Remy and E. Stefan, Universal strategies in research and drug discovery based on protein-fragment complementation assays, *Nat. Rev. Drug Discovery*, 2007, **6**, 569–582.
36. I. Remy and S. W. Michnick, Mapping biochemical networks with protein fragment complementation assays, *Methods Mol. Biol.*, 2015, **1278**, 467–481.
37. M. L. MacDonald, J. Lamerdin, S. Owens, B. H. Keon, G. K. Bilter, Z. Shang, Z. Huang, H. Yu, J. Dias, T. Minami, S. W. Michnick and J. K. Westwick, Identifying off-target effects and hidden phenotypes of drugs in human cells, *Nat. Chem. Biol.*, 2006, **2**, 329–337.
38. H. Yu, M. West, B. H. Keon, G. K. Bilter, S. Owens, J. Lamerdin and J. K. Westwick, Measuring drug action in the cellular context using protein-fragment complementation assays, *Assay Drug Dev. Technol.*, 2003, **1**, 811–822.
39. K. E. Miller, Y. Kim, W.-K. Huh and H.-O. Park, Bimolecular fluorescence complementation (BiFC) analysis: advances and recent applications for genome-wide interaction studies, *J. Mol. Biol.*, 2015, **427**, 2039–2055.
40. T. K. Kerppola, Visualization of molecular interactions using bimolecular fluorescence complementation analysis: characteristics of protein fragment complementation, *Chem. Soc. Rev.*, 2009, **38**, 2876–2886.

41. Y. Kodama and C.-D. Hu, Bimolecular fluorescence complementation (BiFC): a 5-year update and future perspectives, *Biotechniques*, 2012, **53**, 285–298.
42. J. Schoberer and S. W. Botchway, Investigating protein-protein interactions in the plant endomembrane system using multiphoton-induced FRET-FLIM, *Methods Mol. Biol.*, 2014, **1209**, 81–95.
43. F. J. Dekker, O. Rocks, N. Vartak, S. Menninger, C. Hedberg, R. Balamurugan, S. Wetzel, S. Renner, M. Gerauer, B. Schölermann, M. Rusch, J. W. Kramer, D. Rauh, G. W. Coates, L. Brunsveld, P. I. H. Bastiaens and H. Waldmann, Small-molecule inhibition of APT1 affects Ras localization and signaling, *Nat. Chem. Biol.*, 2010, **6**, 449–456.
44. H. Dückert, V. Pries, V. Khedkar, S. Menninger, H. Bruss, A. W. Bird, Z. Maliga, A. Brockmeyer, P. Janning, A. Hyman, S. Grimme, M. Schürmann, H. Preut, K. Hübel, S. Ziegler, K. Kumar and H. Waldmann, Natural product-inspired cascade synthesis yields modulators of centrosome integrity, *Nat. Chem. Biol.*, 2011, **8**, 179–184.
45. G. Zimmermann, B. Papke, S. Ismail, N. Vartak, A. Chandra, M. Hoffmann, S. A. Hahn, G. Triola, A. Wittinghofer, P. I. H. Bastiaens and H. Waldmann, Small molecule inhibition of the KRAS-PDEδ interaction impairs oncogenic KRAS signalling, *Nature*, 2013, **497**, 638–642.
46. J. Bacart, C. Corbel, R. Jockers, S. Bach and C. Couturier, The BRET technology and its application to screening assays, *Biotechnol. J.*, 2008, **3**, 311–324.
47. F. F. Hamdan, M. Audet, P. Garneau, J. Pelletier and M. Bouvier, High-throughput screening of G protein-coupled receptor antagonists using a bioluminescence resonance energy transfer 1-based beta-arrestin2 recruitment assay, *J. Biomol. Screening*, 2005, **10**, 463–475.
48. L. A. Stoddart, E. K. M. Johnstone, A. J. Wheal, J. Goulding, M. B. Robers, T. Machleidt, K. V. Wood, S. J. Hill and K. D. G. Pfleger, Application of BRET to monitor ligand binding to GPCRs, *Nat. Methods*, 2015, **12**, 661–663.
49. M. B. Robers, M. L. Dart, C. C. Woodroofe, C. A. Zimprich, T. A. Kirkland, T. Machleidt, K. R. Kupcho, S. Levin, J. R. Hartnett, K. Zimmerman, A. L. Niles, R. F. Ohana, D. L. Daniels, M. Slater, M. G. Wood, M. Cong, Y.-Q. Cheng and K. V. Wood, Target engagement and drug residence time can be observed in living cells with BRET, *Nat. Commun.*, 2015, **6**, 10091.
50. T. Machleidt, C. C. Woodroofe, M. K. Schwinn, J. Méndez, M. B. Robers, K. Zimmerman, P. Otto, D. L. Daniels, T. A. Kirkland and K. V. Wood, NanoBRET--A novel BRET platform for the analysis of protein-protein interactions, *ACS Chem. Biol.*, 2015, **10**, 1797–1804.
51. L. W. Koblan, D. L. Buckley, C. J. Ott, M. E. Fitzgerald, S. W. J. Ember, J. Zhu, S. Liu, J. M. Roberts, D. Remillard, S. Vittori, W. Zhang, E. Schonbrunn and J. E. Bradner, Assessment of bromodomain target engagement by a series of BI2536 analogues with miniaturized BET-BRET, *ChemMedChem*, 2016, 2575–2581.

52. L. A. Stoddart, L. E. Kilpatrick and S. J. Hill, NanoBRET approaches to study ligand binding to GPCRs and RTKs, *Trends Pharmacol. Sci.*, 2018, **39**, 136–147.
53. N. C. Dale, E. K. M. Johnstone, C. W. White and K. D. G. Pfleger, NanoBRET: the bright future of proximity-based assays, *Front. Bioeng. Biotechnol.*, 2019, **7**, 56.
54. T. A. M. Mocking, E. W. E. Verweij, H. F. Vischer and R. Leurs, Homogeneous, real-time nanoBRET binding assays for the histamine H3 and H4 receptors on living cells, *Mol. Pharmacol.*, 2018, **94**, 1371–1381.
55. Y. J. Shyu and C.-D. Hu, Fluorescence complementation: an emerging tool for biological research, *Trends Biotechnol.*, 2008, **26**, 622–630.
56. E. Wouters, L. Vasudevan, R. A. J. Crans, D. K. Saini and C. P. Stove, Luminescence- and fluorescence-based complementation assays to screen for GPCR oligomerization: current state of the art, *Int. J. Mol. Sci.*, 2019, **20**, 2958.
57. K. Sahlholm, M. Gómez-Soler, M. Valle-León, M. López-Cano, J. J. Taura, F. Ciruela and V. Fernández-Dueñas, Antipsychotic-like efficacy of dopamine D2 receptor-biased ligands is dependent on adenosine A2A receptor expression, *Mol. Neurobiol.*, 2018, **55**, 4952–4958.
58. A. Cannaert, J. Storme, F. Franz, V. Auwärter and C. P. Stove, Detection and activity profiling of synthetic cannabinoids and their metabolites with a newly developed bioassay, *Anal. Chem.*, 2016, **88**, 11476–11485.
59. M. Philpott, C. M. Rogers, C. Yapp, C. Wells, J.-P. Lambert, C. Strain-Damerell, N. A. Burgess-Brown, A.-C. Gingras, S. Knapp and S. Müller, Assessing cellular efficacy of bromodomain inhibitors using fluorescence recovery after photobleaching, *Epigenet. Chromatin*, 2014, **7**, 14.
60. B. Snijder, G. I. Vladimer, N. Krall, K. Miura, A. S. Schmolke, C. Kornauth, O. Lopez de la Fuente, H. S. Choi, E. van der Kouwe, S. Gültekin, L. Kazianka, J. W. Bigenzahn, G. Hoermann, N. Prutsch, O. Merkel, A. Ringler, M. Sabler, G. Jeryczynski, M. E. Mayerhoefer, I. Simonitsch-Klupp, K. Ocko, F. Felberbauer, L. Müllauer, G. W. Prager, B. Korkmaz, L. Kenner, W. R. Sperr, R. Kralovics, H. Gisslinger, P. Valent, S. Kubicek, U. Jäger, P. B. Staber and G. Superti-Furga, Image-based *ex vivo* drug screening for patients with aggressive haematological malignancies: interim results from a single-arm, open-label, pilot study, *Lancet Haematol.*, 2017, e595–e606.
61. Z. E. Perlman, M. D. Slack, Y. Feng, T. J. Mitchison, L. F. Wu and S. J. Altschuler, Multidimensional drug profiling by automated microscopy, *Science*, 2004, **306**, 1194–1198.
62. D. W. Zaharevitz, R. Gussio, M. Leost, A. M. Senderowicz, T. Lahusen, C. Kunick, L. Meijer and E. A. Sausville, Discovery and initial characterization of the paullones, a novel class of small-molecule inhibitors of cyclin-dependent kinases, *Cancer Res.*, 1999, **59**, 2566–2569.
63. T. Glaros, L. Stockwin, M. Mullendore, B. Smith, B. Morrison and D. Newton, The "survivin suppressants" NSC 80467 and YM155 induce a DNA damage response, *Cancer Chemother. Pharmacol.*, 2012, 1–6.

64. J. Lamb, E. D. Crawford, D. Peck, J. W. Modell, I. C. Blat, M. J. Wrobel, J. Lerner, J.-P. Brunet, A. Subramanian, K. N. Ross, M. Reich, H. Hieronymus, G. Wei, S. A. Armstrong, S. J. Haggarty, P. A. Clemons, R. Wei, S. A. Carr, E. S. Lander and T. R. Golub, The Connectivity Map: using gene-expression signatures to connect small molecules, genes, and disease, *Science*, 2006, **313**, 1929–1935.
65. A. Subramanian, R. Narayan, S. M. Corsello, D. D. Peck, T. E. Natoli, X. Lu, J. Gould, J. F. Davis, A. A. Tubelli, J. K. Asiedu, D. L. Lahr, J. E. Hirschman, Z. Liu, M. Donahue, B. Julian, M. Khan, D. Wadden, I. C. Smith, D. Lam, A. Liberzon, C. Toder, M. Bagul, M. Orzechowski, O. M. Enache, F. Piccioni, S. A. Johnson, N. J. Lyons, A. H. Berger, A. F. Shamji, A. N. Brooks, A. Vrcic, C. Flynn, J. Rosains, D. Y. Takeda, R. Hu, D. Davison, J. Lamb, K. Ardlie, L. Hogstrom, P. Greenside, N. S. Gray, P. A. Clemons, S. Silver, X. Wu, W.-N. Zhao, W. Read-Button, X. Wu, S. J. Haggarty, L. V. Ronco, J. S. Boehm, S. L. Schreiber, J. G. Doench, J. A. Bittker, D. E. Root, B. Wong and T. R. Golub, A next generation connectivity map: L1000 platform and the first 1,000,000 profiles, *Cell*, 2017, **171**, 1437–1452.e17.
66. M. Kawatani, H. Takayama, M. Muroi, S. Kimura, T. Maekawa and H. Osada, Identification of a small-molecule inhibitor of dna topoisomerase II by proteomic profiling, *Chem. Biol.*, 2011, **18**, 743–751.
67. V. Ljosa, P. D. Caie, R. Ter Horst, K. L. Sokolnicki, E. L. Jenkins, S. Daya, M. E. Roberts, T. R. Jones, S. Singh, A. Genovesio, P. A. Clemons, N. O. Carragher and A. E. Carpenter, Comparison of methods for image-based profiling of cellular morphological responses to small-molecule treatment, *J. Biomol. Screening*, 2013, **18**, 1321–1329.
68. F. Reisen, A. Sauty de Chalon, M. Pfeifer, X. Zhang, D. Gabriel and P. Selzer, Linking phenotypes and modes of action through high-content screen fingerprints, *Assay Drug Dev. Technol.*, 2015, **13**, 415–427.
69. F. Shah, A. F. Stepan, A. O'Mahony, S. Velichko, A. E. Folias, C. Houle, C. L. Shaffer, J. Marcek, J. Whritenour, R. Stanton and E. L. Berg, Mechanisms of Skin toxicity associated with metabotropic glutamate receptor 5 negative allosteric modulators, *Cell Chem. Biol.*, 2017, **24**, 858–869.e5.
70. F. H. Niesen, H. Berglund and M. Vedadi, The use of differential scanning fluorimetry to detect ligand interactions that promote protein stability, *Nat. Protoc.*, 2007, **2**, 2212–2221.
71. R. Jafari, H. Almqvist, H. Axelsson, M. Ignatushchenko, T. Lundbäck, P. Nordlund and D. M. Molina, The cellular thermal shift assay for evaluating drug target interactions in cells, *Nat. Protoc.*, 2014, **9**, 2100–2122.
72. D. M. Molina, R. Jafari, M. Ignatushchenko, T. Seki, E. A. Larsson, C. Dan, L. Sreekumar, Y. Cao and P. Nordlund, Monitoring drug target engagement in cells and tissues using the cellular thermal shift assay, *Science*, 2013, **341**, 84–87.
73. B. X. Tan, C. J. Brown, F. J. Ferrer, T. Y. Yuen, S. T. Quah, B. H. Chan, A. E. Jansson, H. L. Teo, P. Nordlund and D. P. Lane, Assessing the efficacy of Mdm2/Mdm4-inhibiting stapled peptides using cellular thermal shift assays, *Sci. Rep.*, 2015, **5**, 12116.

74. T.-Y. Xu, S.-L. Zhang, G.-Q. Dong, X.-Z. Liu, X. Wang, X.-Q. Lv, Q.-J. Qian, R.-Y. Zhang, C.-Q. Sheng and C.-Y. Miao, Discovery and characterization of novel small-molecule inhibitors targeting nicotinamide phosphoribosyltransferase, *Sci. Rep.*, 2015, **5**, 10043.
75. B. Seashore-Ludlow, H. Axelsson and T. Lundbäck, Perspective on CETSA literature: toward more quantitative data interpretation, *SLAS Discovery*, 2020, **25**(2), 118–126.
76. D. Zhu, H. Guo, Y. Chang, Y. Ni, L. Li, Z.-M. Zhang, P. Hao, Y. Xu, K. Ding and Z. Li, Cell- and tissue-based proteome profiling and dual imaging of apoptosis markers with probes derived from venetoclax and idasanutlin, *Angew. Chem., Int. Ed.*, 2018, **57**, 9284–9289.
77. A. Alshareef, H.-F. Zhang, Y.-H. Huang, C. Wu, J. D. Zhang, P. Wang, A. El-Sehemy, M. Fares and R. Lai, The use of cellular thermal shift assay (CETSA) to study Crizotinib resistance in ALK-expressing human cancers, *Sci. Rep.*, 2016, **6**, 33710.
78. J. E. Gilda, R. Ghosh, J. X. Cheah, T. M. West, S. C. Bodine and A. V. Gomes, Western blotting inaccuracies with unverified antibodies: need for a western blotting minimal reporting standard (WBMRS), *PLoS One*, 2015, **10**, e0135392.
79. T. A. J. Butler, J. W. Paul, E.-C. Chan, R. Smith and J. M. Tolosa, Misleading westerns: common quantification mistakes in western blot densitometry and proposed corrective measures, *BioMed Res. Int.*, 2019, **2019**, 1–15.
80. N. Rauniyar and J. R. Yates, Isobaric labeling-based relative quantification in shotgun proteomics, *J. Proteome Res.*, 2014, **13**, 5293–5309.
81. M. M. Savitski, F. B. M. Reinhard, H. Franken, T. Werner, M. F. Savitski, D. Eberhard, D. M. Molina, R. Jafari, R. B. Dovega, S. Klaeger, B. Kuster, P. Nordlund, M. Bantscheff and G. Drewes, Tracking cancer drugs in living cells by thermal profiling of the proteome, *Science*, 2014, 1255784.
82. K. V. M. Huber, K. M. Olek, A. C. Müller, C. S. H. Tan, K. L. Bennett, J. Colinge and G. Superti-Furga, Proteome-wide drug and metabolite interaction mapping by thermal-stability profiling, *Nat. Methods*, 2015, **12**, 1055–1057.
83. J. M. Dziekan, H. Yu, D. Chen, L. Dai, G. Wirjanata, A. Larsson, N. Prabhu, R. M. Sobota, Z. Bozdech and P. Nordlund, Identifying purine nucleoside phosphorylase as the target of quinine using cellular thermal shift assay, *Sci. Transl. Med.*, 2019, **11**, eaau3174.
84. I. Becher, T. Werner, C. Doce, E. A. Zaal, I. Togel, C. A. Khan, A. Rueger, M. Muelbaier, E. Salzer, C. R. Berkers, P. F. Fitzpatrick, M. Bantscheff and M. M. Savitski, Thermal profiling reveals phenylalanine hydroxylase as an off-target of panobinostat, *Nat. Chem. Biol.*, 2016, **12**, 908–910.
85. D. E. McNulty, W. G. Bonnette, H. Qi, L. Wang, T. F. Ho, A. Waszkiewicz, L. A. Kallal, R. P. Nagarajan, M. Stern, A. M. Quinn, C. L. Creasy, D.-S. Su, A. P. Graves, R. S. Annan, S. M. Sweitzer and M. A. Holbert, A high-throughput dose-response cellular thermal shift assay for rapid screening of drug target engagement in living cells, exemplified using SMYD3 and IDO1, *SLAS Discovery*, 2018, **23**, 34–46.

86. N. J. Martinez, R. R. Asawa, M. G. Cyr, A. Zakharov, D. J. Urban, J. S. Roth, E. Wallgren, C. Klumpp-Thomas, N. P. Coussens, G. Rai, S.-M. Yang, M. D. Hall, J. J. Marugan, A. Simeonov and M. J. Henderson, A widely-applicable high-throughput cellular thermal shift assay (CETSA) using split Nano Luciferase, *Sci. Rep.*, 2018, **8**, 9472.
87. E. F. Ullman, H. Kirakossian, S. Singh, Z. P. Wu, B. R. Irvin, J. S. Pease, A. C. Switchenko, J. D. Irvine, A. Dafforn and C. N. Skold, Luminescent oxygen channeling immunoassay: measurement of particle binding kinetics by chemiluminescence, *Proc. Natl. Acad. Sci. U. S. A.*, 1994, **91**, 5426–5430.
88. J. Shaw, M. Leveridge, C. Norling, J. Karén, D. M. Molina, D. O'Neill, J. E. Dowling, P. Davey, S. Cowan, M. Dabrowski, M. Main and D. Gianni, Determining direct binders of the Androgen Receptor using a high-throughput Cellular Thermal Shift Assay, *Sci. Rep.*, 2018, **8**, 163.

CHAPTER 11

Chemical Probes in Cellular Assays for Target Validation and Screening in Neurodegeneration

ANA M. SILVA*[a,b,c], ANNA CEDERBALK[a], DANIEL EBNER[b] AND ELENA DI DANIEL[a]

[a]Alzheimer's Research UK Oxford Drug Discovery Institute, Nuffield Department of Medicine, University of Oxford, Old Road Campus, Roosevelt Drive, Oxford OX3 7FZ, UK; [b]Target Discovery Institute, Nuffield Department of Medicine, University of Oxford, Old Road Campus, Roosevelt Drive, Oxford OX3 7FZ, UK; [c]Present address: Medical Research Council Centre for Molecular Bacteriology and Infection, Department of Infectious Disease, Imperial College London, Armstrong Road, London SW7 2AZ, UK
*E-mail: a.ferreira-da-silva@imperial.ac.uk

11.1 Introduction

Therapies to prevent or treat neurodegenerative diseases, due to their increasing prevalence in our ageing population, are in high demand. The majority of neurodegenerative conditions are chronic diseases caused by a combination of genetic, epigenetic and environmental factors rather than a single Mendelian mutation. Common hallmarks of these diseases include an age-dependent accumulation of abnormal intracellular and/or extracellular

Chemical Biology No. 16
The Discovery and Utility of Chemical Probes in Target Discovery
Edited by Paul Brennan and Saleta Vazquez Rodriguez

Published by the Royal Society of Chemistry, www.rsc.org

protein aggregates, synaptic dysfunction and the subsequent death of specific neuronal cell populations, resulting in impaired cognition.[1] For instance, *post-mortem* tissue analysis showed that in Alzheimer's disease (AD), there is an accumulation of extracellular β-amyloid (Aβ) plaques and intracellular phosphorylated tau (p-tau) neurofibrillary tangles (NFTs). Similarly, aggregation of intracellular α-synuclein (α-syn) is characteristic of Parkinson's disease (PD).[2,3]

The selection of druggable targets to develop new therapeutics requires an in-depth understanding of disease biology. Although considerable effort and resources have been employed towards understanding neurodegeneration, the pathological molecular mechanisms remain unclear. Consequently, drug discovery for neurodegenerative diseases is challenging, with an overall success rate for market approval of only 0.4%.[4] This is partly due to the use in screening platforms of genetic or toxin-induced cellular and animal models that often do not recapitulate the pathological phenotype in disease-relevant cells, but also due to the lack of selective and potent compounds to investigate potential drug targets. Therefore, there is a need to establish physiologically relevant disease models amenable to high-throughput screening, thereby facilitating the identification of genes and small-molecule disease modulators.[5] To achieve this, researchers can take advantage of the development of human induced pluripotent stem cell (hiPSC)-based disease models and genome editing techniques. Human adult somatic cells from patients can be reprogrammed into iPSCs[6] and then differentiated to any adult somatic cell, including neurons or glial cells.[7] Many iPSC-derived cells from patients have been shown to recapitulate disease pathology, including AD[8] and PD.[9] Genome editing methods, such as CRISPR/Cas9 nuclease systems, now permit targeted mutagenesis in patient-derived cells to investigate gene function or to create isogenic mutation-corrected controls.[10] It is now possible to generate disease-related cells in numbers necessary for conducting large-scale high-throughput drug screens. In contrast to target-based screening, phenotypic screening in cells aims to uncover biologically active hits relevant to a biological process or expected therapeutic outcome, irrespective of the exact mechanism involved.[5] To link confidently the modulation of a specific target to a relevant disease-modifying phenotype, small-molecule tools or chemical probes should be used. These are well-characterized small-molecule reagents, of well-defined biological potency, selectivity and cell permeability.[11]

This chapter advises on good chemical probes that have emerged from target-based and phenotypic screens, which might serve as tools for deciphering the pathophysiology of neurodegenerative disease and identifying potential therapeutic targets. These compounds were evaluated using the criteria established by the Structural Genomics Consortium (SGC) (https://www.thesgc.org/chemical-probes; see Chapter 1) and focusing on human cell models of AD and PD. Chemical probes according to the SGC criteria are (1) *in vitro* IC_{50} or K_D <100 nM, (2) >30-fold selectivity over proteins in the same family and (3) significant on-target cellular activity at 1 μM.

11.2 Use of Chemical Probes in *In Vitro* Models of Neurodegenerative Diseases

11.2.1 Alzheimer's Disease

AD is the most prevalent cause of dementia and is characterized by progressive memory impairment and acute cognitive dysfunction. Approximately 1% of AD cases [familial AD (fAD)] present with early onset due to rare mutations in one of three genes, β-amyloid precursor protein (*APP*), presenilin 1 (*PSEN1*) and presenilin 2 (*PSEN2*). These genetic mutations lead to amyloid beta (Aβ) 1–42 peptide (Aβ42) production and result in extracellular Aβ aggregation. Sporadic AD (sAD) occurs mostly after the age of 65 years from the combined effects of genetic risk factors and environmental factors, such as ageing.[3] Meta-analysis of genome-wide association studies (GWAS) has identified several genetic variants associated with sAD, including *APOE*, *BIN1*, *PICALM*, *CLU*, *CR1*, *TREM2*, *ABCA7*, *SORL1* and *ADAM10*.[12] The pathways associated with AD are involved in APP metabolism, tau binding proteins, endocytosis, cell trafficking and neuroinflammation.[13] Both genetic and sAD forms share common molecular hallmarks characterized by Aβ plaques and NFTs composed of p-tau, resulting in neurodegeneration[3] due to an imbalance between Aβ production and clearance, leading to the formation of NFTs, cell death, vascular damage and dementia.[14] However, the quantity of Aβ plaques does not seem to correlate with symptom severity. Aβ oligomers, rather than plaques, are thought to be more toxic and to correlate better with cognitive decline.[15] Hyperphosphorylated tau also forms toxic oligomers that ultimately lead to NFTs and neuronal apoptosis. As opposed to Aβ plaques, NFTs strongly correlate with symptom severity and the extent of neuronal loss.[16] Furthermore, Aβ toxicity is thought to be tau dependent, as reducing tau levels protects against Aβ-induced neuronal dysfunction in AD mouse models[17,18].

This section describes small-molecule chemical probes identified in phenotypic screenings or used for target validation aimed at reducing the aggregation and toxicity of Aβ and p-tau and decreasing neuroinflammation.

11.2.1.1 Aβ Modulators

APP, PSEN1 and PSEN2 function in the APP processing pathway. APP is a transmembrane protein with unknown functions present in endosomes and at the cell surface. fAD missense mutations in APP induce changes in APP processing, resulting in increased Aβ42 production. PSEN1 and PSEN2 are alternative subunits of the catalytic site of γ-secretase. The β- and γ-secretases produce multiple Aβ isoforms after sequential cleavages of APP and *PSEN* mutations increase the relative amount of the aggregation-prone Aβ42 and Aβ43 relative to the shorter Aβ38 and Aβ40. The Aβ42/Aβ40 ratio, rather than the total amount of Aβ, inversely correlates with the age of disease onset in both sAD and fAD.[19] Several studies using fAD iPSC-derived neurons showed that these models replicate the altered APP processing and the production and secretion of disease-associated Aβ peptides. These cells exhibited a consistent increase in Aβ42 production and p-tau.[20–23] In addition, iPSC neurons from fAD patients exhibit other cellular phenotypes, including increased susceptibility to oxidative stress, abnormal endosomes, altered axonal transport and impaired mitophagy and autophagy.[22,24–27] Overexpression of *APP* and *PSEN1* mutations in a 3D culture system of human neural stem cells (hNSCs) in matrigel induced both Aβ plaque deposition and aggregates of phosphorylated and filamentous tau, which could be reversed with β-secretase inhibitor (BSI) IV and γ-secretase inhibitor (GSI) XXI/Compound E treatment.[28] Compounds targeting secretases have the potential to inhibit Aβ production and aggregation.[29] In general, Aβ42/40 levels decreased in fAD iPSC-derived neurons after treatment with BSI or GSI, including BSI-IV, LY2886721 and Compound E.[20,21,23,25,28,30,31] BSI-IV is used extensively as a tool compound. It has good cellular and biochemical potency and shows 10-fold selectivity for β-site APP-cleaving enzyme 1 (BACE1) over BACE2.[32] The compound does not inhibit the aspartyl protease renin and no activity for the protease cathepsin D has been reported. The field of BSI has been challenged by the lack of selectivity over cathepsin D and renin, and also between BSI isoforms BACE1 and BACE2. One example is MK-8931/verubecestat, which inhibits Aβ40, Aβ42 and soluble APP β (sAPPβ) production in HEK293-APP$^{swe/lon}$ cells.[33] The compound is a potent BACE1 inhibitor (K_i = 2 nM and APP Aβ cellular IC_{50} = 2 nM), but is an even more potent BACE2 inhibitor (K_i = 0.2 nM).[34] Recently, BSI BACE1-IN-4 was developed by Janssen, showing a >500-fold selectivity for BACE1 over BACE2, K_i = 4 nM and cellular Aβ IC_{50} = 13 nM.[35] The compound has a P-glycoprotein (P-gp) efflux ratio of 20 and corresponding poor blood–brain barrier (BBB) permeability that limits its utility *in vivo*. As no activity towards renin or cathepsin D has been reported, it is unwise to classify this compound as a selective chemical probe for β-secretase.

Another example of a non-selective compound is LY2886721, which is not active towards cathepsin D, renin or pepsin but has a higher potency towards BACE2 than BACE1 (BACE1 IC_{50} = 20 nM, BACE2 IC_{50} = 10 nM).[36] It remains unclear whether it is important to generate BACE1-specific inhibitors. Unlike BACE1, which is highly expressed in the brain, BACE2 is more prominently found in peripheral tissues. In the mouse brain, BACE2 is expressed in subsets of neurons, oligodendrocytes and astrocyte-like cells lining the lateral ventricles and may be an important secretase under inflammatory conditions.[37]

The development of GSI has been challenged by circumventing inhibition of Notch signalling. GSI Compound E exhibits high cellular and biochemical potency (γ-secretase K_i = 0.73 nM, cellular Aβ IC_{50} = 0.3 nM)[38] and is extensively used as a tool compound, but it is an equipotent inhibitor of Notch signalling (IC_{50} = 0.3 nM),[39] making it a less suitable chemical probe for γ-secretase. Whereas GSI BMS-708163 demonstrated 50-fold selectivity for γ-secretase over Notch in a biochemical assay,[40] some sulfonimine pyrazole inhibitors developed by Elan Pharmaceuticals have very good selectivity for γ-secretase over Notch signalling and have excellent biochemical and cellular potency, in addition to good pharmacokinetic profiles. The compounds should therefore be classified as chemical probes for γ-secretase[41] (one example is shown in Table 11.1).

Indirect approaches to inhibit Aβ production have also been explored. For example, Yahata *et al.* used the non-steroidal anti-inflammatory drug (NSAID) sulindac sulfide in iPSCs-derived forebrain neurons and showed that it inhibited Aβ production.[21] Furthermore, high concentrations of other NSAIDs were shown to act as modulators of γ-secretase activity and to reduce Aβ42 secretion.[42] The compound has a weak γ-secretase-related Aβ42 IC_{50} of 20 μM[43] and is more potent towards the human prostaglandin D_2 receptor, hCRTH2 (K_i = 3.6 μM).[44] Additionally, a chemical screen using human embryonic stem cell (hESC)-derived neurons overexpressing the *PSEN1–G378E* mutation showed that both BMS-708163 (GSI) and the tyrosine kinase inhibitor nilotinib were able to decrease Aβ40 levels and restore synaptic dysfunction.[45]

Another screen identified additional γ-secretase-independent modulators of APP processing that would shift the production of Aβ fragments towards shorter and non-toxic forms.[46] Bexarotene, an anticancer drug approved by the US Food and Drug Administration (FDA), selectively targets the primary nucleation step in Aβ42 aggregation and delays the formation of toxic species in neuroblastoma cells.[47] Moreover, treating *PSEN1–M146L* NSCs with bexarotene restored autophagy and mitophagy[48] defects. Bexarotene potently binds to the retinoid X receptors (RXRs: RXRα, RXRβ and RXRγ) in a cotransfection cellular assay (EC_{50} = 28, 25 and 20 nM, respectively[49]) and inhibits Aβ42 aggregation at 3 μM. NMR spectroscopic measurements indicated interaction with Aβ42 primary and secondary oligomers but the exact mode of action of the inhibitor is not clear. Furthermore, there is a distinct difference between EC_{50} values and the concentration needed for inhibition of Aβ42 aggregation. The compound also exhibits a moderate risk of being a promiscuous binder by HitDexter, a program that predicts compounds likely to trigger positive assay readouts either by true promiscuity or by assay interference.[50]

Epigenetic dysregulation, specifically DNA methylation and histone acetylation, is associated with AD neuropathology.[54,55] For example, acetylation of H3K9[56] shows widespread alterations associated with AD, whereas acetylation of H3K27[55] is altered in AD-related pathways involved in the progression of Aβ and tau pathology, such as *APP*, *PSEN1*, *PSEN2* and *MAPT*. In line with this, treatment of HEK293 cells overexpressing the *APP*^{swe} mutation with the HDAC3 inhibitor RGFP-966 (10 μM) increased histone H3 and H4 acetylation, decreased phosphorylation and acetylation of tau, reduced β-secretase cleavage of APP and decreased the Aβ42/Aβ40 ratio while alleviating memory impairment in the triple transgenic AD mouse model.[51] In a cell free activity assay, RGFP-966 is an equipotent HDAC1, -2 and -3 inhibitor with IC_{50} values of 7, 2 and 4 μM, respectively.[51] As RGFP-966 is not selective, it should not be considered as a probe for HDAC3 and it is therefore not certain that

Table 11.1 Recommended chemical probes relevant to Aβ.

Compound	*In vitro* activity	Cellular model	Readout	Recommended as chemical probe
BACE1 inhibitor IV 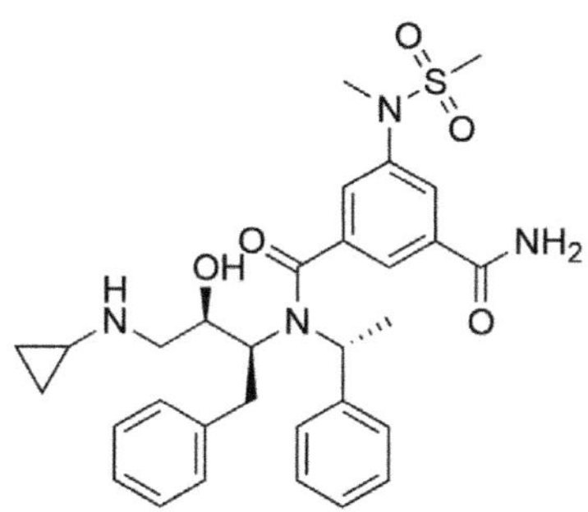	BACE1 IC_{50} = 15 nM BACE2 IC_{50} = 230 nM	iPSC-derived forebrain neurons[21] 3D-differentiated hNPCs with *APP* and *PSEN1* mutations[28] 3D-differentiated APP^{dp} and *PSEN1* mutant organoids[31]	Aβ secretion Aβ aggregation Aβ aggregation	No
LY2886721 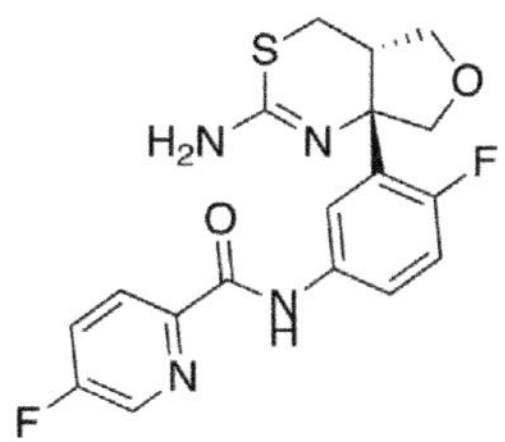	BACE1 IC_{50} = 20 nM BACE2 IC_{50} = 10 nM	hiPSC-3D neurospheroids from AD patients' blood[30]	Aβ aggregation	No
MK-8931/verubecestat 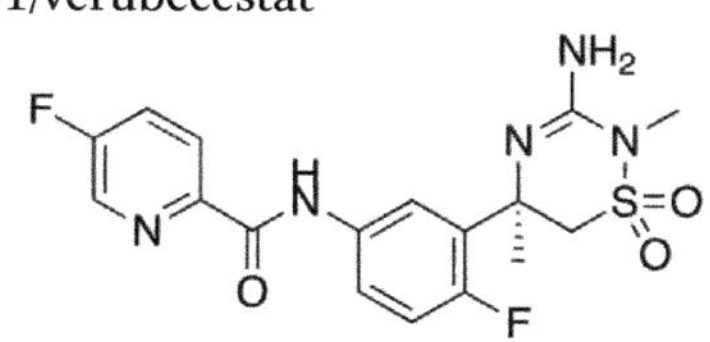	BACE1 K_i = 2 nM APP Aβ cellular IC_{50} = 2 nM BACE2 K_i = 0.2 nM	HEK293-$APP^{swe/lon}$ cells[33]	Aβ production	No

(*continued*)

Table 11.1 (*continued*)

Compound	*In vitro* activity	Cellular model	Readout	Recommended as chemical probe
Compound E 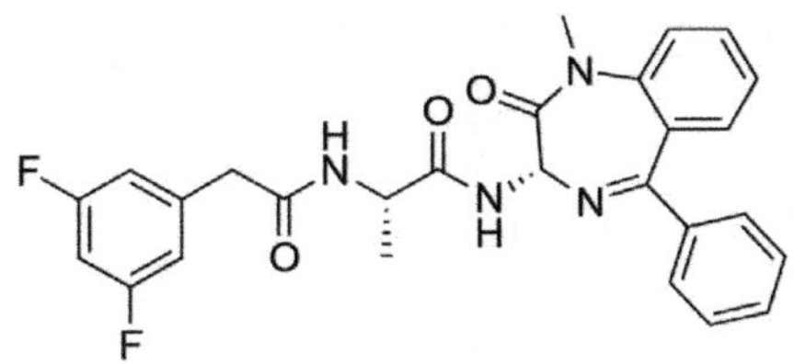	γ-Secretase K_i = 0.73 nM Cellular Aβ IC_{50} = 0.3 nM	iPSC-derived forebrain neurons[21]	Aβ secretion	No
		hiPSC-3D neurospheroids from AD patients' blood[30]	Aβ aggregation	
		3D-differentiated hNPCs with *APP* and *PSEN1* mutations[28]	Aβ aggregation	
		3D-differentiated *APP*^{dp} and *PSEN1* mutant organoids[31]	Aβ aggregation	
		PSEN1–A246E and *PSEN2–N141I* iPSC-derived CNs[20]	Aβ secretion	
BACE1-IN-4 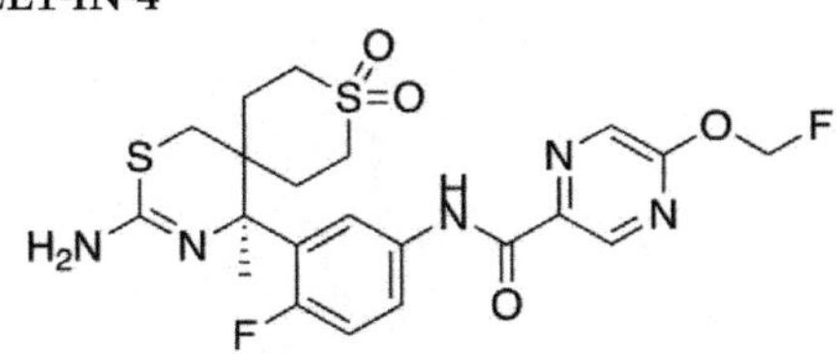	BACE1 IC_{50} = 3.8 nM BACE2 IC_{50} = 2090 nM Aβ IC_{50} = 13 nM	Human neuroblastoma SKNBE2 cells expressing wild-type APP[35]	Aβ42 secretion	No
Avagacestat/BMS-708163 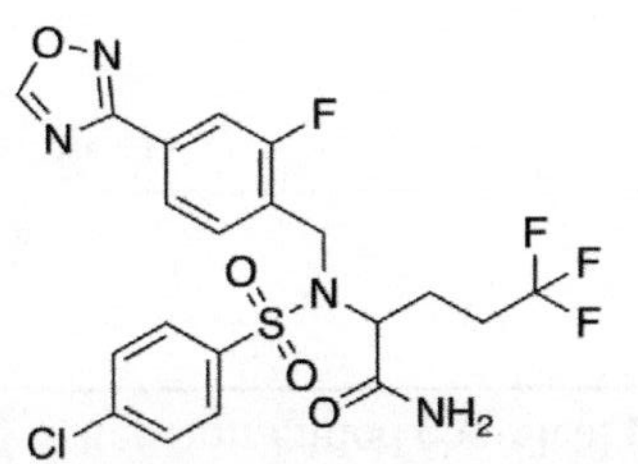	γ-Secretase-related Aβ42 IC_{50} = 0.27 nM CYP2C19 IC_{50} = 20 μM	hESC-derived neurons overexpressing the *PSEN1–G378E* mutation[45]	Aβ aggregation	Yes

Sulfonimine pyrazole compound 3a 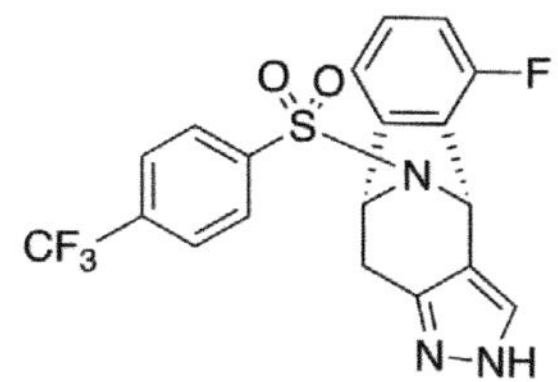	γ-Secretase IC_{50} = 0.4 nM Cellular ED_{50} = 2 nM			
Sulindac sulfide 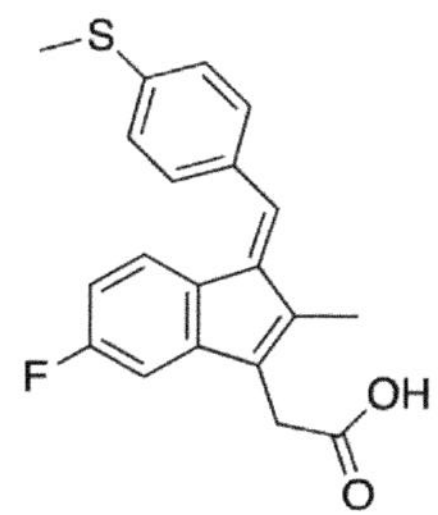	γ-Secretase-related Aβ42 IC_{50} = 20 μM hCRTH2 K_i = 3.6 μM	iPSC-derived forebrain neurons[21]	Aβ secretion	No
Nilotinib	Bcr-Abl	*PSEN1–G378E* overexpression in hESC-derived neurons[45]	Aβ production and secretion	No
Bexarotene 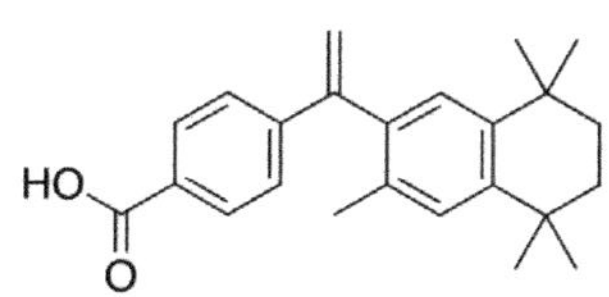	RXRα EC_{50} = 28 nM RXRβ EC_{50} = 25 nM RXRγ EC_{50} = 20 nM	SH-SY5Y cells incubated with Aβ42[47]	Aβ42 aggregation	No
		hiPSC-derived hippocampal *PSEN1–M146L* NSCs[48]	Mitophagy/autophagy	No

(*continued*)

Table 11.1 (*continued*)

Compound	*In vitro* activity	Cellular model	Readout	Recommended as chemical probe
RGFP-966 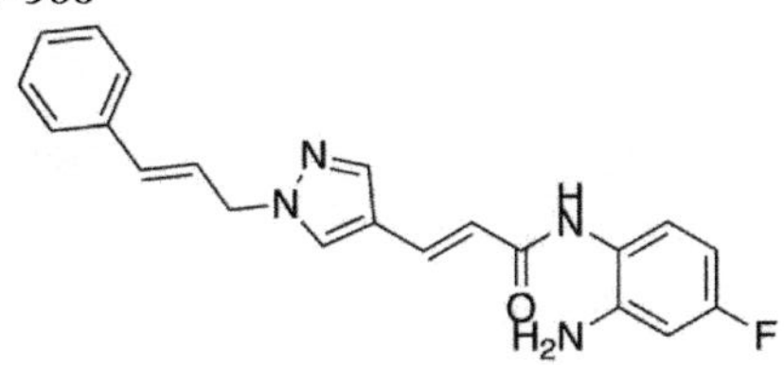	HDAC1 IC_{50} = 7 µM HDAC 2 IC_{50} = 2 µM HDAC3 IC_{50} = 4 µM	HEK293–APP[swe] cells[51]	Aβ42:40 production	No
JQ1 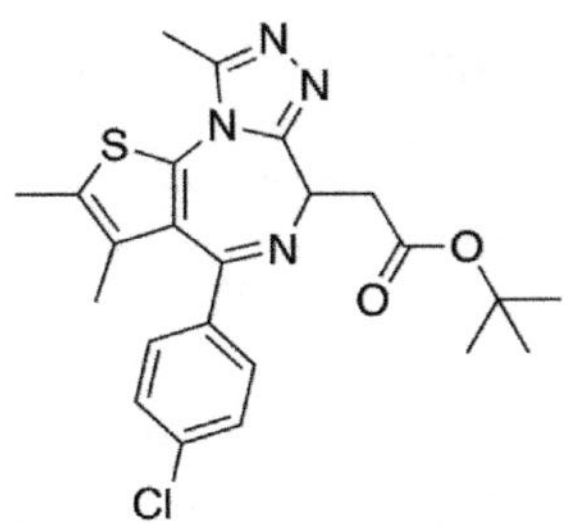	BRD4 K_D = 50 nM	Hippocampal mouse cultures; WT and APP/PSEN1 mice[52]	Restoration of physiological expression of amyloid-associated genes	Yes
CM-414 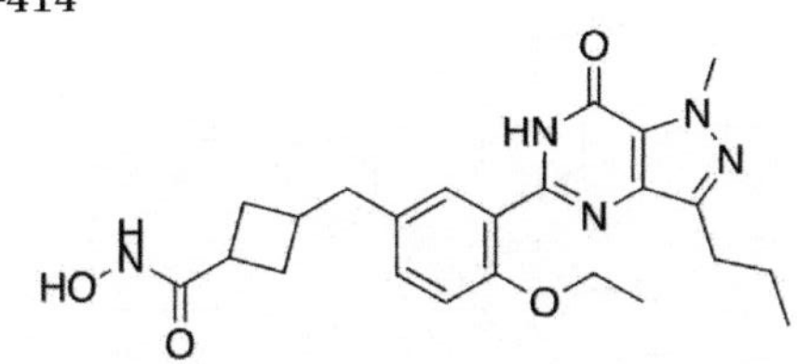	PDE3 IC_{50} = 1.8 µM PDE5 IC_{50} = 60 nM PDE6 IC_{50} = 3 nM HDAC1 IC_{50} = 310 nM HDAC2 IC_{50} = 490 nM HDAC3 IC_{50} = 322 nM HDAC6 IC_{50} = 91 nM	Hippocampal mouse cultures; *APP/PSEN1* mice[53]	Aβ production	Unclear

the reduction of Aβ42 is due solely to HDAC3 inhibition. In another study, JQ1 (Table 11.1) displaced BRD4 at 250 nM in hippocampal mouse cultures and enhanced cognitive performance and long-term potentiation (LTP) in wild-type and APP/PSEN1 AD mouse models.[52] This probe also restored the physiological expression of a subgroup of amyloid-associated genes in APP/PSEN1 mice.[52] JQ1 is an excellent and well-developed probe for BET bromodomains with $K_D = 50$ nM for BRD4. The compound potently binds to BRDT, BRD2, BRD3 and BRD4, whereas little binding was found to other proteins outside the bromodomain family. This is in contrast to the negative control JQ1(−) (enantiomer of JQ1).[57] JQ1 is a good example of a nicely developed chemical probe, as it has a very closely related chemical analogue as negative control, has been well studied in cellular assays and is highly selective and potent for the BET BRD family. In another study, a new first-in-class small molecule, CM-414, which acts as a dual inhibitor of PDE5 and HDACs, was shown to decrease brain levels of Aβ and p-tau, increase the inactive form of GSK3β and reverse the cognitive deficits of mice carrying the APP^{swe} mutation.[53] CM-414 is an interesting compound with one part of the molecule having a hydroxamic acid, typical of HDAC inhibitors, and the pyrimidine pyrazolo scaffold, typical of PDE inhibitors. It is a moderate potency inhibitor of PDE5, HDAC1, HDAC2, HDAC3 and HDAC6 with IC_{50} values of 60, 310, 490, 322 and 91 nM, respectively.[58] Although the authors showed that the compound is not active towards PDE9 and has moderate activity towards PDE3 ($IC_{50} = 1.8$ μM), they did not investigate the activity towards PDE1, an enzyme that is a close homologue to PDE5 [59] and PDE1 inhibitors are known to induce LTP.[60] In addition, CM-414 also potently inhibits PDE6 ($IC_{50} = 3$ nM). The compound has good physiochemical properties and crosses the BBB,[58] making it an attractive tool compound. However, owing to the limited selectivity data on CM-414 in additional pharmacologically relevant assays, it is unclear whether or not its effects are due to its annotated HDAC and PDE targets.

11.2.1.2 Tau

In healthy neurons, tau has a key role in the regulation of microtubule dynamics, axonal transport and neurite outgrowth. In AD, tau becomes hyperphosphorylated owing to increased activity of kinases and decreased activity of phosphatases and loses its ability to bind microtubules. This leads to increased cytoplasmic tau levels, tau aggregation and polymerization into NFTs.[61] Given that tau pathology correlates better with cognitive impairments than do Aβ plaques, chemical screens aim to identify compounds and targets that are able to prevent aggregation *via* inhibiting tau–tau interactions or by either inhibiting kinases that phosphorylate tau or activating phosphatases.[62]

Targeting and modulating heat shock proteins (HSPs) show potential to decrease tau aggregation. Methylene blue (MB) and its derivative LMTX showed tau aggregation inhibitor activity *in vitro* and *in vivo* by trapping tau monomers in an aggregation-incompetent conformation, preventing tau–tau binding.[63–65] MB modulates Hsp70 activity, which was shown to correlate with reduced tau levels in cells and transgenic animals. Furthermore, MB reduced tau levels by inducing autophagy in tau-expressing CHO cells, primary neurons, organotypic slice cultures and frontal temporal dementia (FTD) transgenic animals.[64] MB is a dye and should be used with caution as a tool compound. It is known to be a promiscuous compound (highest score on HitDexter), it is a possible pan-assay interfering compound (PAIN) and is a photosensitizer. The Hsp90 inhibitor 17-DMAG/alvespimycin was shown to reverse altered chromatin structure induced by tau overexpression in iPSC-derived cortical neurons.[56] It potently inhibits Hsp90 (EC_{50} = 62 nM;[66] Hsp90α K_i = 2 nM[67]) and is even more potent in MDA-MB231 cells (IC_{50} = 17 nM). 17-DMAG is a quinone, which is a known PAIN motif,[68] and it is recommended that further investigations should be performed prior to use. The HSP90 inhibitor SNX-0723/PF-04924868 (reviewed in Section 11.2.2.1) is an improved inhibitor, although the compound lacks kinase selectivity and has not been tested in the MDA-MB231 cell model.

Kinases known to mediate tau phosphorylation include GSK3β, mitogen-activated protein (MAP) kinases, AKT serine/threonine kinases, protein kinase A (PKA), calcium/calmodulin-dependent protein kinase II (CAMKII), Fyn, cyclin-dependent kinase-5 (CDK5) and thousand-and-one amino acid kinases (TAOKs).[69] In addition to p-tau, enhanced kinase activity can promote neurodegeneration through other mechanisms. CDK5 activity is implicated in the deposition of Aβ, indirect reduction of nerve growth factor, exacerbation of oxidative stress, promotion of aberrant cell-cycle re-entry and activation of c-JUN N-terminal kinase (JNK).[70] To the best of our knowledge, no selective inhibitors for CDK5 meet chemical probe criteria. Significant efforts have been made to develop kinase inhibitors that alter tau pathology. The GSK3β inhibitors 1-azakenpaullone (1-AZA) and SB415286 decreased p-tau levels in 3D-differentiated human neural progenitor cells (NPCs) with *APP* and *PSEN1* mutations without changing the total tau levels or Aβ levels.[28] 1-AZA has an IC_{50} value of 18 nM.[71] The compound is 100-fold selective for GSK3β over CDK1 and CDK5, but complete kinase selectivity panel data have not been reported. AZA-1 decreased p-tau levels in 3D-differentiated hNPCs at 2.5 μM. SB415286 has the highest possible score in the promiscuous hit finder software, possibly owing to its phthalimide motif. It is therefore recommended to investigate whether this compound is a PAIN compound. The compound is potent towards GSK3β with a biochemical IC_{50} = 77 nM and stimulated glycogen synthesis with a cellular EC_{50} of 3 μM. Only a minor kinase selectivity scan has been performed with 25 other kinases, where SB415286 inhibits MAPK and RSK-2 >50% at 10 μM.[72] Another GSK3 inhibitor, CHIR99021, was shown to modulate Wnt/GSK3 signalling in human iPSC–NPCs[73] and to decrease p-tau in neuroblastoma cells.[74] CHIR99021 potently inhibits GSK3β and GSK3α with IC_{50} values of 7 and 10 nM, respectively, and had a significant effect on glycogen deposition, with an EC_{50} of 763 nM.[75] The compound hit 7% of 311 kinases at 10 μM in a kinase selectivity panel and showed a >800 fold selectivity over a selected set of pharmacological relevant receptors. Therefore, CHIR99021 can be classified as a chemical probe for GSK3. An additional GSK3 inhibitor identified in a high-throughput screen and extensively optimized is ML320 (Table 11.2). ML320 inhibited GSK3β-mediated p-tau in SH-SY5Y cells at low micromolar concentrations.[74] The compound potently inhibits GSK3β and GSK3α (GSK3β IC_{50} = 20 nM, at 10 μM GSK3α/β inhibition is 100% whereas CDK5 IC_{50} = 9 μM. The inactive enantiomer of ML320 did not show any activity in a cellular assay targeting GSK3β and was inactive towards CDK5.[74] ML320 fulfils the criteria of a chemical probe for GSK3β/α and is superior to CHIR99021 (Table 11.2) as it shows a better kinase selectivity profile.

TAOKs 1 and 2 also phosphorylate tau at residues Ser262, Ser324 and Ser356, which are associated with the microtubule-binding repeat domain on tau and are specific for TAOK. The TAOK inhibitor Compound 43 reduced p-tau in HEK293T cells overexpressing htau, primary rat cortical neurons and FTD iPSC-derived human cortical neurons.[76] Compound 43 is highly potent in a kinase biochemical assay (IC_{50} = 11 nM) and has good selectivity for TAOK1-3 at 1 μM over 62 kinases.[77] A significant discrepancy between biochemical and cellular potency (cellular IC_{50} > 15 μM) was reported. This might be due to strong ATP competition, but could also be due to the compound hitting other targets at high micromolar compound concentrations.

11.2.1.3 Neuroinflammation

Aβ plaques lead to activation of microglia and astrocytes, which secrete proinflammatory cytokines and try to phagocytize the debris from plaques and injured cells.[78] In a 3D microfluidic-based system of AD, overexpression of *APP* and *PSEN1* mutations in hNSC-derived neurons and astrocytes cultured with immortalized microglia cells exhibited neuroinflammation, in addition to Aβ aggregation and p-tau accumulation. Recruited microglia induced neuronal loss *via* a TNF-γ- and TLR4-dependent mechanism. However, knockdown of TLR4 in microglia reduced neuron and astrocyte loss.[79] Resatorvid/TAK-242 is a TLR4 inhibitor and inhibits NO, TNF-α and IL-6 production in cells with IC_{50} values of 3.2, 1.3 and 1.3 nM, respectively.[38] TAK-242, having a Michael acceptor motif, is a covalent inhibitor and binds to the C747A cysteine of TLR4. No inhibition of TLR2-, -3-, -5- or -9-related proteins was observed, highlighting the selectivity of TAK-242 for TLR4. Furthermore, the enantiomer and negative control of TAK-242 did not show any activity towards any of the TLRs.[80] TAK-242 is a selective and potent TLR4 inhibitor in cells but no cell-free binding data have been reported. TAK-242 is a high-quality tool compound for cellular use, although a lack of full selectivity characterization means that it cannot be considered a high-quality chemical probe.

Table 11.2 Recommended chemical probes relevant to tau.

Compound	*In vitro* activity	Cellular model	Readout	Recommended as chemical probe
Methylene blue 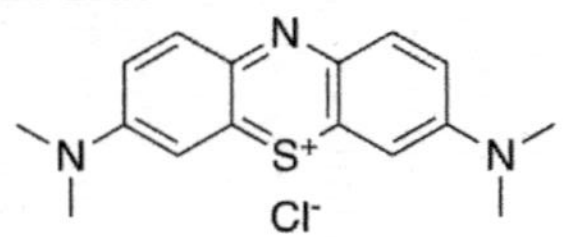	*In vitro* activity unknown	Stable, tau-expressing CHO cells and primary mice neurons[64]	Tau aggregation	No
LMTX 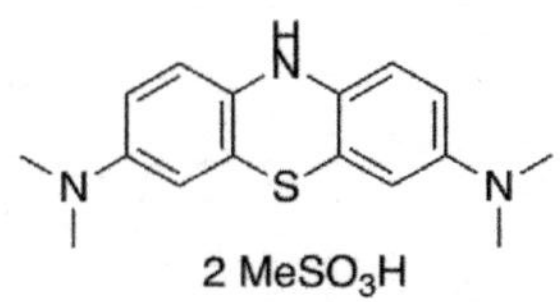	*In vitro* activity unknown	Inducible expression of htau40 in 3T6H cells[65]	Tau aggregation	No
17-DMAG/alvespimycin 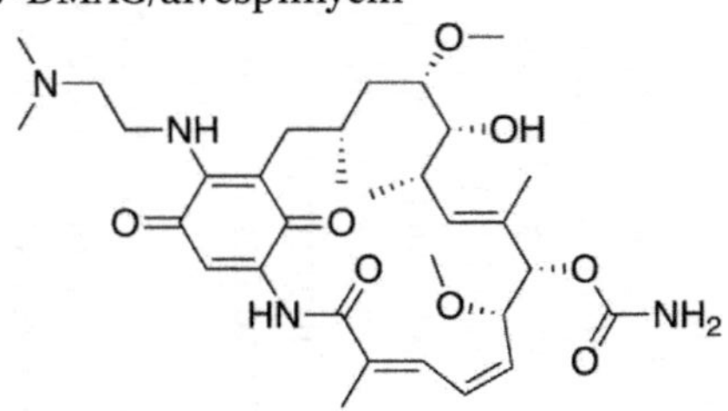	Hsp90 IC_{50} = 28 nM Hsp90α K_i = 2 nM MEC_{50} = 62 nM	*MAPT* iPSC-derived fore-brain neurons[56]	Tau-induced chro-matin marks	No
1-AZA 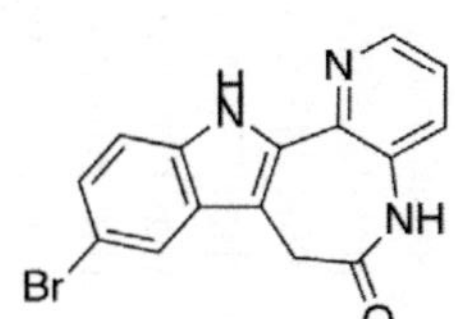	GSK3β IC_{50} = 18 nM	3D-differentiated hNPCs with *APP* and *PSEN1* mutations[28]	p-Tau	No

SB415286 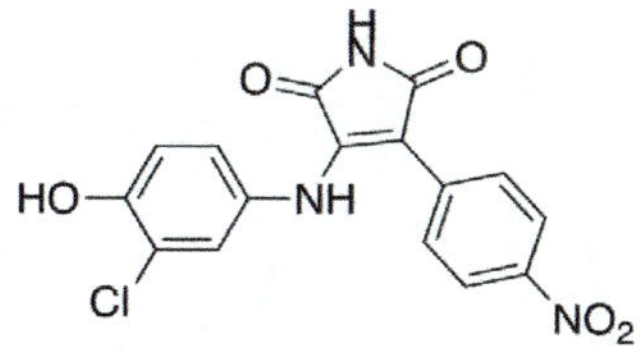	GSK3β IC_{50} = 77 nM	3D-differentiated hNPCs with *APP* and *PSEN1* mutations[28]	p-Tau	No
CHIR99021 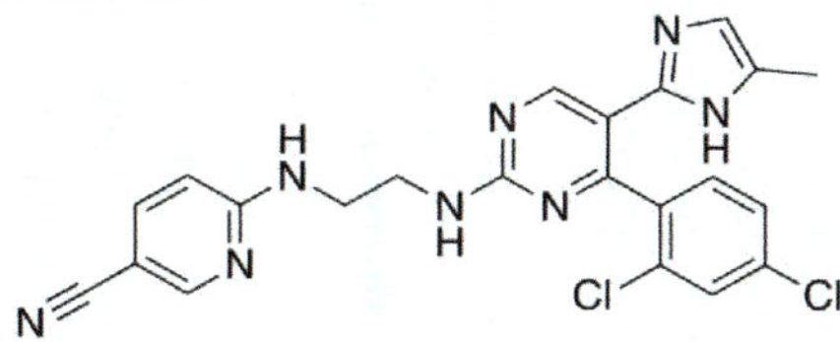	GSK3α IC_{50} = 10 nM GSK3β IC_{50} = 7 nM	iPSC–NPCs;[73] SH-SY5Y cells[74]	p-Tau (Ser199)	Yes
ML320 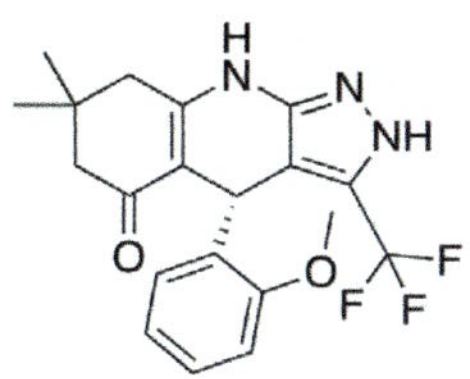	GSK3α % inhibition = 100 at 10 μM GSK3β IC_{50} = 20 nM	SH-SY5Y cells[74]	p-Tau (Ser199)	Yes
Compound 43 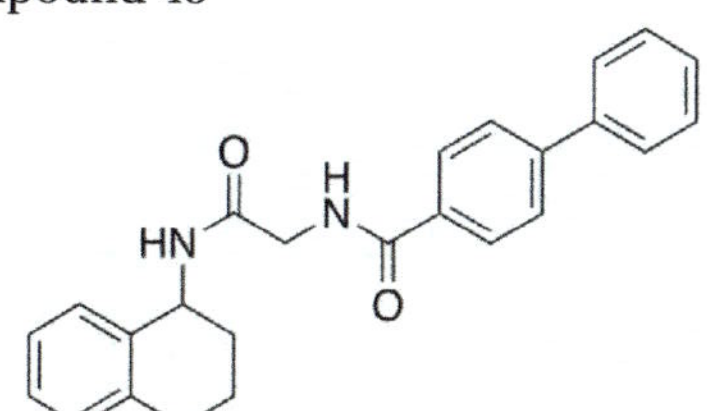	TAOK1 IC_{50} = 11 nM TAOK2 IC_{50} = 15 nM	Overexpression of htau (2N4R) and TAOK2 in HEK293T cells and primary rat CNs[76]	p-Tau	No

Another microglial target expressed in human AD is RIPK1.[81] RIPK1 mediates the microglial expression of CH25H and Cst7, the latter encoding an endosomal/lysosomal cathepsin inhibitor, leading to microglial dysfunction by blocking lysosomal trafficking.[82] Inhibiting RIPK1 using necrostatin-1 (Nec-1) promoted microglial Aβ degradation *in vitro*.[82] The development of Nec-1 as a drug was halted owing to low metabolic stability; however, it has been used extensively as a tool compound. It has a moderate biochemical potency (IC_{50} > 1 μM) with an EC_{50} < 1 μM in a TNF-induced cell assay. Nec-1 shows good selectivity in a kinase selectivity panel. The negative control Nec1i (demethylated analogue) showed a very low activity in the kinase panel but it was equipotent to Nec-1 at 10 μM in the TNF-induced assay.[83,84] Furthermore, Nec-1 has also been referred to as MTH-Trp, an indoleamine 2,3-dioxygenase (IDO) inhibitor.[84] RIPK1 inhibitor compounds with better potency, selectivity and physiochemical properties have been developed. An example is Compound 22 from Takeda, which exhibited an RIPK1 $pK_i = 9$ and significantly reduced necroptisis in cells (IC_{50} = 2 nM) and was potent in a direct target engagement CETSA assay for RIPK1 (EC_{50} = 6.5 nM).[85] It has a good central nervous system penetration and good selectivity in a kinase (400 kinases, tested at 10 μM) and non-kinase selectivity panel.[86] In our opinion, Compound 22 can therefore be qualified as a chemical probe for RIPK1 (Table 11.3). It should be noted that this compound has not been evaluated in AD-related models.

ApoE encodes a cholesterol-transporting lipoprotein and generates three major isoforms, ApoEε2, -ε3 and -ε4. The *ApoEε4* allele is the strongest genetic risk factor for sAD. In the human brain, ApoE is predominantly expressed in astrocytes and to a lesser degree in microglia. The *ApoEε4* allele increases the density of Aβ deposits and p-tau levels, decreases Aβ clearance, impairs mitophagy and enhances neuroinflammation.[27,78,91] ApoE4 alters the transcriptional profiles in different cell types, with the most affected genes involved in synaptic function (neurons), lipid metabolism (astrocytes) and immune response (microglia-like cells).[92] Relative to ApoE3 cells, ApoE4 neurons exhibited an increased number of synapses, early endosomes and Aβ42 secretion; ApoE4 astrocytes displayed impaired Aβ uptake and cholesterol accumulation; ApoE4 microglia-like cells showed altered morphologies and reduced Aβ phagocytosis.[92] Another study showed that ApoE4 hiPSC-derived forebrain neurons had high levels of tau phosphorylation, Aβ42 secretion and GABAergic neuron degeneration.[88,92]

Strategies to target ApoE include increasing its physiological functions and inhibiting the detrimental effects of ApoE4. U18666A and AY-9944 target two key enzymes involved in cholesterol biosynthesis, DHCR24 and DHCR7, respectively, which increase ApoE secretion *in vitro* and *in vivo* in primary astrocytes.[87] These compounds provide an alternative to known promoters of ApoE secretion and lipidation, such as agonists of nuclear liver X receptors (LXRs) or retinoid X receptors (RXRs) that induce expression of several genes other than ApoE and have widespread physiological effects.[93,94] Although being known as DHCR24 and DHCR7 inhibitors, the literature reveals that

Table 11.3 Recommended chemical probes relevant to neuroinflammation.

Compound	*In vitro* activity	Cellular model	Readout	Recommended as chemical probe
Resatorvid/TAK-242 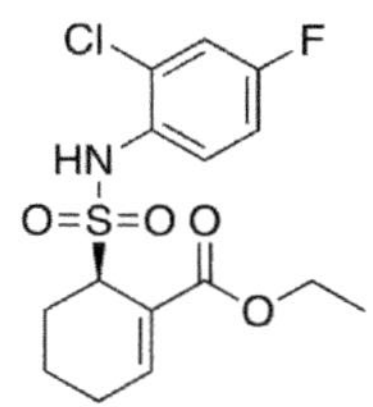	Inhibition of production in cells: NO IC_{50} = 1.8 nM TNF-α IC_{50} = 1.9 nM IL-6 IC_{50} = 1.3 nM	LPS-treated mouse macrophage cell line RAW264.7 cells[80]	TLR4 signalling	No
Nec-1 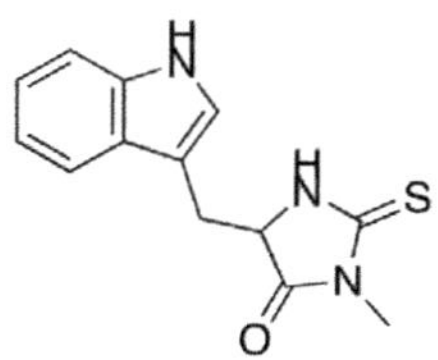	TNF-induced cell assay EC_{50} < 1 μM	Aβ42-treated primary microglia[82]	Aβ degradation	No
Compound 22 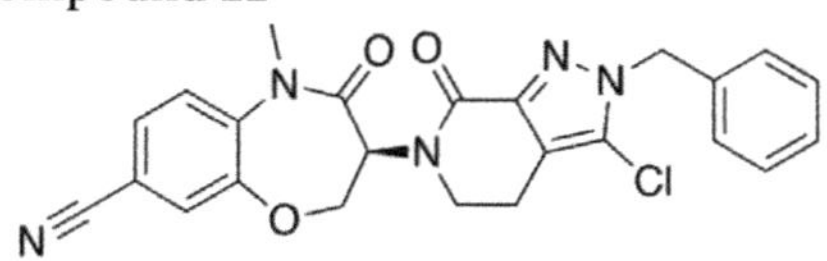	RIPK1 pK_i = 9 RIPK1 EC_{50} = 6.5 nM	Necroptotic cell death in HT-29 cells	Suppression of necroptotic cell death and phosphorylation of MLKL	Yes
U18666A 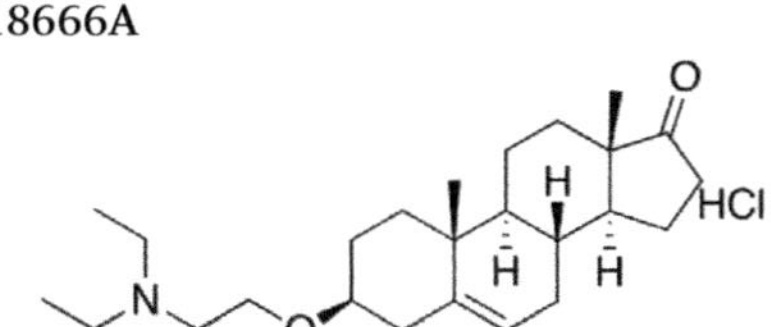	ERG2 K_i = 0.2 nM EBP K_i = 2 nM Sigma receptor-1 K_i = 1 nM	Primary human astrocytes[87]	sApoE levels	No

(*continued*)

Table 11.3 (continued)

Compound	*In vitro* activity	Cellular model	Readout	Recommended as chemical probe
AY-9944 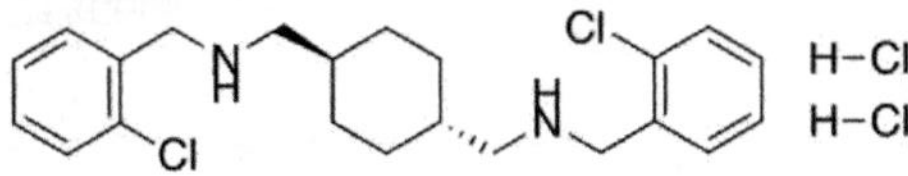	ERG2 K_i = 65 nM EBP K_i = 4 nM Sigma receptor-1 K_i = 0.5 nM	Primary human astrocytes[87]	sApoE levels	No
PH002 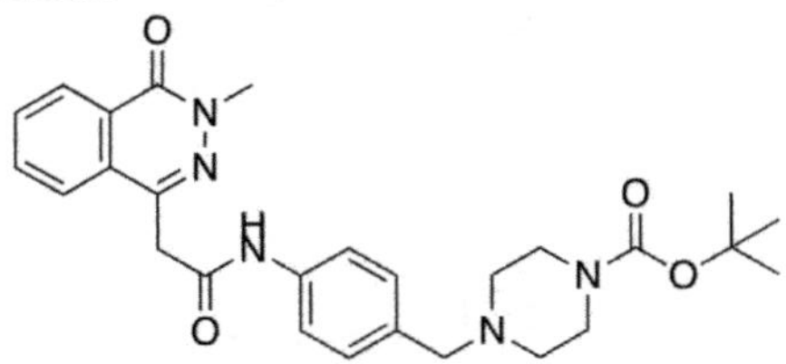	Cellular ApoE4 IC_{50} = 116 nM	ApoE3/3 or ApoE4/4 iPSC-derived GABAergic neurons[88]	ApoE, Aβ42, Aβ40	No
3AC 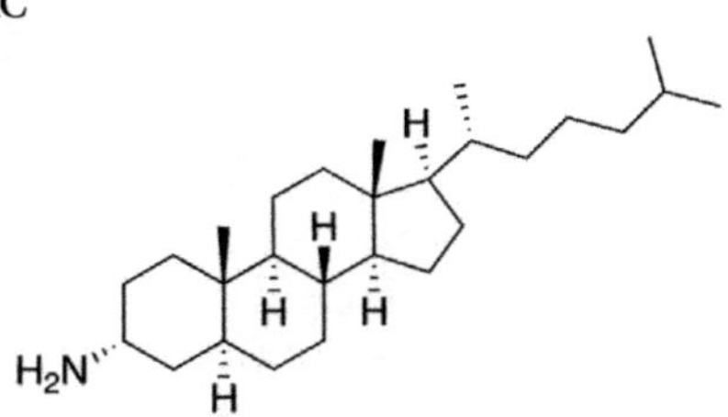	SHIP1 IC_{50} = 2.5 μM	Haematopoietic cancer cells[89]	SHIP1 activity	No
K118 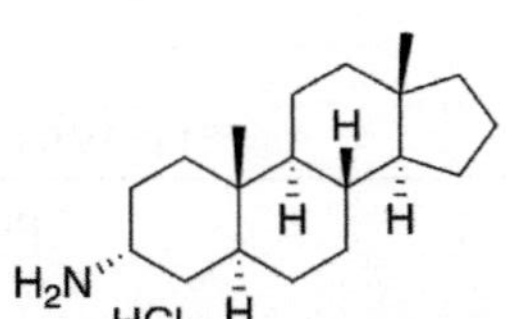	SHIP1 IC_{50} = 14 μM	Mouse mesenchymal stem cells[90]	SHIP1 activity	No

the compounds are potent modulators of ERG2 (K_i = 0.2 nM for U18666A and 65 nM for AY-9944), EBP (K_i = 2 nM for U18666A and 4 nM for AY-9944) and Sigma receptor-1 (K_i = 1 nM for U18666A and 0.5 nM for AY-9944).[95] Therefore, owing to the poor selectivity, these compounds cannot be classified as chemical probes.

Altering the ApoE4 structure to form an ApoE3-like molecule has attracted great interest and shown high potential.[88,96] Small-molecule structure correctors can bind to ApoE4 and interfere with domain interactions between the N- and C-termini. Treating ApoE4 iPSC-derived neurons with submicromolar concentrations of the ApoE4 structure corrector PH002 significantly decreased ApoE4 fragment levels, reduced p-tau levels and decreased Aβ40 and Aβ42 production and/or secretion in a dose-dependent manner.[88] The compound is potent in cells, having an ApoE4 IC_{50} = 116 nM in a FRET assay.[97] It should also be noted that the compound has a moderate predicted promiscuity score and therefore it should not be classified as a chemical probe.

Several microglia-expressed genes have emerged from GWAS studies, with many of them pointing to a TREM2 signalling pathway, including PLCγ2 and SHIP1. TREM2 is a transmembrane receptor of the IgG superfamily, which in the brain is predominantly expressed in microglia. Upon ligand binding (cell debris, Aβ, ApoE), TREM2 activates a signalling cascade that promotes metabolism, survival proliferation and phagocytosis.[98]

Loss of function homozygous mutations in TREM2 causes Nasu–Hakola disease with early-onset dementia, whereas heterozygous variants that lead to a partial loss of function confer an increased risk for AD.[99] The most common TREM2 variant R47H increases the risk for AD by two-to threefold. It prevents the normal folding of TREM2 and leads to defective ligand binding and signalling, reduced phagocytosis and reduced migration of microglia to Aβ plaques.[100] Another variant, H157H, has been reported to decrease TREM2 signalling by enhancing TREM2 shedding in HEK293 cells. Elevated shedding reduced cell-surface full-length TREM2 and lowered TREM2-dependent phagocytosis.[101]

There are no known small-molecule activators of TREM2. However, an alternative strategy to modulate TREM2 expression and secretion would be to target other proteins of the TREM2–DAP12 signalling pathway, such as partially activating PLCγ2[102] or inhibiting SHIP1.[98] U73122 has been used as a PLCγ2 activator and inhibitor.[103] 3α-Aminocholestane (3AC) was identified in a high-throughput screen for SHIP1 inhibitors. The compound is selective for SHIP1 (IC_{50} = 2.5 μM) and shows no inhibition of SHIP2. However, it shows poor aqueous solubility.[89] The water-soluble 3AC derivative K118 was then developed that has an approximate SHIP1 IC_{50} = 14 μM. Although K118 targets SHIP1, it may also inhibit SHIP2.[90] However, neither compound has been fully characterized for selectivity in AD cellular models.

11.2.2 Parkinson's Disease

PD is the second most prevalent neurodegenerative disease, affecting 1–2% of individuals above the age of 65 years. It is characterized by the selective loss of dopaminergic neurons (DNs) located in the *substantia nigra pars compacta*, leading to motor deficits such as tremor, rigidity, akinesia and postural instability.[104]

The pathological hallmarks of PD are the presence of intra-neuronal Lewy bodies (LBs) and Lewy neurites primarily composed of α-syn.[105] LBs follow a predictable pattern of development in the brain as the disease progresses, distinguishable by six stages of α-syn deposition.[104] At the cellular level, PD pathology leads to impaired clearance of protein aggregates, endoplasmic reticulum (ER) stress, mitochondria and lysosomal dysfunction, increased oxidative stress, dysfunctional intracellular trafficking and neuroinflammation, ultimately converging towards neurodegeneration.[106]

Approximately 10% of PD cases are caused by inherited genetic mutations in genes, including *LRRK2*, *SNCA* (α-synuclein, α-syn), *PARK2* (parkin), *PINK1*, *PARK7* (DJ-1) and *ATP13A2*. However, most PD cases are sporadic (sPD), likely involving both genetic and environmental factors. GWAS data have identified gene variants, such as in *GBA1* and *MAPT* (tau), as highly associated with sPD.[107] Environmental factors associated with PD pathogenesis include exposures to certain pesticides, herbicides and heavy metals.[2]

This section describes small-molecule chemical probes identified in phenotypic screenings aimed at reducing levels of α-syn, inhibiting LRRK2 kinase activity, modulating β-glucocerebrosidase (GBA), increasing mitophagy or rescuing cytotoxicity.

11.2.2.1 α-syn

Although the pathology of LBs is not fully understood, α-syn plays a key role in PD pathogenesis supported by the fact that rare *SNCA* mutations, duplications or triplications cause autosomal dominant familial PD. In its native form, α-syn is mainly present in presynaptic terminals and associated with synaptic vesicles. How α-syn converts to amyloid aggregates that cause neuronal toxic species of prefibrillar oligomeric α-syn[108] is unclear, while the presence of aggregates in the brains of patients suggests that proteostasis of α-syn is disrupted in PD. Studies have shown that elevated levels of α-syn disrupt intracellular trafficking pathways, including those involving mitochondria, ER and Golgi, the early and late endosomes and the lysosome.[109–111]

The α-syn is released from cells through secretion or exocytosis of lysosomes, spreading the pathology to neighbouring cells. This may then trigger inflammatory microglia towards damaged neurons, further enhancing neuronal damage.[112] Indeed, a clear link between α-syn aggregation and inflammasome activation has been observed in the brain microglia of PD animal models.[113]

In order to modulate *SNCA* expression, a large-scale screen looked for both activators and inhibitors of expression of α-syn 5′UTR luciferase construct expressed in neuroglioma H4 cells.[114,115] The probes ML163 and ML150 (Table 11.4) were able to activate or inhibit *SNCA* expression levels, respectively, and were designed to target selectively the 5′-untranslated region (5′UTR) stem–loop of α-syn. ML150 had an IC_{50} of 1.8 μM and ML163 an EC_{50} of 2 μM. The compounds were not tested in a biophysical assay but the binding modes were confirmed by testing the compounds in an α-syn expression cell assay lacking the α-syn 5′UTR. Both compounds showed 100-fold selectivity over cells lacking the 5′UTR.[115] To the best of our knowledge, other selectivity screens towards other targets have not been reported. Although the selectivity profile of the compounds could be further examined, ML150 and ML163 compounds may be considered as chemical probes for studying α-syn expression in cells.

Reducing α-syn cytotoxicity is a common assay endpoint in screenings. A chemical screen in yeasts identified an *N*-arylbenzimidazole (NAB) compound that reduced α-syn toxicity by promoting the activity of the E3 ubiquitin–protein ligase RSP5/NEDD4. For example, NAB2 was then tested in iPSC-derived A53T α-syn cortical neurons and shown to rescue the accumulation of ER-associated degradation substrates and nitrosative stress[111] at 2.5 μM. Interestingly, another study showed that Nedd4 overexpression in human cells enhanced α-syn ubiquitination and clearance *via* the endosomal–lysosomal pathway, whereas Nedd4 knockdown increased the α-syn content.[116] NAB2 has a moderate cellular potency (α-syn EC_{50} = 5 μM) and its selectivity has not been evaluated in detail. NAB2 is proposed potentially to target Rsp5, but more data are needed to confirm this claim.[111] Although the data with NAB2 activator mimic the effect of Nedd4 overexpression, the compound lacks biochemical and cellular potency, hence further studies should be carried out to determine the role of Nedd4 in α-syn regulation.

An alternative approach to decrease α-syn levels is to enhance its clearance. XCT 790 is an inverse agonist of the orphan nuclear receptor estrogen-related receptor α (ERRα), with good cellular potency (IC_{50} = 370 nM).[117] The compound was tested in SH-SY5Y cells overexpressing A53T α-syn and shown to increase α-syn clearance at 5 μM concentration[118,119] in an autophagy-dependent MTOR-independent manner. It should also be noted that the compound is a possible covalent modulator as it has a Michael acceptor motif. No biochemical data have been reported, although the compound was highly potent in a cellular ERRα assay. Furthermore, the compound showed good selectivity towards ERRα over ERRγ and the other estrogen receptors ERα and ERβ but had activity towards the receptor PPARγ (EC_{50} = 1.3 μM). Although detailed investigation is recommended prior to using XCT 790, the compound can be classified as a chemical probe for ERRα.

Pharmacological reduction of pathological forms of α-syn may alter the course of PD. To achieve this, several high-throughput screens have targeted α-syn in order to reduce its oligomerization or *SNCA* gene expression. Expression of α-syn from separate constructs with a split GFP or

Table 11.4 Recommended chemical probes relevant to α-syn.

Compound	*In vitro* activity	Cellular model	Readout	Recommended as chemical probe
ML163 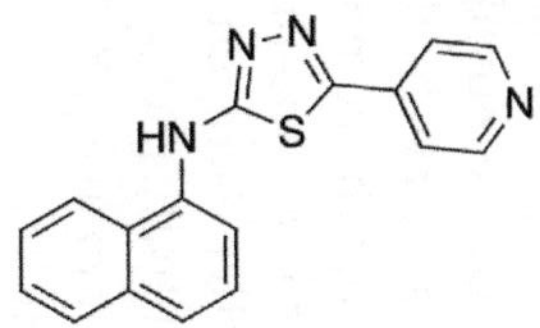	α-syn EC_{50} = 2 μM	α-syn 5′UTR luciferase reporter in H4 cells[114,115]	α-syn expression	Yes
ML150 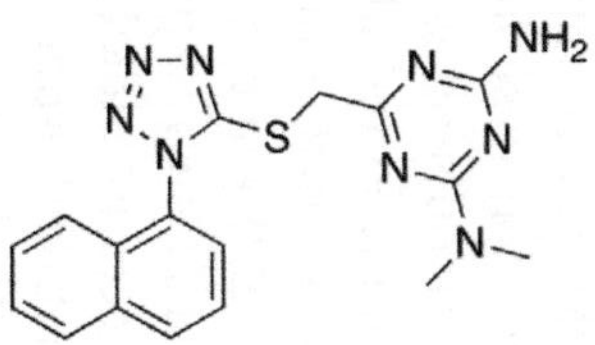	α-syn IC_{50} = 1.8 μM	α-syn 5′UTR luciferase reporter in H4 cells[114,115]	α-syn expression	Yes
NAB2 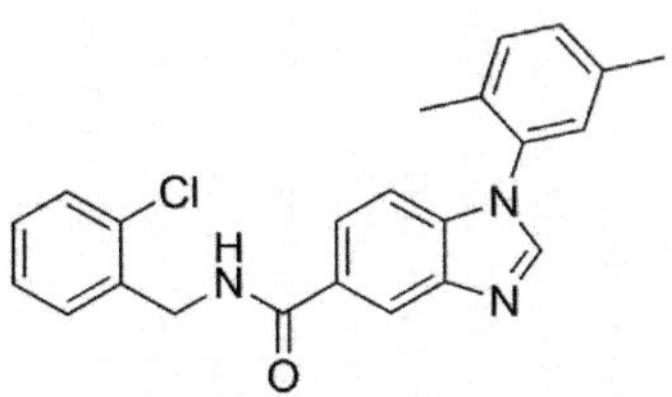	NEDD4 data not known α-syn EC_{50} = 5 μM	iPSC-derived A53T α-syn CNs[111]	ER stress	No
XCT 790 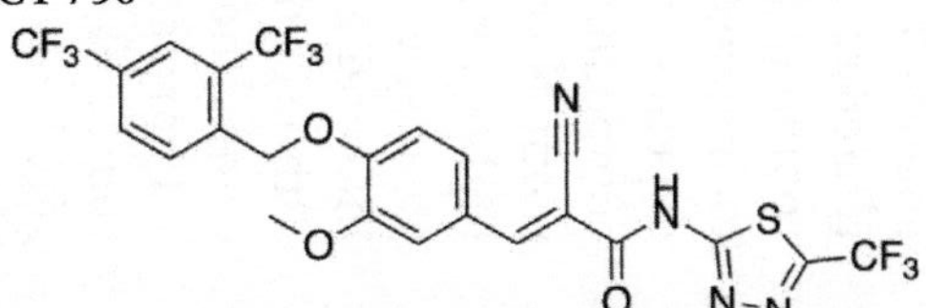	ERRα cellular IC_{50} = 370 nM	Expression of GFP–SNCA in SK-N-MC cells[118,119]	α-syn clearance	Yes

AGK2 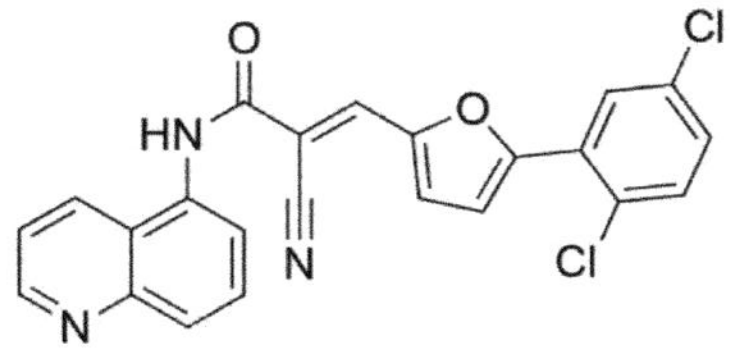	SIRT2 IC_{50} = 2.5 µM	Human neuroglioma H4 cells[120]	α-syn inclusions	No
SNX-0723/PF-04924868 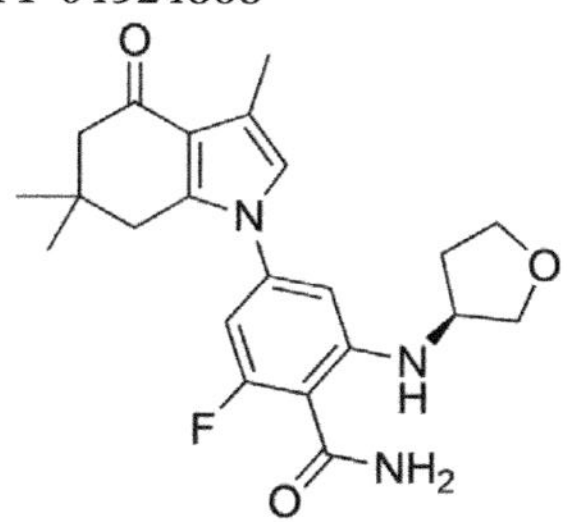	Hsp90 IC_{50} = 14 nM Hsp70 IC_{50} = 31 nM α-syn oligomer EC_{50} = 48 nM	Stable expression of α-syn fused to split luciferase in H4 cells[121]	Luciferase activity	No
NPT100-18A 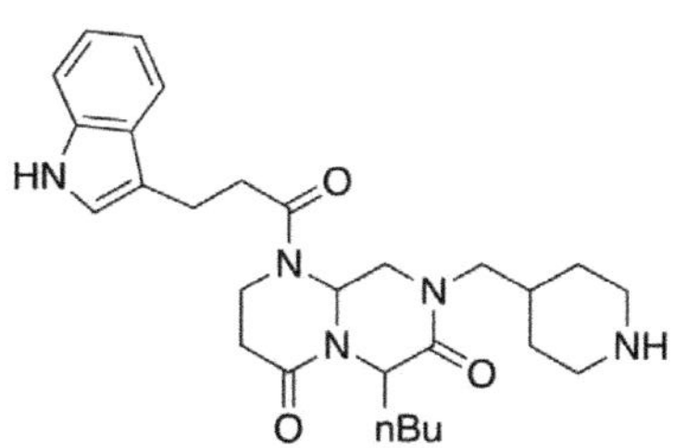	Unclear mode of action	Epoxomicin or MG-132-treated iPSC-derived A53T α-syn dopaminergic, GABAergic and glutamatergic neurons[124]	α-syn toxicity	No
		Lentiviral transduction of wild-type or E83R α-syn in primary mouse CNs[123]	α-syn oligomerization	

(*continued*)

Table 11.4 (*continued*)

Compound	*In vitro* activity	Cellular model	Readout	Recommended as chemical probe
NPT100-14A	Unclear mode of action	Epoxomicin or MG-132-treated iPSC-derived A53T α-syn dopaminergic, GABAergic and glutamatergic neurons[124]	α-syn toxicity	No
Structure unknown		Lentiviral transduction of wild-type or E83R α-syn in primary mouse CNs[123]	α-syn oligomerization	
MCC950 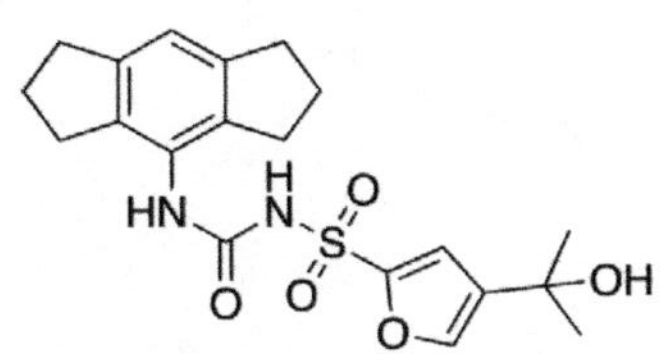	NLRP3 K_D = 224 nM Cellular IC_{50} = 7.5 nM	Primary mouse microglia incubated with α-syn PFF[113] Decrease of α-syn-mediated inflammasome activation	Yes	

luciferase results in cellular α-syn oligomerization and reconstitution of a functional GFP or luciferase, respectively. These methods allowed the identification of hit compounds, such as AGK2, a SIRT2 inhibitor[120] that reduces α-syn oligomerization and cytotoxicity.[121] This compound has low biochemical (IC_{50} = 2.5 μM) and cellular potency (IC_{50} > 15 μM), it has been scarcely characterized and it is potentially a covalent inhibitor. More potent SIRT2 inhibitor compounds have been developed but none qualifies as a chemical probe,[122] therefore preventing the ability to determine if SIRT2 inhibition plays a role in α-syn oligomerization. An HSP90 inhibitor, SNX-0723/PF-04924868, also reduces α-syn oligomerization. It shows excellent biochemical (Hsp90 IC_{50} = 14 nM) and cellular potency (EC_{50} = 48 nM in an α-syn oligomerization assay) but lacks selectivity. It is nearly equipotent for Hsp70 (IC_{50} = 31 nM) and has only been tested against a panel of 36 different kinases.[121] The small-molecule NPT100-18A was shown to decrease α-syn oligomerization by interfering with the interaction between α-syn and lipid membranes.[123,124] The authors demonstrated binding of NPT100-18A to α-syn by NMR studies and postulated that the molecule interacts with the C-terminus of the protein. *In vitro* studies showed that NPT100-18A reduced the formation of wild-type α-syn oligomers in membranes, reduced the neuronal accumulation of α-syn and decreased markers of cell toxicity in primary neuronal cells expressing wild-type α-syn.[123] NPT100-18A, together with two other aggregation inhibitors, NPT100-14A and ELN484228,[125] was tested further using iPSC-derived A53T α-syn dopaminergic, GABAergic and glutamatergic neurons.[124] All three compounds reverted the degenerative phenotype under both basal and induced stress conditions. The limitation of using NPT100-18A and NPT100-14 as chemical probes is the lack of data regarding selectivity and the unclear mode of action.[123–125]

Recently, a link between α-syn aggregation and inflammasome activation in preclinical PD models was demonstrated.[113] Preformed fibrils (PFFs) of α-syn added to mouse microglial cells resulted in a delayed but robust activation of the NLRP3 inflammasome, inducing extracellular ASC release. This was also shown *in vivo* as treatment with the selective NLRP3 inhibitor MCC950 (CRID3) rescued α-syn-mediated inflammasome activation, mitigating motor deficits, nigrostriatal dopaminergic degeneration and accumulation of α-syn aggregates in mice treated with 6-OHDA, α-syn PFFs or containing an inactivated *Tfam* gene.[113] MCC950 has been used extensively as a tool compound. It is an NLRP3 direct inhibitor binding to the Walker B motif within the NLRP3 NACHT domain.[126] The compound shows good binding to NLRP3 in SPR (K_D = 224 nM)[126] and inhibits IL-1β in cells (IC_{50} = 7.5 nM).[127] MCC950 has a good selectivity profile *versus* the other inflammasomes AIM2, NLRC4 and NLRP1. The compound is cell permeable, metabolically stable and bioavailable. Therefore, MCC950 can be classified as a chemical probe for NLRP3 (Table 11.4).

11.2.2.2 LRRK2

Autosomal dominant mutations in the *LRRK2* gene are the most prevalent cause of familial PD. *LRRK2* encodes a large, multidomain protein that contains a kinase, a GTPase and other protein-interaction domains. The G2019S *LRRK2* mutation is responsible for 2–7% of genetic cases, but it is also observed in 1% of sPD cases. This mutation leads to an increase in LRRK2 kinase activity, whereas other mutations, such as R1441C, reduce GTPase activity.[2] Wild-type LRRK2 has also been shown to play a role in sPD as its kinase activity was selectively enhanced in DNs in the *substantia nigra* of *post-mortem* brain tissue from sPD patients.[128] LRRK2 interacts with many key proteins involved in PD, playing a central role in vesicular membranes, autophagy and lysosome functions.

Increased levels of α-syn are measured in *LRRK2–G2019S* iPSC-derived DNs,[129,130] suggesting that α-syn might be a cargo relevant for LRRK2-associated trafficking or accumulation of α-syn might be the result of a more general endosome and lysosome dysfunction.[2] Indeed, mutant LRRK2 has been associated with altered mitochondrial function, defective endosome-to-lysosome trafficking, accumulation of abnormal lysosomal structures and a reduction in the number of neurite processes in neurons.[9]

LRRK2 kinase activity inhibitors may serve as a useful tool to examine the function of LRRK2 in PD cellular models and several high-quality, well-characterized LRRK2 inhibitors have been developed.[131] An example is the inhibitor PF-06447475, which demonstrates excellent biochemical and cellular potency with IC_{50} = 11 and 25 nM, respectively. The compound has much increased biochemical and cellular selectivity compared with previously discovered LRKK2 inhibitors but hits several other kinases in a selectivity panel.[132] The LRRK2 inhibitor MLi-2 (Table 11.5) also has excellent biochemical and cellular potency (IC_{50} = 0.76 and 4.8 nM, respectively) and is superior to PF-06447475 in terms of selectivity (>295-fold selective over 306 kinases at 1 μM).[133] Therefore, MLi-2 qualifies as a chemical probe for LRRK2. Wallings *et al.* used MLi-2 and PF-06447475 to elucidate the mechanisms by which LRRK2 affects autophagy in PD.[134] Treatment with these LRKK2 inhibitors showed that defects in autolysosome maturation in *LRRK2-R1441C* primary neurons were independent of LRRK2 kinase activity, whereas the reduction of autophagosome formation in *LRRK2–G2019S* and *hWT-LRRK2* cortical neurons was kinase dependent.

11.2.2.3 GBA

Autosomal recessive mutations in the *GBA* gene cause the early-onset lysosomal storage disorder Gaucher's disease, whereas heterozygous mutations are the most common genetic risk factor for PD, accounting for 7–10% of sPD patients.[136] *GBA* encodes the lysosomal enzyme β-glucocerebrosidase (GCase), which is involved in the breakdown of glycosphingolipid glucoceramide

Table 11.5 Recommended chemical probes relevant to LRRK2 and GBA.

Compound	*In vitro* activity	Cellular model	Readout	Recommended as chemical probe
PF-06447475 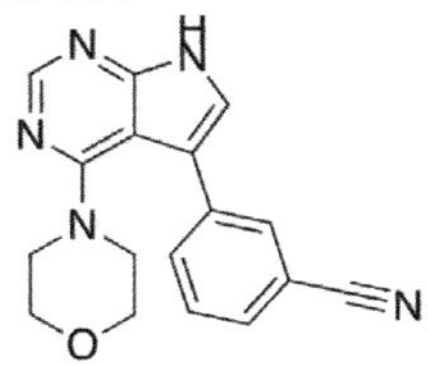	LRRK2 IC_{50} = 11 nM Cellular IC_{50} = 25 nM	BAC transgenic expression of *LRRK2–G2019S*, *LRRK2–R1441C*, or *WT-LRRK2* in primary rat CNs[134]	Autophagosome maturation and autophagosome/lysosome fusion	No
MLi-2 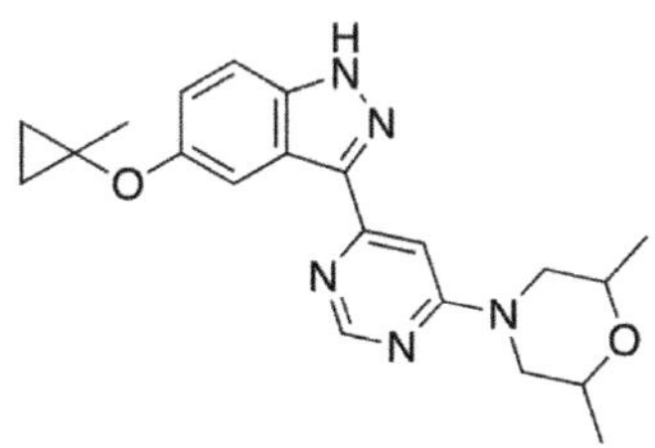	LRRK2 IC_{50} = 0.76 nM	*WT-LRRK2*, *LRRK2–G2019S*, *LRRK2* knockout in HEK293 cells[128]	LRRK2 phosphorylation state (pSer1292)	Yes
	Cellular IC_{50} = 4.8 nM	BAC transgenic expression of *LRRK2–G2019S*, *LRRK2–R1441C* or *WT-LRRK2* primary rat cortical neurons[134]	Autophagy flux	
GCSase modulator 758 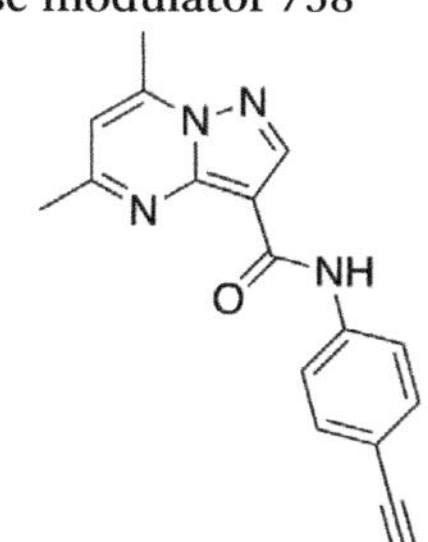	GCase AC_{50} = 0.4–4 μM	iPSC-derived dopaminergic neurons with *SNCA* triplication; inducible α-syn overexpression in H4 cells[135]	Lysosomal GCase activity and α-syn expression	No

(GlcCer) into ceramide and glucose. Mutations in *GBA* (N370S and L444P) result in lowered GCase activity and increased α-syn accumulation in PD iPSC-derived DNs.[135,137] Furthermore, brain tissue from sPD patients confirms that GCase activity is also reduced without *GBA* mutations.[138] GCase deficiency seems to lead to neurodegeneration either through endosomal/lysosomal dysfunction or through a link between the accumulation of GlcCer and α-syn.[137]

Molecules able to increase lysosomal GCase activity therefore have the potential to be disease-modifying therapies for PD. A modulator of GCase (GCSase modulator 758) was shown to specifically enhance GCase activity within lysosomal compartments regardless of the PD-causing mutation, reducing GCase substrates and clearing pathological α-syn in inducible α-syn H4 cells and in both iPSC-derived *GBA1–N370S* DNs or carrying an *SNCA* triplication[135] (Table 11.5). The compound is a moderately potent GCase modulator in both biochemical (AC_{50} = 0.4–4 μM) and cellular (AC_{50} = 9 μM) assays[139] and its selectivity towards other proteins has not been reported. It therefore does not qualify as a chemical probe and better GCase enhancers are greatly needed.

11.2.2.4 Mitochondrial Homeostasis

Early-onset hereditary forms of PD can be caused by loss of function mutations in *PINK1* or *PARK2* (parkin).[140] PINK1 is a mitochondrial kinase that phosphorylates parkin and ubiquitin, playing a role in maintaining mitochondrial homeostasis. Parkin is an E3 ubiquitin–protein ligase that promotes the ubiquitination of several substrates and targets them for degradation by the proteasome or lysosome. Deficits in mitophagy, the process of turning over mitochondria, are prevalent in PD patients. *PINK1* and *PARK2* mutant iPSC-derived DNs exhibit a reduced ability to remove damaged mitochondria, enlarged mitochondria and mitochondrial stress, increased levels of α-syn and abnormal dopamine homeostasis.[141]

Cell vulnerability associated with mitochondrial dysfunction in *PINK1(Q456X)* DNs could be rescued by rapamycin.[142] The macrocycle rapamycin is an FDA-approved immunosuppressant. The molecule is known to bind potently to FKBP12 and the resulting complex in turn inhibits mTOR complex 1.[143] Rapamycin is very selective, it has shown beneficial effects in several ageing, AD and PD mouse models[142,144] and it is a potent mTOR inhibitor (cellular IC_{50} = 0.1 nM).[145] However, owing to the complex interactions of mTORC1 in the regulation of growth and metabolism, rapamycin treatment will likely generate varied effects. It should be noted that suppression of mTOR leads to cell death.[146]

Another way to regulate the PINK1/parkin pathway is to target the corresponding deubiquitinase, USP30. USP30 deubiquitinates mitochondrial proteins and inhibits mitophagy.[147] A high-throughput screen for USP30 inhibitors identified a racemic phenylalanine derivative (Compound 1). Compound 1 has a biochemical $IC_{50} < 1$ μM and does not inhibit the ubiquitin-specific proteases (USP)-1, −8 and −9 at 10 μM. Subsequently, a series of analogues with low nanomolar IC_{50} values in a biochemical assay were described as non-covalent inhibitors of USP30. Although Compound 31 was not the most potent inhibitor with an IC_{50} of 120 nM, it was evaluated further. This compound and a closely related inactive analogue were screened in a panel of 22 USP assays and both compounds showed <30% inhibition at 10 μM.[148] Evaluating the two compounds in a cellular assay ranging from 0.2 to 5 μM led to inhibition of USP30 and mitophagy induction in a dose-responsive manner for Compound 31 (Table 11.6), whereas the inactive analogue did not. Compound 31 and other potent USP30 analogues have been tested using non-PD cellular assays. Therefore, further evaluation is needed in a PD cellular context.

Environmental factors such as mitochondrial toxins and heavy metals are known to cause PD. MPTP (1-methyl-4-phenyl-1,2,3,6-tetrahydropyrid ine) induces a rapid PD-like syndrome. MPTP crosses the BBB and is subsequently metabolized to MPP+ (1-methyl-4-phenylpyridinium), which is taken up by dopamine transporters into DNs. MPP+ inhibits mitochondrial complex I, leads to oxidative stress and energetic impairment and triggers α-syn accumulation in DNs. Paraquat (1,1′-dimethyl-4,4′-bipyri dinium dichloride), a commonly used herbicide, shares structural similarity with MPP+. Paraquat crosses the BBB, generates reactive oxygen and nitrogen species and causes loss of DNs in animal models. Additional pesticides, including the fungicide maneb [manganese ethylene bis(dithiocarbamate)] and the insecticide rotenone, can induce neuronal death in PD models. Rotenone is a mitochondrial complex I inhibitor and leads to the production of oxidative stress. Like MPTP, rotenone exposure upregulates α-syn and triggers its oligomerization.[151] Hence these toxins are used as tools in disease models to induce mitochondrial dysfunction and subsequent cell injury. Yoon *et al.* screened a chemical library containing 1120 bioactive compounds and identified carnosic acid as a suppressor of rotenone-induced toxicity *via* induction of HSP70 expression.[149] Mitochondrial toxins also induce cell vulnerability in PD iPSC-derived neurons carrying the A53T α-syn mutation. Treatment with isoxazole rescued neurons from apoptosis induced by rotenone or paraquat/manganese and was reported to validate the MEF2C–PGC1α pathway as a new drug target for PD.[150] One should be cautious when using isoxazole in cellular assays as it has a molecular weight of only 69, has limited selectivity data and ought to be considered more as a small fragment than a tool compound.

Table 11.6 Recommended chemical probes relevant to mitochondrial homeostasis.

Compound	*In vitro* activity	Cellular model	Readout	Recommended as chemical probe
Rapamycin 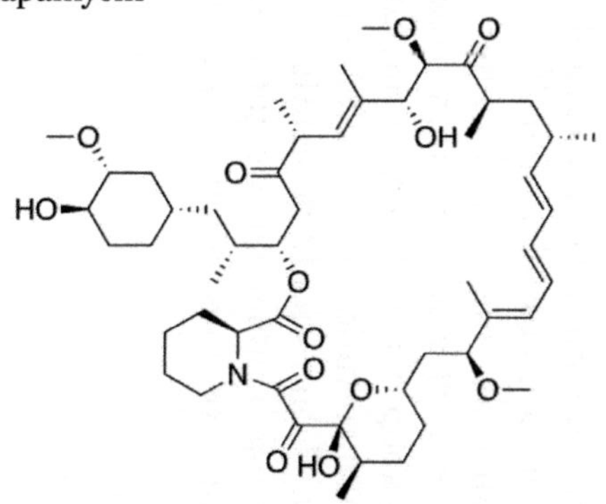	mTOR cellular $IC_{50} = 0.1$ nM	Valinomycin- or concanamycin A-treated iPSC-derived neural cells with *PINK1–Q456X*[57]	Cellular vulnerability	No
Compound 1 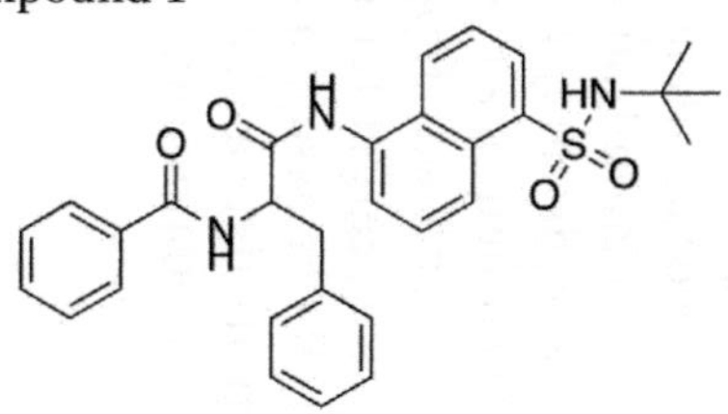	USP30 $IC_{50} < 1$ μM	C2C12 cells[148]	Mitophagy induction	No

Compound 31 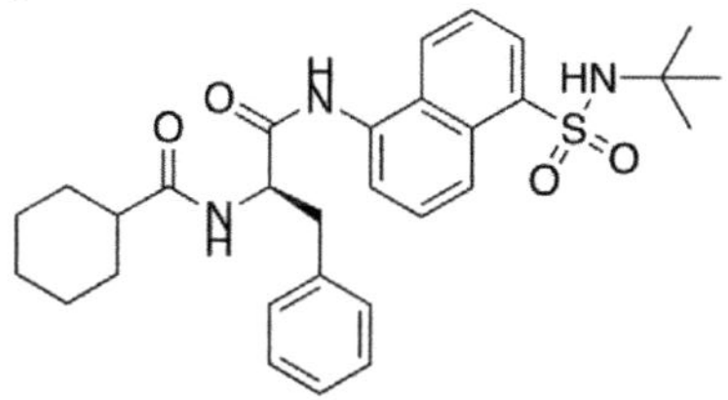	USP30 IC_{50} = 120 nM	C2C12 cells[148]	Mitophagy induction	Yes
Carnosic acid 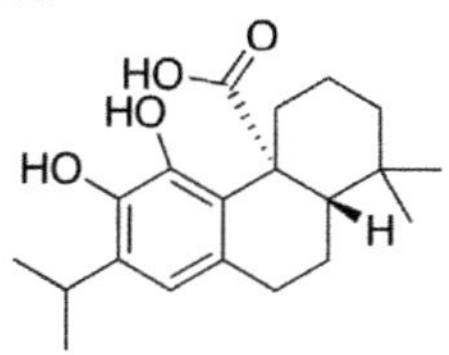	Unclear mode of action	Rotenone-treated SK-N-SH cells[149]	Cytotoxicity	No
Isoxazole 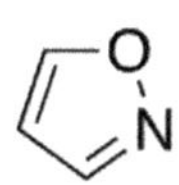	Unclear mode of action	iPSC-derived A53T α-syn neurons[150]	*MEF2* transcription	No

11.3 Conclusions and Future Directions

Neurodegenerative diseases have generally proven to be therapeutically intractable for a number of reasons; among them is the poor availability of well-characterized chemical probes to evaluate candidate targets and the lack of physiologically relevant disease models amenable to high-throughput screening. To address this, considerable effort has been dedicated to developing iPSC neurons and glial cells that recapitulate key features of neurodegenerative pathology. With these *in vitro* models, a more thorough investigation of disease initiation and progression can be made and, as the field of iPSCs matures, it is to be expected that the physiological relevance of these models, through the application of co-cultures, extracellular matrices, medias/environmental controls and differentiation procedures, will continue to develop towards a more robust and biologically accurate platform. These advances are crucial as it is the limited relevance of current models that has severely hindered the development of small-compound probes to investigate AD and PD. Indeed, probes developed using yeast models, immortalized or surrogate cell models or biochemical targets have proven to be mostly disappointing as tools for target discovery and mechanistic understanding of disease biology. Currently, as reviewed in this chapter, there are a limited number of small-compound probes of sufficient quality to help decipher the complex pathophysiology of neurodegeneration. This limited tool set has clearly delayed the pace of drug discovery and, although powerful genome editing tools exist, such as CRISPR/Cas9, a knockout of a protein does not mimic the physiological effects of a small compound binding to a protein. If clinical history is a guide, these drugs will most likely be a small-compound modulator of the disease biology. With this in mind, an in-depth interrogation of multiple cellular pathways implicated in neurodegeneration can only be accurately investigated by using the high-quality small-compound probes highlighted in this review, and poor-quality probes should be avoided as their weak potency and uncharacterized selectivity will only confound target identification. In addition, there is a clear need to expand the number of quality chemical probes to cover ideally the entire proteome, allowing target validation on a larger scale in cell-specific disease models.

Abbreviations

1-AZA	1-Azakenpaullone
3AC	3α-Aminocholestane
6-OHDA	6-Hydroxydopamine
Aβ	β-Amyloid
ABCA7	ATP-binding cassette subfamily A member 7 gene
AC50	half maximal activation concentration
AD	Alzheimer's disease
ADAM10	ADAM metallopeptidase domain 10 gene
AIM2	Absent in melanoma-2

APOE	Apolipoprotein E
APP	β-Amyloid precursor protein
ASC	Apoptosis-associated speck-like protein containing a caspase recruit domain
ATP13A2	ATPase cation transporting 13A2 gene
BACE	β-Site APP-cleaving enzyme
BBB	Blood–brain barrier
BET	Bromodomain and extra-terminal protein
BIN1	Bridging integrator 1 gene
BRD4	Bromodomain-containing protein 4
BSI	β-Secretase inhibitor
CDK5	Cyclin-dependent kinase-5
CETSA	Cellular thermal shift assay
CN	Cortical neuron
CLU	Clusterin gene
CR1	Complement receptor 1 gene
CRISPR	Clustered regularly interspaced short palindromic repeats
DHCR	Dehydrocholesterol reductase
DN	Dopaminergic neuron
ER	Endoplasmic reticulum
ERG2	Early growth response 2
ERRα	Estrogen-related receptor α
fAD	Familial Alzheimer's disease
FDA	Food and Drug Administration
FKBP12	FK506 binding protein 12
FTD	Frontal temporal dementia
GBA	Glucocerebrosidase gene
GCase	β-Glucocerebrosidase
GFP	Green fluorescent protein
GlcCer	Glycosphingolipid glucoceramide
GSI	γ-Secretase inhibitor
GSK3β	Glycogen synthase kinase 3 beta
GSM	γ-Secretase modulator
GWAS	Genome-wide association studies
H3K9	Histone 3 lysine 9
H3K27	Histone 3 lysine 27
HDAC	Histone deacetylase
HEK293	Human embryonic kidney cells 293
hiPSC	Human induced pluripotent stem cell
hNSC	Human neural stem cell
HSP	Heat shock protein
IDO	Indoleamine 2,3-dioxygenase
IL-6	Interleukin 6
iPSC	Induced pluripotent stem cell
JNK	c-JUN N-terminal kinase

LBs	Lewy bodies
LRRK2	Leucine-rich repeat kinase 2
LPS	Lipopolysaccharide
LTP	Long-term potentiation
LXR	Liver X receptor
MAP	Mitogen-activated protein
MAPT	Microtubule-associated protein tau gene
MB	Methylene blue
MEF2	Myocyte enhancer factor-2 gene
MPP+	1-Methyl-4-phenylpyridinium
MPTP	1-Methyl-4-phenyl-1,2,3,6-tetrahydropyridine
MTOR	Mammalian target of rapamycin
NAB	*N*-arylbenzimidazole
NACHT	NAIP (neuronal apoptosis inhibitory protein), CIITA (MHC class II transcription activator), HET-E (incompatibility locus protein from *Podospora anserina*) and TP1 (telomerase-associated protein)
NFTs	Neurofibrillary tangles
NLRC4	NLR family CARD domain containing 4
NLRP3	NOD-, LRR- and pyrin domain-containing protein 3
NMR	Nuclear magnetic resonance
NPC	Neural progenitor cell
NSC	Neural stem cell
NSAID	Non-steroidal anti-inflammatory drug
PAIN	Pan-assay interfering compound
PD	Parkinson's disease
PDE5	Phosphodiesterase type 5
PFF	Preformed fibril
PGC1α	Peroxisome proliferator-activated receptor (PPARγ) coactivator 1α
P-gp	P-glycoprotein
PICALM	Phosphatidylinositol-binding clathrin assembly protein
PINK1	PTEN-induced kinase 1
PKA	Protein kinase A
PLCγ2	Phosphoinositide-specific phospholipase C
PSEN1	Presenilin 1 gene
PSEN2	Presenilin 2 gene
p-tau	Phosphorylated tau
RIPK1	Receptor interacting serine/threonine kinase 1
RXR	Retinoid X receptor
sAD	Sporadic Alzheimer's disease
SGC	Structural Genomics Consortium
SHIP	SH2-containing inositol 5′-phosphatase
SH-SY5Y	Human neuroblastoma cell line
SIRT2	Sirtuin 2
SNCA	α-Synuclein gene
SORL1	Sortilin-related receptor 1 gene

sPD	Sporadic Parkinson's disease
SPR	Surface plasmon resonance
α-syn	α-Synuclein
TAOK	Thousand-and-one amino acid kinase
TLR4	Toll-like receptor 4
TNF-γ	Tumour necrosis factor gamma
TREM2	Triggering receptor expressed on myeloid cells 2
USP	Ubiquitin-specific protease
UTR	Untranslated region

References

1. T. Wyss-Coray, *Nature*, 2016, **539**, 180–186.
2. A. Abeliovich and A. D. Gitler, *Nature*, 2016, **539**, 207–216.
3. R. G. Canter, J. Penney and L. H. Tsai, *Nature*, 2016, **539**, 187–196.
4. J. Cummings, G. Lee, T. Mortsdorf, A. Ritter and K. Zhong, *Alzheimer's Dementia*, 2017, **3**, 367–384.
5. V. Khurana, D. F. Tardiff, C. Y. Chung and S. Lindquist, *Nat. Rev. Neurol.*, 2015, **11**, 339–350.
6. K. Takahashi and S. Yamanaka, *Cell*, 2006, **126**, 663–676.
7. Y. Shi, H. Inoue, J. C. Wu and S. Yamanaka, *Nat. Rev. Drug Discovery*, 2017, **16**, 115–130.
8. T. C. W. Julia, *Neurosci. Lett.*, 2019, **699**, 31–40.
9. M. M. Cobb, A. Ravisankar, G. Skibinski and S. Finkbeiner, *Cell Tissue Res.*, 2018, **373**, 61–77.
10. R. M. Gupta and K. Musunuru, *J. Clin. Invest.*, 2014, **124**, 4154–4161.
11. J. Blagg and P. Workman, *Cancer Cell*, 2017, **32**, 9–25.
12. C. Reitz, *Future Neurol.*, 2015, **10**, 547–558.
13. B. De Strooper and E. Karran, *Cell*, 2016, **164**, 603–615.
14. J. A. Hardy and G. A. Higgins, *Science*, 1992, **256**, 184LP–185.
15. J. Hardy and D. J. Selkoe, *Science*, 2002, **297**, 353LP–356.
16. H. Braak and E. Braak, *Acta Neuropathol.*, 1991, **82**, 239–259.
17. E. D. Roberson, K. Scearce-Levie, J. J. Palop, F. Yan, I. H. Cheng, T. Wu, H. Gerstein, G.-Q. Yu and L. Mucke, *Science*, 2007, **316**, 750 LP–754.
18. L. M. Ittner, Y. D. Ke, F. Delerue, M. Bi, A. Gladbach, J. van Eersel, H. Wölfing, B. C. Chieng, M. J. Christie, I. A. Napier, A. Eckert, M. Staufenbiel, E. Hardeman and J. Götz, *Cell*, 2010, **142**, 387–397.
19. S. Kumar-Singh, J. Theuns, B. Van Broeck, D. Pirici, K. Vennekens, E. Corsmit, M. Cruts, B. Dermaut, R. Wang and C. Van Broeckhoven, *Hum. Mutat.*, 2006, **27**, 686–695.
20. T. Yagi, D. Ito, Y. Okada, W. Akamatsu, Y. Nihei, T. Yoshizaki, S. Yamanaka, H. Okano and N. Suzuki, *Hum. Mol. Genet.*, 2011, **20**, 4530–4539.
21. N. Yahata, M. Asai, S. Kitaoka, K. Takahashi, I. Asaka, H. Hioki, T. Kaneko, K. Maruyama, T. C. Saido, T. Nakahata, T. Asada, S. Yamanaka, N. Iwata and H. Inoue, *PLoS One*, 2011, **6**, e25788.

22. M. A. Israel, S. H. Yuan, C. Bardy, S. M. Reyna, Y. Mu, C. Herrera, M. P. Hefferan, S. Van Gorp, K. L. Nazor, F. S. Boscolo, C. T. Carson, L. C. Laurent, M. Marsala, F. H. Gage, A. M. Remes, E. H. Koo and L. S. B. Goldstein, *Nature*, 2012, **482**, 216–220.
23. T. Kondo, K. Imamura, M. Funayama, K. Tsukita, M. Miyake, A. Ohta, K. Woltjen, M. Nakagawa, T. Asada, T. Arai, S. Kawakatsu, Y. Izumi, R. Kaji, N. Iwata and H. Inoue, *Cell Rep.*, 2017, **21**, 2304–2312.
24. E. M. Rodrigues, A. M. Weissmiller and L. S. B. Goldstein, *Hum. Mol. Genet.*, 2012, **21**, 4587–4601.
25. T. Kondo, M. Asai, K. Tsukita, Y. Kutoku, Y. Ohsawa, Y. Sunada, K. Imamura, N. Egawa, N. Yahata, K. Okita, K. Takahashi, I. Asaka, T. Aoi, A. Watanabe, K. Watanabe, C. Kadoya, R. Nakano, D. Watanabe, K. Maruyama, O. Hori, S. Hibino, T. Choshi, T. Nakahata, H. Hioki, T. Kaneko, M. Naitoh, K. Yoshikawa, S. Yamawaki, S. Suzuki, R. Hata, S. I. Ueno, T. Seki, K. Kobayashi, T. Toda, K. Murakami, K. Irie, W. L. Klein, H. Mori, T. Asada, R. Takahashi, N. Iwata, S. Yamanaka and H. Inoue, *Cell Stem Cell*, 2013, **12**, 487–496.
26. L. Li, J. H. Roh, E. H. Chang, Y. Lee, S. Lee, M. Kim, W. Koh, J. W. Chang, H. J. Kim, M. Nakanishi, R. A. Barker, D. L. Na and J. Song, *Exp. Neurobiol.*, 2018, **27**, 350–364.
27. E. F. Fang, Y. Hou, K. Palikaras, B. A. Adriaanse, J. S. Kerr, B. Yang, S. Lautrup, M. M. Hasan-Olive, D. Caponio, X. Dan, P. Rocktäschel, D. L. Croteau, M. Akbari, N. H. Greig, T. Fladby, H. Nilsen, M. Z. Cader, M. P. Mattson, N. Tavernarakis and V. A. Bohr, *Nat. Neurosci.*, 2019, **22**, 401–412.
28. S. H. Choi, Y. H. Kim, M. Hebisch, C. Sliwinski, S. Lee, C. D'Avanzo, H. Chen, B. Hooli, C. Asselin, J. Muffat, J. B. Klee, C. Zhang, B. J. Wainger, M. Peitz, D. M. Kovacs, C. J. Woolf, S. L. Wagner, R. E. Tanzi and D. Y. Kim, *Nature*, 2014, **515**, 274–278.
29. B. De Strooper, R. Vassar and T. Golde, *Nat. Rev. Neurol.*, 2010, **6**, 99.
30. H.-K. Lee, C. Velazquez Sanchez, M. Chen, P. J. Morin, J. M. Wells, E. B. Hanlon and W. Xia, *PLoS One*, 2016, **11**, e0163072.
31. W. K. Raja, A. E. Mungenast, Y.-T. Lin, T. Ko, F. Abdurrob, J. Seo and L.-H. Tsai, *PLoS One*, 2016, **11**, e0161969.
32. S. J. Stachel, C. A. Coburn, T. G. Steele, K. G. Jones, E. F. Loutzenhiser, A. R. Gregro, H. A. Rajapakse, M. T. Lai, M. C. Crouthamel, M. Xu, K. Tugusheva, J. E. Lineberger, B. L. Pietrak, A. S. Espeseth, X. P. Shi, E. Chen-Dodson, M. K. Holloway, S. Munshi, A. J. Simon, L. Kuo and J. P. Vacca, *J. Med. Chem.*, 2004, **47**, 6447–6450.
33. M. E. Kennedy, A. W. Stamford, X. Chen, K. Cox, J. N. Cumming, M. F. Dockendorf, M. Egan, L. Ereshefsky, R. A. Hodgson, L. A. Hyde, S. Jhee, H. J. Kleijn, R. Kuvelkar, W. Li, B. A. Mattson, H. Mei, J. Palcza, J. D. Scott, M. Tanen, M. D. Troyer, J. L. Tseng, J. A. Stone, E. M. Parker and M. S. Forman, *Sci. Transl. Med.*, 2016, **8**, 363ra150.
34. J. D. Scott, S. W. Li, A. P. J. Brunskill, X. Chen, K. Cox, J. N. Cumming, M. Forman, E. J. Gilbert, R. A. Hodgson, L. A. Hyde, Q. Jiang, U. Iserloh,

I. Kazakevich, R. Kuvelkar, H. Mei, J. Meredith, J. Misiaszek, P. Orth, L. M. Rossiter, M. Slater, J. Stone, C. O. Strickland, J. H. Voigt, G. Wang, H. Wang, Y. Wu, W. J. Greenlee, E. M. Parker, M. E. Kennedy and A. W. Stamford, *J. Med. Chem.*, 2016, **59**, 10435.

35. K. Fujimoto, E. Matsuoka, N. Asada, G. Tadano, T. Yamamoto, K. Nakahara, K. Fuchino, H. Ito, N. Kanegawa, D. Moechars, H. J. M. Gijsen and K.-I. Kusakabe, *J. Med. Chem.*, 2019, **62**, 5080–5095.
36. P. C. May, B. A. Willis, S. L. Lowe, R. A. Dean, S. A. Monk, P. J. Cocke, J. E. Audia, L. N. Boggs, A. R. Borders, R. A. Brier, D. O. Calligaro, T. A. Day, L. Ereshefsky, J. A. Erickson, H. Gevorkyan, C. R. Gonzales, D. E. James, S. S. Jhee, S. F. Komjathy, L. Li, T. D. Lindstrom, B. M. Mathes, F. Martényi, S. M. Sheehan, S. L. Stout, D. E. Timm, G. M. Vaught, B. M. Watson, L. L. Winneroski, Z. Yang, X. Dustin and J. Mergott, *Neurobiol. Dis.*, 2015, **35**, 1199–1210.
37. I. Voytyuk, S. A. Mueller, J. Herber, A. Snellinx, D. Moechars, G. van Loo, S. F. Lichtenthaler and B. De Strooper, *Life Sci. Alliance*, 2018, **1**, e201800026.
38. M. Yamada, T. Ichikawa, M. Ii, M. Sunamoto, K. Itoh, N. Tamura and T. Kitazaki, *J. Med. Chem.*, 2005, **48**, 7457–7467.
39. D. Beher, J. D. J. Wrigley, A. Nadin, G. Ve Evin, C. L. Masters, T. Harrison, J. L. Castro and M. S. Shearman, *J. Biol. Chem.*, 2001, **48**, 45394–45402.
40. K. W. Gillman, J. E. Starrett, M. F. Parker, K. Xie, J. J. Bronson, L. R. Marcin, K. E. Mcelhone, C. P. Bergstrom, R. A. Mate, R. Williams, J. E. Meredith, C. R. Burton, D. M. Barten, J. H. Toyn, S. B. Roberts, K. A. Lentz, J. G. Houston, R. Zaczek, C. F. Albright, C. P. Decicco, J. E. Macor and R. E. Olson, *ACS Med. Chem. Lett.*, 2010, **1**, 120–124.
41. X. M. Ye, A. W. Konradi, M. Sun, S. Yuan, D. L. Aubele, M. Dappen, D. Dressen, A. W. Garofalo, J. J. Jagodzinski, L. Latimer, G. D. Probst, H. L. Sham, D. Wone, Y. Xu, D. Ness, E. Brigham, G. T. Kwong, C. Willtis, G. Tonn, E. Goldbach, K. P. Quinn, H. H. Zhang, J.-M. Sauer, M. Bova and G. S. Basi, *Bioorg. Med. Chem. Lett.*, 2013, **23**, 996–1000.
42. J. Mertens, K. Stüber, P. Wunderlich, J. Ladewig, J. C. Kesavan, R. Vandenberghe, M. Vandenbulcke, P. Van Damme, J. Walter, O. Brüstle and P. Koch, *Stem Cell Rep.*, 2013, **1**, 491–498.
43. Y. Takahashi, I. Hayashi, Y. Tominari, K. Rikimaru, Y. Morohashi, T. Kan, H. Natsugari, T. Fukuyama, T. Tomita and T. Iwatsubo, *J. Biol. Chem.*, 2003, **278**, 18664–18670.
44. N. Sawyer, E. Cauchon, A. Chateauneuf, R. P. G. Cruz, D. W. Nicholson, K. M. Metters, G. P. O'Neill and F. G. Gervais, *Br. J. Pharmacol.*, 2002, **137**, 1163–1172.
45. H. Nishioka, N. Tooi, T. Isobe, N. Nakatsuji and K. Aiba, *Sci. Rep.*, 2016, **6**, 1–9.
46. P. W. Brownjohn, J. Smith, E. Portelius, L. Serneels, H. Kvartsberg, B. De Strooper, K. Blennow, H. Zetterberg and F. J. Livesey, *Stem Cell Rep.*, 2017, **8**, 870–882.

47. J. Habchi, P. Arosio, M. Perni, A. R. Costa, M. Yagi-Utsumi, P. Joshi, S. R. Chia, S. I. A. Cohen, M. B. D. Müller, S. Linse, E. A. A. Nollen, C. M. Dobson, T. P. J. Knowles and M. Vendruscolo, *Sci. Adv.*, 2016, e1501244.
48. P. Martín-Maestro, A. Sproul, H. Martinez, D. Paquet, M. Gerges, S. Noggle and A. A. Starkov, *Mol. Neurobiol.*, 2019, 8220–8236, DOI: 10.1007/s12035-019-01665-y.
49. M. F. Boehm, L. Zhang, L. Zhi, M. R. Mcclurg, E. Berger, M. Wagoner, D. E. Mais, C. M. Suto, P. J. A. Davies, R. A. Heyman and A. M. Nadzan, Design and synthesis of potent retinoid X receptor selective ligands that induce apoptosis in leukemia cells, *J. Med. Chem.*, 1995, **38**, 3146–3155.
50. C. Stork, Y. Chen, M. Šícho and J. Kirchmair, *ChemMedChem*, 2018, **13**, 564–571.
51. K. J. Janczura, C.-H. Volmar, G. C. Sartor, S. J. Rao, N. R. Ricciardi, G. Lambert, S. P. Brothers and C. Wahlestedt, *Proc. Natl. Acad. Sci. U. S. A.*, 2018, **115**, E11148–E11157.
52. E. Benito, B. Ramachandran, H. Schroeder, G. Schmidt, H. Urbanke, S. Burkhardt, V. Capece, C. Dean and A. Fischer, *Transl. Psychiatry*, 2017, **7**, e1239.
53. M. Cuadrado-Tejedor, C. Garcia-Barroso, J. A. Sánchez-Arias, O. Rabal, M. Pérez-González, S. Mederos, A. Ugarte, R. Franco, V. Segura, G. Perea, J. Oyarzabal and A. Garcia-Osta, *Neuropsychopharmacology*, 2017, **42**, 524–539.
54. P. L. De Jager, G. Srivastava, K. Lunnon, J. Burgess, L. C. Schalkwyk, L. Yu, M. L. Eaton, B. T. Keenan, J. Ernst, C. McCabe, A. Tang, T. Raj, J. Replogle, W. Brodeur, S. Gabriel, H. S. Chai, C. Younkin, S. G. Younkin, F. Zou, M. Szyf, C. B. Epstein, J. A. Schneider, B. E. Bernstein, A. Meissner, N. Ertekin-Taner, L. B. Chibnik, M. Kellis, J. Mill and D. A. Bennett, *Nat. Neurosci.*, 2014, **17**, 1156.
55. S. J. Marzi, S. K. Leung, T. Ribarska, E. Hannon, A. R. Smith, E. Pishva, J. Poschmann, K. Moore, C. Troakes, S. Al-Sarraj, S. Beck, S. Newman, K. Lunnon, L. C. Schalkwyk and J. Mill, *Nat. Neurosci.*, 2018, **21**, 1618–1627.
56. H.-U. Klein, C. McCabe, E. Gjoneska, S. E. Sullivan, B. J. Kaskow, A. Tang, R. V. Smith, J. Xu, A. R. Pfenning, B. E. Bernstein, A. Meissner, J. A. Schneider, S. Mostafavi, L.-H. Tsai, T. L. Young-Pearse, D. A. Bennett and P. L. De Jager, *Nat. Neurosci.*, 2019, **22**, 37–46.
57. P. Filippakopoulos, J. Qi, S. Picaud, Y. Shen, W. B. Smith, O. Fedorov, E. M. Morse, T. Keates, T. T. Hickman, I. Felletar, M. Philpott, S. Munro, M. R. Mckeown, Y. Wang, A. L. Christie, N. West, M. J. Cameron, B. Schwartz, T. D. Heightman, N. La Thangue, C. A. French, O. Wiest, A. L. Kung, S. Knapp and J. E. Bradner, *Nature*, 2010, **468**, 1067.
58. O. Rabal, J. A. Sánchez-Arias, M. Cuadrado-Tejedor, I. de Miguel, M. Pérez-González, C. García-Barroso, A. Ugarte, A. Estella-Hermoso de Mendoza, E. Sáez, M. Espelosin, S. Ursua, H. Tan, W. Wu, M. Xu, A. Pineda-Lucena, A. Garcia-Osta and J. Oyarzabal, *ACS Chem. Neurosci.*, 2019, **10**, 4076–4101.
59. J. M. Humphrey, in *Phosphodiesterases and their Inhibitors*, 2014, pp. 155–164.

60. A. E. Medina, *Front. Neurosci.*, 2011, **5**, 1–5.
61. M. G. Spillantini and M. Goedert, *Lancet Neurol.*, 2013, **12**, 609–622.
62. C. Li and J. Götz, *Nat. Rev. Drug Discovery*, 2017, **16**, 863–883.
63. V. Melis, M. Magbagbeolu, J. E. Rickard, D. Horsley, K. Davidson, K. A. Harrington, K. Goatman, E. A. Goatman, S. Deiana, S. P. Close, C. Zabke, K. Stamer, S. Dietze, K. Schwab, J. M. D. Storey, C. R. Harrington, C. M. Wischik, F. Theuring and G. Riedel, *Behav. Pharmacol.*, 2015, **26**, 353–368.
64. E. E. Congdon, J. W. Wu, N. Myeku, Y. H. Figueroa, M. Herman, P. S. Marinec, J. E. Gestwicki, C. A. Dickey, W. H. Yu and K. E. Duff, *Autophagy*, 2012, **8**, 609–622.
65. C. R. Harrington, J. M. D. Storey, S. Clunas, K. A. Harrington, D. Horsley, A. Ishaq, S. J. Kemp, C. P. Larch, C. Marshall, S. L. Nicoll, J. E. Rickard, M. Simpson, J. P. Sinclair, L. J. Storey and C. M. Wischik, *J. Biol. Chem.*, 2015, **290**, 10862–10875.
66. J. Ge, E. Normant, J. R. Porter, J. A. Ali, M. S. Dembski, Y. Gao, A. T. Georges, L. Grenier, R. H. Pak, J. Patterson, J. R. Sydor, T. T. Tibbitts, J. K. Tong, J. Adams and V. J. Palombella, *J. Med. Chem.*, 2006, **49**(15), 4606–4615.
67. J. T. Ernst, M. Liu, H. Zuccola, T. Neubert, K. Beaumont, A. Turnbull, A. Kallel, B. Vought and D. Stamos, *Bioorg. Med. Chem. Lett.*, 2014, **24**, 204–208.
68. J. B. Baell and J. M. Willem Nissink, *ACS Chem. Biol.*, 2010, **13**, 44.
69. E. E. Congdon and E. M. Sigurdsson, *Nat. Rev. Neurol.*, 2018, **14**, 399–415.
70. S. L. Liu, C. Wang, T. Jiang, L. Tan, A. Xing and J. T. Yu, *Mol. Neurobiol.*, 2016, **53**, 4328–4342.
71. C. Kunick, K. Lauenroth, M. Leost, L. Meijer and T. Lemcke, *Bioorg. Med. Chem. Lett.*, 2004, **14**, 413–416.
72. M. P. Coghlan, A. A. Culbert, D. A. E. Cross, S. L. Corcoran, J. W. Yates, N. J. Pearce, O. L. Rausch, G. J. Murphy, P. S. Carter, L. Roxbee Cox, D. Mills, M. J. Brown, D. Haigh, R. W. Ward, D. G. Smith, K. J. Murray, A. D. Reith and J. C. Holder, *Chem. Biol.*, 2000, **7**, 793.
73. A. M. Hossini, M. Megges, A. Prigione, B. Lichtner, M. R. Toliat, W. Wruck, F. Schröter, P. Nuernberg, H. Kroll, E. Makrantonaki, C. C. Zoubouliss and J. Adjaye, *BMC Genomics*, 2015, **16**, 84.
74. W. F. An, A. R. Germain, J. A. Bishop, P. P. Nag, S. Metkar, J. Ketterman, M. Walk, M. Weiwer, X. Liu, D. Patnaik, Y.-L. Zhang, J. Gale, W. Zhao, T. Kaya, D. Barker, F. F. Wagner, E. B. Holson, S. Dandapani, J. Perez, B. Munoz, M. Palmer, J. Q. Pan, S. J. Haggarty and S. L. Schreiber, *Probe Reports from NIH Molecular Libraries Program*, 2014, pp. 1–115.
75. A. S. Wagman, R. S. Boyce, S. P. Brown, E. Fang, D. Goff, J. M. Jansen, V. P. Le, B. H. Levine, S. C. Ng, Z.-J. Ni, J. M. Nuss, K. B. Pfister, S. Ramurthy, P. A. Renhowe, D. B. Ring, W. Shu, S. Subramanian, X. A. Zhou, C. M. Shafer, S. D. Harrison, K. W. Johnson and D. E. Bussiere, *J. Med. Chem.*, 2017, **60**, 8282–8515.

76. C. Giacomini, C. Y. Koo, N. Yankova, I. A. Tavares, S. Wray, W. Noble, D. P. Hanger and J. D. H. Morris, *Acta Neuropathol. Commun.*, 2018, **6**, 1–16.
77. C.-Y. Koo, C. Giacomini, M. Reyes-Corral, Y. Olmos, I. A. Tavares, C. M. Marson, S. Linardopoulos, A. N. Tutt and J. D. H. Morris, *Mol. Cancer Ther.*, 2017, 1–13.
78. M. T. Heneka, M. J. Carson, J. El Khoury, G. E. Landreth, F. Brosseron, D. L. Feinstein, A. H. Jacobs, T. Wyss-Coray, J. Vitorica, R. M. Ransohoff, K. Herrup, S. A. Frautschy, B. Finsen, G. C. Brown, A. Verkhratsky, K. Yamanaka, J. Koistinaho, E. Latz, A. Halle, G. C. Petzold, T. Town, D. Morgan, M. L. Shinohara, V. H. Perry, C. Holmes, N. G. Bazan, D. J. Brooks, S. Hunot, B. Joseph, N. Deigendesch, O. Garaschuk, E. Boddeke, C. A. Dinarello, J. C. Breitner, G. M. Cole, D. T. Golenbock and M. P. Kummer, *Lancet Neurol.*, 2015, **14**, 388–405.
79. J. Park, I. Wetzel, I. Marriott, D. Dréau, C. D'Avanzo, D. Y. Kim, R. E. Tanzi and H. Cho, *Nat. Neurosci.*, 2018, **21**, 941–951.
80. K. Takashima, N. Matsunaga, M. Yoshimatsu, K. Hazeki, T. Kaisho, M. Uekata, O. Hazeki, S. Akira, Y. Iizawa and M. Ii, *Br. J. Pharmacol.*, 2009, **157**, 1250–1262.
81. A. Degterev, D. Ofengeim and J. Yuan, *Proc. Natl. Acad. Sci. U. S. A.*, 2019, **116**, 9714 LP–9722.
82. D. Ofengeim, S. Mazzitelli, Y. Ito, J. P. DeWitt, L. Mifflin, C. Zou, S. Das, X. Adiconis, H. Chen, H. Zhu, M. A. Kelliher, J. Z. Levin and J. Yuan, *Proc. Natl. Acad. Sci. U. S. A.*, 2017, **114**, E8788–E8797.
83. A. Degterev, Z. Huang, M. Boyce, Y. Li, P. Jagtap, N. Mizushima, G. D. Cuny, T. J. Mitchison, M. A. Moskowitz and J. Yuan, *Nat. Chem. Biol.*, 2005, **1**(2), 112–119.
84. N. Takahashi, L. Duprez, S. Grootjans, A. Cauwels, W. Nerinckx, J. B. Duhadaway, V. Goossens, R. Roelandt, F. Van Hauwermeiren, C. Libert, W. Declercq, N. Callewaert, G. C. Prendergast, A. Degterev, J. Yuan and P. Vandenabeele, *Cell Death Discovery*, 2012, e437.
85. T. Ishii, T. Okai, M. Iwatani-Yoshihara, M. Mochizuki, S. Unno, M. Kuno, M. Yoshikawa, S. Shibata, M. Nakakariya, T. Yogo and T. Kawamoto, *Sci. Rep.*, 2017, **7**, 1–14.
86. M. Yoshikawa, M. Saitoh, T. Katoh, T. Seki, S. V Bigi, Y. Shimizu, T. Ishii, T. Okai, M. Kuno, H. Hattori, E. Watanabe, K. S. Saikatendu, H. Zou, M. Nakakariya, T. Tatamiya, Y. Nakada, T. Yogo and † Research, *J. Med. Chem.*, 2018, **61**, 2384.
87. G. M. Finan, R. Realubit, S. Chung, D. Lütjohann, N. Wang, J. R. Cirrito, C. Karan and T. W. Kim, *Cell Chem. Biol*, 2016, **23**, 1526–1538.
88. C. Wang, R. Najm, Q. Xu, D. E. Jeong, D. Walker, M. E. Balestra, S. Y. Yoon, H. Yuan, G. Li, Z. A. Miller, B. L. Miller, M. J. Malloy and Y. Huang, *Nat. Med.*, 2018, **24**, 647–657.
89. R. Brooks, G. M. Fuhler, S. Iyer, M. J. Smith, M.-Y. Park, K. H. T. Paraiso, R. W. Engelman and W. G. Kerr, *J. Immunol.*, 2010, **184**, 3582 LP–3589.
90. R. Brooks, S. Iyer, H. Akada, S. Neelam, C. M. Russo, J. D. Chisholm and W. G. Kerr, *Stem Cells*, 2015, **33**, 848–858.

91. C.-C. Liu, T. Kanekiyo, H. Xu and G. Bu, *Nat. Rev. Neurol.*, 2013, **9**, 106.
92. Y. T. Lin, J. Seo, F. Gao, H. M. Feldman, H. L. Wen, J. Penney, H. P. Cam, E. Gjoneska, W. K. Raja, J. Cheng, R. Rueda, O. Kritskiy, F. Abdurrob, Z. Peng, B. Milo, C. J. Yu, S. Elmsaouri, D. Dey, T. Ko, B. A. Yankner and L. H. Tsai, *Neuron*, 2018, **98**, 1141–1154.e7.
93. M. I. Dawson and Z. Xia, *Biochim. Biophys. Acta, Mol. Cell Biol. Lipids*, 2012, **1821**, 21–56.
94. C. Gabbi, M. Warner and J.-Å. Gustafsson, *Biochem. Biophys. Res. Commun.*, 2014, **446**, 647–650.
95. C. Laggner, C. Schieferer, B. Fiechtner, G. Poles, R. D. Hoffmann, H. Glossmann, T. Langer and F. F. Moebius, *J. Med. Chem.*, 2005, **48**, 4754–4764.
96. S. Ye, Y. Huang, K. Müllendorff, L. Dong, G. Giedt, E. C. Meng, F. E. Cohen, I. D. Kuntz, K. H. Weisgraber and R. W. Mahley, *Proc. Natl. Acad. Sci. U. S. A.*, 2005, **102**, 18700–18705.
97. J. Brodbeck, J. McGuire, Z. Liu, A. Meyer-Franke, M. E. Balestra, D. E. Jeong, M. Pleiss, C. McComas, F. Hess, D. Witter, S. Peterson, M. Childers, M. Goulet, N. Liverton, R. Hargreaves, S. Freedman, K. H. Weisgraber, R. W. Mahley and Y. Huang, *J. Biol. Chem.*, 2011, **286**, 17217–17226.
98. Q. Peng, S. Malhotra, J. A. Torchia, W. G. Kerr, K. M. Coggeshall and M. B. Humphrey, *Sci. Signaling*, 2010, **3**, ra38.
99. R. Guerreiro, A. Wojtas, J. Bras, M. Carrasquillo, E. Rogaeva, E. Majounie, C. Cruchaga, C. Sassi, J. S. K. Kauwe, S. Younkin, L. Hazrati, J. Collinge, J. Pocock, T. Lashley, J. Williams, J.-C. Lambert, P. Amouyel, A. Goate, R. Rademakers, K. Morgan, J. Powell, P. St. George-Hyslop, A. Singleton and J. Hardy, *N. Engl. J. Med.*, 2012, **368**, 117–127.
100. Y. Shi and D. M. Holtzman, *Nat. Rev. Immunol.*, 2018, **18**, 759–772.
101. K. Schlepckow, G. Kleinberger, A. Fukumori, R. Feederle, S. F. Lichtenthaler, H. Steiner and C. Haass, *EMBO Mol. Med.*, 2017, **9**, 1356–1365.
102. L. Magno, C. B. Lessard, M. Martins, V. Lang, P. Cruz, Y. Asi, M. Katan, J. Bilsland, T. Lashley, P. Chakrabarty, T. E. Golde and P. J. Whiting, *Alzheimer's Res. Ther.*, 2019, **11**, 16.
103. L. F. Horowitz, W. Hirdes, B.-C. Suh, D. W. Hilgemann, K. Mackie and B. Hille, *J. Gen. Physiol.*, 2005, **126**, 243–262.
104. M. Goedert, M. G. Spillantini, K. Del Tredici and H. Braak, *Nat. Rev. Neurol.*, 2013, **9**, 13–24.
105. M. G. Spillantini, M. L. Schmidt, V. M.-Y. Lee, J. Q. Trojanowski, R. Jakes and M. Goedert, *Nature*, 1997, **388**, 839–840.
106. M. Cherubini and R. Wade-Martins, *Cell Tissue Res.*, 2018, **373**, 79–90.
107. K. J. Billingsley, S. Bandres-Ciga, S. Saez-Atienzar and A. B. Singleton, *Cell Tissue Res.*, 2018, **373**, 9–20.
108. B. Winner, R. Jappelli, S. K. Maji, P. A. Desplats, L. Boyer, S. Aigner, C. Hetzer, T. Loher, M. Vilar, S. Campioni, C. Tzitzilonis, A. Soragni, S. Jessberger, H. Mira, A. Consiglio, E. Pham, E. Masliah, F. H. Gage and R. Riek, *Proc. Natl. Acad. Sci. U. S. A.*, 2011, **108**, 4194 LP–4199.

109. A. A. Cooper, A. D. Gitler, A. Cashikar, C. M. Haynes, K. J. Hill, B. Bhullar, K. Liu, K. Xu, K. E. Strathearn, F. Liu, S. Cao, K. A. Caldwell, G. A. Caldwell, G. Marsischky, R. D. Kolodner, J. LaBaer, J.-C. Rochet, N. M. Bonini and S. Lindquist, *Science*, 2006, **313**, 324LP–328.
110. J. R. Mazzulli, F. Zunke, O. Isacson, L. Studer and D. Krainc, *Proc. Natl. Acad. Sci. U. S. A.*, 2016, **113**, 1931–1936.
111. C. Y. Chung, V. Khurana, P. K. Auluck, D. F. Tardiff, J. R. Mazzulli, F. Soldner, V. Baru, Y. Lou, Y. Freyzon, S. Cho, A. E. Mungenast, J. Muffat, M. Mitalipova, M. D. Pluth, N. T. Jui, B. Schüle, S. J. Lippard, L.-H. Tsai, D. Krainc, S. L. Buchwald, R. Jaenisch and S. Lindquist, *Science*, 2013, **342**, 983–987.
112. S. Wang, C.-H. Chu, T. Stewart, C. Ginghina, Y. Wang, H. Nie, M. Guo, B. Wilson, J.-S. Hong and J. Zhang, *Proc. Natl. Acad. Sci. U. S. A.*, 2015, **112**, E1926 LP–E1935.
113. R. Gordon, E. A. Albornoz, D. C. Christie, M. R. Langley, V. Kumar, S. Mantovani, A. A. B. Robertson, M. S. Butler, D. B. Rowe, L. A. O. Neill, A. G. Kanthasamy, K. Schroder, M. A. Cooper and T. M. Woodruff, *Sci. Transl. Med.*, 2018, **10**, eaah4066.
114. N. Ross, S. Metkar, H. Le, J. Burbank, C. Cahill, J. Germain, A. MacPherson, L. Bittker, M. Palmer, J. Rogers and S. L. Schreiber, in *Probe Reports from the NIH Molecular Libraries Program [Internet]*, National Center for Biotechnology Information (US), Bethesda (MD), 2011.
115. N. Ross, S. Metkar, H. Le, J. Burbank, C. Cahill, J. Germain, A. MacPherson, L. Bittker, M. Palmer, J. Rogers and S. L. Schreiber, in *Probe Reports from the NIH Molecular Libraries Program [Internet]*, National Center for Biotechnology Information (US), Bethesda (MD), 2011.
116. G. K. Tofaris, H. T. Kim, R. Hourez, J.-W. Jung, K. P. Kim and A. L. Goldberg, *Proc. Natl. Acad. Sci. U. S. A.*, 2011, **108**, 17004–17009.
117. B. B. Busch, W. C. Stevens, R. Martin, P. Ordentlich, S. Zhou, D. W. Sapp, R. A. Horlick and R. Mohan, *J. Med. Chem.*, 2004, 5593–5596.
118. S. N. Suresh, A. K. Chavalmane, D. J. Vidyadhara, H. Yarreiphang, S. Rai, A. Paul, J. P. Clement, P. A. Alladi and R. Manjithaya, *Autophagy*, 2017, **13**, 1221–1234.
119. S. N. Suresh, A. K. Chavalmane, M. Pillai, V. Ammanathan, D. J. Vidyadhara, H. Yarreiphang, S. Rai, A. Paul, J. P. Clement, P. A. Alladi and R. Manjithaya, *Front. Mol. Neurosci.*, 2018, **11**, 109.
120. T. F. Outeiro, E. Kontopoulos, S. M. Altmann, I. Kufareva, K. E. Strathearn, A. M. Amore, C. B. Volk, M. M. Maxwell, J.-C. Rochet, P. J. McLean, A. B. Young, R. Abagyan, M. B. Feany, B. T. Hyman and A. G. Kazantsev, *Science*, 2007, **317**, 516–519.
121. P. Putcha, K. M. Danzer, L. R. Kranich, A. Scott, M. Silinski, S. Mabbett, C. D. Hicks, J. M. Veal, P. M. Steed, B. T. Hyman and P. J. McLean, *J. Pharmacol. Exp. Ther.*, 2010, **332**, 849–857.
122. S. Sébastien, M. Forgione, A. Lucidi, G. S. Hailu, A. Nebbioso, V. Carafa, F. Baratta, L. Altucci, N. Giacché, D. Passeri, R. Pellicciari, A. Mai, C. Steegborn and D. Rotili, *J. Med. Chem.*, 2017, **60**, 2344–2360.

123. W. Wrasidlo, I. F. Tsigelny, D. L. Price, G. Dutta, E. Rockenstein, T. C. Schwarz, K. Ledolter, D. Bonhaus, A. Paulino, S. Eleuteri, Å. A. Skjevik, V. L. Kouznetsova, B. Spencer, P. Desplats, T. Gonzalez-Ruelas, M. Trejo-Morales, C. R. Overk, S. Winter, C. Zhu, M. F. Chesselet, D. Meier, H. Moessler, R. Konrat and E. Masliah, *Brain*, 2016, **139**, 3217–3236.
124. G. Kouroupi, E. Taoufik, I. S. Vlachos, K. Tsioras, N. Antoniou, F. Papastefanaki, D. Chroni-Tzartou, W. Wrasidlo, D. Bohl, D. Stellas, P. K. Politis, K. Vekrellis, D. Papadimitriou, L. Stefanis, P. Bregestovski, A. G. Hatzigeorgiou, E. Masliah and R. Matsas, *Proc. Natl. Acad. Sci. U. S. A.*, 2017, **114**, E3679–E3688.
125. G. Tóth, S. J. Gardai, W. Zago, C. W. Bertoncini, N. Cremades, S. L. Roy, M. A. Tambe, J.-C. Rochet, C. Galvagnion, G. Skibinski, S. Finkbeiner, M. Bova, K. Regnstrom, S.-S. Chiou, J. Johnston, K. Callaway, J. P. Anderson, M. F. Jobling, A. K. Buell, T. A. Yednock, T. P. J. Knowles, M. Vendruscolo, J. Christodoulou, C. M. Dobson, D. Schenk and L. McConlogue, *PLoS One*, 2014, **9**, e87133.
126. R. C. Coll, J. R. Hill, C. J. Day, A. Zamoshnikova, D. Boucher, N. L. Massey, J. L. Chitty, J. A. Fraser, M. P. Jennings, A. A. B. Robertson and K. Schroder, *Nat. Chem. Biol.*, 2019, **15**, 556–559.
127. R. C. Coll, A. A. B. Robertson, J. J. Chae, S. C. Higgins, R. Muñoz-Planillo, M. C. Inserra, I. Vetter, L. S. Dungan, B. G. Monks, A. Stutz, D. E. Croker, M. S. Butler, M. Haneklaus, C. E. Sutton, G. Núñez, E. Latz, D. L. Kastner, K. H. G. Mills, S. L. Masters, K. Schroder, M. A. Cooper and L. A. J. O'Neill, *Nat. Med.*, 2015, **21**, 248–257.
128. R. Di Maio, E. K. Hoffman, E. M. Rocha, M. T. Keeney, L. H. Sanders, B. R. De Miranda, A. Zharikov, A. Van Laar, A. F. Stepan, T. A. Lanz, J. K. Kofler, E. A. Burton, D. R. Alessi, T. G. Hastings and J. T. Greenamyre, *Sci. Transl. Med.*, 2018, **10**, eaar5429.
129. A. Sánchez-Danés, Y. Richaud-Patin, I. Carballo-Carbajal, S. Jiménez-Delgado, C. Caig, S. Mora, C. Di Guglielmo, M. Ezquerra, B. Patel, A. Giralt, J. M. Canals, M. Memo, J. Alberch, J. López-Barneo, M. Vila, A. M. Cuervo, E. Tolosa, A. Consiglio and A. Raya, *EMBO Mol. Med.*, 2012, **4**, 380 LP–395.
130. G. Skibinski, V. Hwang, D. M. Ando, A. Daub, A. K. Lee, A. Ravisankar, S. Modan, M. M. Finucane, B. A. Shaby and S. Finkbeiner, *Proc. Natl. Acad. Sci. U. S. A.*, 2017, **114**, 1165–1170.
131. Y. Zhao and N. Dzamko, *Drugs*, 2019, 1037–1051.
132. J. L. Henderson, B. L. Kormos, M. M. Hayward, K. J. Coffman, J. Jasti, R. G. Kurumbail, T. T. Wager, P. R. Verhoest, G. Stephen Noell, Y. Chen, E. Needle, Z. Berger, S. J. Steyn, C. Houle, W. D. Hirst and P. Galatsis, *J. Med. Chem.*, 2015, **58**, 16.
133. J. D. Scott, D. E. DeMong, T. J. Greshock, K. Basu, X. Dai, J. Harris, A. Hruza, S. W. Li, S.-I. Lin, H. Liu, M. K. Macala, Z. Hu, H. Mei, H. Zhang, P. Walsh, M. Poirier, Z.-C. Shi, L. Xiao, G. Agnihotri, M. A. S. Baptista, J. Columbus, M. J. Fell, L. A. Hyde, R. Kuvelkar, Y. Lin, C. Mirescu, J. A. Morrow, Z. Yin, X. Zhang, X. Zhou, R. K. Chang, M. W. Embrey, J. M.

Sanders, H. E. Tiscia, R. E. Drolet, J. T. Kern, S. M. Sur, J. J. Renger, M. T. Bilodeau, M. E. Kennedy, E. M. Parker, A. W. Stamford, R. Nargund, J. A. McCauley and M. W. Miller, *J. Med. Chem.*, 2017, **60**, 2983–2992.
134. R. Wallings, N. Connor-Robson and R. Wade-Martins, *Hum. Mol. Genet.*, 2019, 2696–2710.
135. J. R. Mazzulli, F. Zunke, T. Tsunemi, N. J. Toker, S. Jeon, L. F. Burbulla, S. Patnaik, E. Sidransky, J. J. Marugan, C. M. Sue and D. Krainc, *J. Neurosci.*, 2016, **36**, 7693–7706.
136. A. H. V. Schapira, *Mol. Cell. Neurosci.*, 2015, **66**, 37–42.
137. D. C. Schöndorf, M. Aureli, F. E. McAllister, C. J. Hindley, F. Mayer, B. Schmid, S. P. Sardi, M. Valsecchi, S. Hoffmann, L. K. Schwarz, U. Hedrich, D. Berg, L. S. Shihabuddin, J. Hu, J. Pruszak, S. P. Gygi, S. Sonnino, T. Gasser and M. Deleidi, *Nat. Commun.*, 2014, **5**, 4028.
138. K. E. Murphy, A. M. Gysbers, S. K. Abbott, N. Tayebi, W. S. Kim, E. Sidransky, A. Cooper, B. Garner and G. M. Halliday, *Brain*, 2014, **137**, 834–848.
139. S. Patnaik, W. Zheng, J. H. Choi, O. Motabar, N. Southall, W. Westbroek, W. A. Lea, A. Velayati, E. Goldin, E. Sidransky, W. Leister and J. J. Marugan, *J. Med. Chem.*, 2012, **55**, 5734–5748.
140. T. M. Dawson and V. L. Dawson, *Mov. Disord.*, 2010, **25**(Suppl 1), S32–S39.
141. S. Y. Chung, S. Kishinevsky, J. R. Mazzulli, J. Graziotto, A. Mrejeru, E. V. Mosharov, L. Puspita, P. Valiulahi, D. Sulzer, T. A. Milner, T. Taldone, D. Krainc, L. Studer and J. w. Shim, *Stem Cell Rep.*, 2016, **7**, 664–677.
142. O. Cooper, H. Seo, S. Andrabi, C. Guardia-laguarta, J. Graziotto, M. Sundberg, J. R. Mclean, L. Carrillo-reid, Z. Xie, T. Osborn, G. Hargus, M. Deleidi, T. Lawson, H. Bogetofte, E. Perez-torres, L. Clark, C. Moskowitz, J. Mazzulli, L. Chen, L. Volpicelli-daley, N. Romero, H. Jiang, R. J. Uitti, Z. Huang, G. Opala, L. A. Scarffe, V. L. Dawson, C. Klein, J. Feng, O. A. Ross, J. Q. Trojanowski, V. M. Lee, K. Marder, D. J. Surmeier, Z. K. Wszolek, S. Przedborski, D. Krainc, T. M. Dawson and O. Isacson, *Sci. Transl. Med.*, 2012, **4**, 141ra90.
143. R. Sedrani, L. H. Jones, A. M. Jutzi-Eme, W. Schuler and S. Cottens, *Bioorg. Med. Chem. Lett.*, 1999, **9**, 459.
144. M. Kaeberlein and V. Galvan, *Science*, 2019, **2**, 1–6.
145. S. Patnaik, W. Zheng, J. H. Choi, O. Motabar, N. Southall, W. Westbroek, W. A. Lea, A. Velayati, E. Goldin, E. Sidransky, W. Leister and J. J. Marugan, *J. Med. Chem.*, 2012, **55**, 5734–5748.
146. C. Malagelada, Z. H. Jin, V. Jackson-Lewis, S. Przedborski and L. A. Greene, *J. Neurosci.*, 2010, **30**, 1166–1175.
147. B. Bingol, J. S. Tea, L. Phu, M. Reichelt, C. E. Bakalarski, Q. Song, O. Foreman, D. S. Kirkpatrick and M. Sheng, *Nature*, 2014, **510**, 370–375.
148. A. F. Kluge, B. R. Lagu, P. Maiti, M. Jaleel, M. Webb, J. Malhotra, A. Mallat, P. A. Srinivas and J. E. Thompson, *Bioorg. Med. Chem. Lett.*, 2018, **28**, 2655–2659.
149. I. S. Yoon, Q. Au, J. R. Barber, S. C. Ng and B. Zhang, *Anal. Biochem.*, 2010, **407**, 205–210.

150. S. D. Ryan, N. Dolatabadi, S. F. Chan, X. Zhang, M. W. Akhtar, J. Parker, F. Soldner, C. R. Sunico, S. Nagar, M. Talantova, B. Lee, K. Lopez, A. Nutter, B. Shan, E. Molokanova, Y. Zhang, X. Han, T. Nakamura, E. Masliah, J. R. Yates III, N. Nakanishi, A. Y. Andreyev, S. Okamoto, R. Jaenisch, R. Ambasudhan and S. A. Lipton, *Cell*, 2013, **155**, 1351–1364.
151. A. Moretto and C. Colosio, *Toxicology*, 2013, **307**, 24–34.

Subject Index